C000150101

The Rinderpest Campaigns

Amanda Kay McVety has written the first history of the international effort to eradicate rinderpest – a devastating cattle disease – which began in the 1940s and ended in 2011. Rinderpest is the only other disease besides smallpox to have been eradicated, but very few people in the United States know about it, because it did not infect humans and never broke out in North America. In other parts of the world, however, rinderpest was a serious economic and social burden and the struggle against it was a critical part of the effort to fight poverty and hunger globally. McVety's history of the campaigns against the virus expands our understanding of development, internationalism, and national security by showing how these concepts were framed not only by economic and political concerns, but also by biological and environmental ones.

Amanda Kay McVety is Associate Professor of History at Miami University. She is the author of *Enlightened Aid: U.S. Development as Foreign Policy in Ethiopia* and has published articles in the journals *Diplomatic History* and *The Journal of the Gilded Age and Progressive Era*.

Global and International History

Series Editors

Erez Manela, *Harvard University*
John McNeill, *Georgetown University*
Aviel Roshwald, *Georgetown University*

The Global and International History series seeks to highlight and explore the convergences between the new International History and the new World History. Its editors are interested in approaches that mix traditional units of analysis such as civilizations, nations and states with other concepts such as transnationalism, diasporas, and international institutions.

Titles in the Series

Antoine Acker, *Volkswagen in the Amazon: The Tragedy of Global Development in Modern Brazil*

Christopher R. W. Dietrich, *Oil Revolution: Anti-Colonial Elites, Sovereign Rights, and the Economic Culture of Decolonization*

Stefan Rinke, *Latin America and the First World War*

Nathan J. Citino, *Envisioning the Arab Future: Modernization in U.S.-Arab Relations, 1945–1967*

Timothy Nunan, *Humanitarian Invasion: Global Development in Cold War Afghanistan*

Michael Goebel, *Anti-Imperial Metropolis: Interwar Paris and the Seeds of Third World Nationalism*

Stephen J. Macekura, *Of Limits and Growth: The Rise of Global Sustainable Development in the Twentieth Century*

Michele L. Louro, *Comrades against Imperialism*

The Rinderpest Campaigns

A Virus, Its Vaccines, and Global Development in the Twentieth Century

AMANDA KAY MCVETY

Miami University

CAMBRIDGE
UNIVERSITY PRESS

CAMBRIDGE
UNIVERSITY PRESS

University Printing House, Cambridge CB2 8BS, United Kingdom

One Liberty Plaza, 20th Floor, New York, NY 10006, USA

477 Williamstown Road, Port Melbourne, VIC 3207, Australia

314-321, 3rd Floor, Plot 3, Splendor Forum, Jasola District Centre, New Delhi - 110025, India

103 Penang Road, #05-06/07, Visioncrest Commercial, Singapore 238467

Cambridge University Press is part of the University of Cambridge.

It furthers the University's mission by disseminating knowledge in the pursuit of education, learning and research at the highest international levels of excellence.

www.cambridge.org
Information on this title: www.cambridge.org/9781108434065
DOI: 10.1017/9781108381673

First published 2018
First paperback edition 2021

A catalogue record for this publication is available from the British Library

ISBN 978-1-108-42274-1 Hardback
ISBN 978-1-108-43406-5 Paperback

Cambridge University Press has no responsibility for the persistence or accuracy of URLs for external or third-party internet websites referred to in this publication, and does not guarantee that any content on such websites is, or will remain, accurate or appropriate.

For
Harvey and Bonnie Vernell Franks

Contents

List of Figures		*page* viii
Acknowledgments		ix
List of Abbreviations		xii
	Introduction: Grosse Île, September 1942	1
1	Rinderpest and the Origins of International Cooperation for Disease Control	13
2	GIR-1: Rinderpest in World War II	47
3	"Freedom from Want": UNRRA's Rinderpest Campaigns	86
4	The Machinery of Development: FAO's Rinderpest Campaigns	121
5	Back to Grosse Île: Biological Warfare in the Postwar World	164
6	"Freedom from Rinderpest"	207
	Conclusion	250
Select Bibliography		254
Index		278

Figures

0.1 Concrete pier and former immigrant inspection
building on Grosse Île where researchers worked on
rinderpest during World War II.   *page* 3

1.1 "God strikes the Netherlands with rinderpest."   19

2.1 Junji Nakamura and Keikichi Yamawaki at the Pasteur
Institute in Paris.   49

2.2 Richard Shope during World War II.   65

3.1 Field veterinarian in village office near Taipei.   98

3.2–3.5 Four photographs demonstrating the manufacture of
lapinized rinderpest vaccine in China in 1949.   103–105

3.6 Huang Sung-ling of the Provincial Bureau of
Agriculture and Forestry, Robert C. Reisinger of FAO
(working as a JCRR veterinary consultant), and others
investigate a case of suspected rinderpest during an
outbreak in Taiwan.   116

6.1 Walter Plowright and Kazuya Yamanouchi in
Tokyo in 1994.   212

6.2 Dr. William Taylor looks for the characteristic
rinderpest lesions in the mouth of a cow in Sudan in
1987.   224

6.3 President Mwai Kibaki of Kenya and other dignitaries
at the unveiling of a buffalo statue at Meru National
Park in 2010.   237

6.4 Tom Olaka, a community animal worker in Karamoja,
Uganda, using thermostable rinderpest vaccine to
vaccinate local cattle in 1994.   241

Acknowledgments

I owe thanks to many people who helped me with this project. Drew Cayton read the first draft of this manuscript. It was in very rough shape, to say the least, but Drew, as he had done for countless others before, offered insightful feedback that helped me turn it into the book it is today. He was an amazing colleague and friend. I am one of many who miss him. Daniel Immerwahr graciously read the second draft of this manuscript and he too offered insightful feedback that helped me tighten the narrative and vastly improve the overall structure. I could not have written this book without the assistance of Kazuya Yamanouchi who shared a wealth of knowledge and sources with me. His generosity knew no bounds and I am very much in his debt. When I visited FAO, Felix Njeumi spoke with me at length about rinderpest and helped me to secure some valuable documents. The head of the FAO archives, Fabio Ciccarello, could not have been more gracious or more helpful. Peter Roeder, Jeff Mariner, and Michael Baron all kindly agreed to be interviewed for this project and offered valuable insights. Richard Shope spoke with me about his grandfather and cheerfully shared articles and images. Many others helped along the way in conversations at conferences and workshops, including, but not limited to: Tatiana Seijas, Nick Cullather, Julia Irwin, Stephen Macekura, Tom Robertson, Jenny Smith, Erez Manela, Thadeus Sunseri, Ed Miller, Megan Black, Jessica Wang, Karen Rader, Jim Webb, Prakash Kumar, Karl Appuhn, Paul Adler, Sheyda Jahanbani, and Steve Porter. Special thanks to Stephen, Erez, Tom, and Jenny for organizing two of those excellent workshops. George Sweat translated some material for me. Sheila Sparks secured innumerable sources through interlibrary loan. My anonymous readers at Cambridge University Press provided excellent advice for improving the manuscript. Debbie Gershenowitz was enthusiastic about the project from the first time I spoke with her, as were Erez Manela, John McNeill, and Aviel Roshwald. Along with Debbie, Ruth Boyes, Krishna Prasath Ganesan, Faye Roberts, Tad Cook, and Kristina Deusch turned the manuscript into a book. I am very grateful for all of the assistance. Any errors are my own.

Luckily for me, I work with some of the best people in the business at Miami University: they are smart and funny and make work a source of joy. I want to thank everyone in the history department for their friendship and support, with particular nods to Steve Norris, Renée Baernstein, Charlotte Goldy, Wietse de Boer, Allan Winkler, Elena Albarrán, Dan Prior, Peggy Shaffer, and Steve Conn for their encouragement along the way. I owe special thanks to Erik Jensen, who, by traveling with me on several research trips, turned them into adventures. He also graciously translated some of Robert Koch's writing from the German. José Amador, with his enthusiasm and support, helped bring this project to life. As Altman Fellows at the Miami University Humanities Center, we co-organized the program "The Human and the Nonhuman: Exploring Intersections between Science and the Humanities" during the 2012–2013 academic year. That collaboration convinced me that I could turn my fascination with rinderpest into a book. This is just one of many projects that the Humanities Center, under the direction of the supremely talented Tim Melley, has nourished. We are lucky to have it and him.

I could not have written this book without the assistance of archivists at FAO, the Harry S. Truman Presidential Library, the National Archives at College Park, Library and Archives Canada, the National Archives at Kew, the United Nations Archives, the Herbert Hoover Presidential Library, the Rockefeller Archive Center, the Schlesinger Library, the Hoover Institution, and Oklahoma State University Library Special Collections and University Archives. The Harry S. Truman Library Institute and the Ewing Marion Kauffman Foundation generously helped fund my research and writing through the Truman-Kauffman Research Fellowship Program. Miami University also provided funding for some of my research trips. I am grateful for all the support.

My final thanks go to my family. This book is dedicated to my grandparents, Harvey and Bonnie Vernell Franks, who dedicated their lives to making sure that their children and grandchildren had access to far more opportunities than had been available to them. My grandfather passed away in 2015 at the age of 90. I miss him every day. My grandmother continues to bring love and joy to everyone who knows her. She and my mother both traveled with me to Geneva while I was researching this book and I have wonderful memories of the adventures that we had together there. My parents, Bruce and Elaine McVety, have provided unending love and support from the beginning. I owe them a great deal. The same is true for my sister and brother-in-law, Christy and David Rowe, and my many wonderful aunts, uncles, and cousins. While writing this book, I met and married Brian Hogan, gaining, in the process, both a marvelous husband and a new pack of wonderful relatives. We now have two sons, William Harvey and Eamon Bruce, who fill the house with laughter and chaos. Joy abounds.

Abbreviations

CBAHW	Community-Based Animal Health Workers
CCTA	Commission for Technical Cooperation in Africa South of the Sahara
CDC	Centers for Disease Control and Prevention
CLARA	Communist Liberated Areas Relief Administration
CNRRA	China National Relief and Rehabilitation Administration
CO	Colonial Office
CWS	Chemical Warfare Service
DRB	Defense Research Board
EC	European Commission
ECA	Economic Cooperation Administration
EFO	Economic and Financial Organization (League of Nations)
EMPRES	Emergency Prevention System for Transboundary Animal and Plant Pests and Diseases
FAO	Food and Agriculture Organization
GIR-1	Grosse Île Rinderpest (WWII research program)
GREP	Global Rinderpest Eradication Programme
IAEA	International Atomic Energy Agency
IAH	Institute for Animal Health
IBAH	Interafrican Bureau for Animal Health
IBAR	Interafrican Bureau for Animal Resources
IEG	Imperial Ethiopian Government
IIA	International Institute of Agriculture
ILO	International Labour Organization
JCRR	Joint Commission on Rural Reconstruction
JP15	Joint Programme 15

KAG	Kabete attenuated goat vaccine
LNHO	League of Nations Health Organization
MEP	Malaria Eradication Programme
MeV	Measles virus
OAU	Organization for African Unity
OFRRO	Office of Foreign Relief and Rehabilitation Operations
OIE	Office International des Epizooties (now World Animal Health Organization)
PANVAC	Pan African Veterinary Vaccine Centre
PARC	Pan African Rinderpest Campaign
PE	Participatory Epidemiology
PENAPH	Participatory Epidemiology Network for Animal and Public Health
PPR	Peste des petits ruminants
RPV	Rinderpest virus
RVCM	Rinderpest virus-containing materials
SAREC	South Asian Rinderpest Eradication Campaign
SEP	Smallpox Eradication Programme
STRC	Scientific Technical and Research Commission
UNESCO	United Nations Educational, Scientific and Cultural Organization
UNICEF	United Nations International Children's Emergency Fund
UNRRA	United Nations Relief and Rehabilitation Administration
USAID	United States Agency for International Development
USBWC	United States Biological Warfare Committee
USDA	United States Department of Agriculture
USOM	United States Operations Mission
WAREC	West Asia Rinderpest Eradication Campaign
WBC	War Bureau of Consultants Committee
WFB	World Food Board
WHA	World Health Assembly
WHO	World Health Organization
WRS	War Research Service

Grosse Île, September 1942

"Modern bacteriology has brought about a state of affairs which may exert profound influence upon the future economic and political history of the world."

Hans Zinsser, *Rats, Lice and History*, 1935

Despite its name, it is not big. It is one mile long and a half mile wide, a small island in the middle of the Saint Lawrence River, thirty miles north of Quebec City. Its geography had determined its destiny long before American scientists arrived in the fall of 1942. Grosse Île – Canada's main quarantine station for 105 years – evoked sorrow, thoughts of disease and death. If the Americans had not known the story before they arrived, they would have been told it by the Canadian military personnel and workers who met them on the dock. Moreover, they would have seen it in the cemetery of thousands of coffins marked not by stones with names, but by indentions in the ground itself that had settled into valleys between hills of the dead. The land bore witness to suffering, and to the power of the nonhuman.

Grosse Île was just the right thing (an island), just the right size (large enough), in just the right location (in the middle of the river with a three-mile-wide harbor between it and a neighboring island), and surrounded by just the right depths of water ("7 to 19 fathoms") for the task assigned to it by the Assembly of Lower Canada in 1832. The island was to guard the province from cholera, which had recently broken out of India and onto the global stage. On the recommendation of a local captain who was surveying the river, the assembly rented Grosse Île from its owner. Following the spring thaw, officials made their way to the island, kicked out its sole resident farmer, and started building a quarantine station.

Ships hoping to unload passengers in Quebec had to first get a certificate of health from the officials at Grosse Île. They would drop anchor near its dock and wait for inspection; if there were sick passengers on board, all would have to be rowed to the island for further observation. It was a good idea, but in 1832 it did not work; *Vibrio Cholerae* slipped through the quarantine. Over 3,000 people died of cholera in Quebec that summer. Officials buried twenty-eight victims on Grosse Île; they would be far from its last burials. The year of 1847 was the worst. That year, typhus, courtesy of *Rickettsia Prowazekii*, claimed over 5,000 Irish men, women, and children in flight from the famine. In 1909, the Ancient Order of Hibernians erected a massive stone memorial to them on the highest point on the island, 140 feet above the river. The Celtic cross on the top ever after warned all approaching that the island was a cemetery.[1]

Times changed. Quarantine islands went out of fashion as humans gained more power over microbes. People were now treated instead of isolated. The Canadian government ordered the quarantine station closed in 1937; sick ship passengers now went directly to hospitals in Quebec City and Montreal. Responsibility for the island moved from the Department of Pensions and National Health to the Department of Public Works, and there it stayed until it was transferred to the Department of Defense in 1942. It was a fitting home. The government's involvement with the island had always really been about defense, about humans trying to protect themselves from dangerous nonhumans.

In 1942, several of the same geographical attributes that had made Grosse Île an ideal quarantine station would make it an ideal research station for scientists charged by their governments to find a defense against a pathogen that had never yet reached North America. They called it rinderpest, the German word for "cattle plague." It was a fierce virus that attacked not humans, but their cattle. Strict cattle quarantines had long denied it entry into North America, but they feared that the quarantines were no longer enough, because now they had enemies. They worried that the virus would arrive not in a host who would reveal its unwanted passenger with sickness and death but in a glass vial of frozen infected tissue that could be thawed and unleashed with a single injection. They were preparing to fight a biological attack. They needed a defense.

---

[1] O'Gallagher, Marianna. *Grosse Ile: Gateway to Canada, 1832–1937.* Quebec City: Livres Carraig Books, 1984, 15–26, 47–58, 85. For more on the island's history, see also, Renaud, Anne. *Island of Hope and Sorrow: The Story of Grosse Île.* Montreal: Lobster Press, 2007.

FIGURE 0.1 Concrete pier and former immigrant inspection building on Grosse Île where researchers worked on rinderpest during World War II. *Author photograph.*

They needed a vaccine. To make it, they would have to import the very thing they feared: a glass vial of frozen infected tissue. And that is why they needed Grosse Île.

There, in quickly constructed laboratories erected in rooms which had previously witnessed medical examinations and forced "disinfecting" showers, they gave new life to the virus, freeing it from its confines. They had to use hosts to do it, for they could not keep it alive without them. They brought their victims – hundreds of them from the cattle market in Montreal – to the island in boats. They needed the virus to flourish so that they could figure out how to destroy it. Their efforts would be successful. Within nineteen months, the scientists on the island not only produced 100,000 doses of a known vaccine, they created a better one. The United States and Canada ended up not needing either. The feared biological attack never came. The war ended with the vaccines, the virus, and the surviving hosts still on the island. The futures of the virus, the vaccine and the hosts were uncertain, but some of the scientists involved had plans for them, plans that involved shipping the virus off an island designed to contain it. They had to act quickly, for the virus's days were numbered.

On January 19, 1946, Dominion Animal Pathologist Charles "Chas." Alexander Mitchell, quickly composed a heated memorandum about the imminent closure of the research facility at Grosse Île on behalf of the Deputy Minister at Canada's Department of Agriculture. "When hostilities ceased," Mitchell wrote, "a directive was given in the United States that all war projects would terminate within a month. An exception was made to the Grosse Isle[2] project because of its biological nature, it being self-evident that time is required to inactivate living material." But that time had run out. The United States Department of War, he warned, had announced that it "will no longer be associated with the project" as of midnight on February 28. "Unless the Dominion Department of Agriculture takes over the project immediately, decontamination, that is killing of the virus, will commence in the last week of January." This would, Mitchell insisted, be a grievous mistake. The virus was too precious to lose because it was so dangerous.[3]

Humans still had a great deal left to learn about the rinderpest virus in 1946, but they already knew quite a bit about rinderpest the disease – they had been fighting it for centuries. In 1902, Duncan Hutcheon, Colonial Veterinary Surgeon to the Government of Cape Colony, described it as "a specific malignant and highly contagious fever, characterized by congestion, and a peculiar form of inflammation of the mucous membranes, more particularly of the digestive tract." He charted its progress in its bovine victims. The first symptom was "a rise in temperature," usually "between the third and fourth day, often a little earlier." By the fifth day, "the animal is visibly dull ... and the appetite less." Sometimes this came with "twitching" and "a mild but rough-sounding cough." On the sixth day, "the characteristic symptoms appear": inflamed mucous membranes turned red and mucous started flowing from the eyes, mouth, and nostrils. Diarrhea followed on the seventh or eighth day. The animal stopped eating, but became desperately thirsty, and "when allowed to get to water ... will stand and sip continuously, and frequently die in the water," poisoning it with "secretions and excretions." Animals who

---

[2] At the time, officials referred to the island as Grosse Isle.

[3] Chas. A. Mitchell to the Deputy Minister, Department of Agriculture (January 19, 1946), Diseases of Animals, Rinderpest Control Vaccination Project, Grosse Isle, QC, Record Group 17, Volume 3029, File 37–23, LAC.

could not make it to water simply laid down and died in "a semi-comatose condition." Entire herds could be wiped out in a few days. Many were.[4] Hutcheon labeled the disease behind the devastation as "rinderpest, bovine pest, or cattle plague." He preferred the first term, as did most of his fellow English-speakers in Cape Colony. The French called it *peste bovine*. Other languages used other names, but most meant the same thing: killer of cattle.[5]

The disease spread quickly and terribly: sometimes only claiming 20 percent of its victims, sometimes 100 percent, depending on the strain and the host species. In the outbreak Hutcheon witnessed – the Great African Rinderpest Panzootic – the mortality rate was over 90 percent for the entire continent. Rinderpest could infect many animals, but was most dangerous to cattle, yaks, wild African buffalos, and Asian water buffalos.[6] It devastated humans by depriving them of their most valuable livestock: the source of their food, their labor, their wealth, their security. It was a dangerous enemy.

The virus Mitchell was terrified of losing to decontamination was a strain of rinderpest that had proven amenable to being turned into a live vaccine: a living weapon whose presence in a host would render it immune to attack by one of the other deadly strains that roamed freely throughout Africa and Asia. It was not enough just to save the vaccine, Mitchell warned; the strain it had come from also needed to be preserved because it was "the only one which has been proved capable of propagation and its loss would be a major catastrophe having regard to the future protection of food supplies of the world." The United States and Canada no longer had any need of it, Mitchell acknowledged, but other nations did. Mitchell begged for a stay of execution: six months "for the purpose of permitting representatives from other parts of the world ... to visit the Station and train in the methods of vaccine production, also so that the particular strain which has been propagated will not be lost."[7] He got it and, in the process, the world got a new weapon against a deadly foe.

---

[4] Hutcheon, Duncan. "Rinderpest in South Africa." *Journal of Comparative Pathology and Therapeutics* 15:4 (December 31, 1902): 300–305.

[5] Ibid., 300–305; Scott, Gordon R. "Rinderpest," in *Advances in Veterinary Science*, v. 9, ed. C. A. Brandly and E. L. Jungherr. New York: Academic Press, 1964) 114–115.

[6] OIE, Rinderpest Technical Disease Card, available at http://www.oie.int/fileadmin/Home /eng/Animal_Health_in_the_World/docs/pdf/Disease_cards/RINDERPEST.pdf.

[7] Chas. A. Mitchell to The Deputy Minister, Department of Agriculture (January 19, 1946), Diseases of Animals, Rinderpest Control Vaccination Project, Grosse Isle, QC, Record Group 17, Volume 3029, File 37-23, LAC.

This book is in part a history of that weapon and others like it: the vaccines that fought rinderpest. On one level, it is the story of how humans manipulated a virus to make cattle and buffalo immune to it, but it is also the story of how the vaccines that were the product of that manipulation affected human behavior. The technology exerted influence that reached far beyond the virus and its bovine victims, because the virus had always reached beyond them as well.[8] Mitchell wanted to save the virus for scientific purposes, but also for political ones. He wrote openly about his concern regarding global food security. He did not mention his concern about biological warfare but in the fall of 1946, when they shut down the facility at Grosse Île, they left seed virus and vaccine in a freezer, in case they needed them again. Mitchell and his team sent vaccines off the island and they kept some on it; both decisions were the product of a recognition of the biological connections that bound humans, bovines, and the virus together. It was an act of global consciousness – the idea, as Akira Iriye defined it, "that there is a wider world over and above separate states and national societies, and that individuals and groups, no matter where they are, share certain interests and concerns in that wider world."[9]

Rinderpest had played a role in the creation of that consciousness of connection. During the nineteenth century, the virus had used the expansion of global trade to spread further and faster than ever before, bringing devastation in its wake. Outbreaks in virgin territories often produced mortality rates of 90 percent. The virus sparked fear and encouraged international cooperation for resistance. That cooperation took a highly nationalistic form. European nations worked together to drive rinderpest out. North and South American nations worked together to keep it out altogether. (Their efforts were successful with the exception of a quickly contained outbreak in Brazil in 1920.) Meanwhile, the same European governments that organized to control the disease at home actually furthered its spread abroad through the construction of imperial networks of trade and

---

[8] For broader thoughts on this connection between technology and politics, see Hecht, Gabrielle. "Introduction," in *Entangled Geographies: Empire and Technopolitics in the Global Cold War*. ed. Gabrielle Hecht. Cambridge, MA: MIT Press, 2011), 1–12.

[9] Iriye, Akira. *Global Community.* Berkeley: University of California Press, 2004, 8.

conquest. The quest for empire helped to make rinderpest endemic in much of Asia and Africa.

In the aftermath, rinderpest became a largely imperial disease, existing primarily in parts of the world that had been claimed by nations in other parts of the world. Its existence there was largely tolerated until a 1920 outbreak, in a quarantine stable in Antwerp from infected Indian cattle, reminded European officials that the virus's existence anywhere was a potential threat to cattle in their nations; they began organizing against it. Researchers waged war against it in imperial laboratories and moved vaccines along imperial networks, but they also published the results of their work in open journals and visited each other's laboratories, sharing the technical knowledge of their creations outside imperial boundaries. The vaccines encouraged an inter-imperialism made tangible with the establishment of the Office International des Épizooties, or OIE, in 1924, to collect and publicize information about epizootic diseases and how to control them, to support research about those diseases, and "to prepare and stimulate international agreements on sanitary regulations, and to assist governments in the pursuit and enforcement of such measures."[10]

World War II, with all its devastations, only heightened this sense of global interconnectedness. The United States and Canada sent their scientists to Grosse Île in 1942 to prepare for a biological attack. If the enemy had access to rinderpest, then it had access to a potentially devastating weapon; it would not have to do anything to it except unleash it in a North America that had never had an outbreak. Such a threat was particularly ominous at that moment because North America was the Allied world's primary source of food. Cattle were part of the war effort. When the war was over, the new vaccine created at Grosse Île opened the door to a different way of thinking about interconnectedness. The United States and Canada had previously not been involved in the global struggle against rinderpest, but now they had something to contribute. They also had the machinery through which to do so.

---

[10] FAO Standing Advisory Committee on Agriculture, Minutes of Meeting of Subcommittee on Animal Health (March 31–April 4, 1947), Sub-Committee on Animal Health, 1946–1947, Animal Production and Health Division, 10AGA407, FAO. On "inter-imperialism," see Akami, Tomoko. "A Quest to be Global: The League of Nations Health Organization and Inter-Colonial Regional Governing Agendas of the Far Eastern Association of Tropical Medicine 1910–25." *The International History Review* 38:1 (2016): 1–23.

While the scientists had been at work on Grosse Île building the vaccine, their governments, struck by a sense of both economic and biological interconnectedness, had been constructing international machinery for not only winning the war, but for making the postwar world a better, safer place, calling for freedom from want and freedom from fear. Their efforts initially focused on the immediate relief of liberated areas and here the scientists at Grosse Île found the machinery that allowed them to share their vaccine, sending it via the United Nations Relief and Rehabilitation Administration to China to fight hunger by fighting rinderpest. The success of the subsequent rinderpest campaign there encouraged those involved to expand their vision of what the struggle to secure freedom from want could look like in the postwar world. Growing confidence in technological innovation helped to broaden confidence that the international machinery created during the war could provide more than relief, more even than rehabilitation: that it could provide development. Here, once again, the rinderpest vaccines played an important, yet now largely forgotten, role.

Starting in the late 1940s, the Food and Agriculture Organization of the United Nations (FAO) made the fight against rinderpest a core part of its mission to fight hunger around the world. It did so because the vaccines convinced its leadership that it was a fight that humanity could win. Its struggle was against more than just the virus, it was also a struggle for the United Nations system and for the sense of a global community. Leadership at FAO hoped to use the rinderpest campaign to demonstrate the point that the world's biological interconnectedness required human political interconnectedness. It was not an easy fight. Throughout the 1950s FAO was both helped and hindered at different turns by the Cold War and by imperialism, but it persevered, determined to eradicate the virus and, in the process, to provide undeniable proof of the new international system's value.

The postwar recognition of global interconnectedness encouraged FAO to fight rinderpest, but that was not humankind's only response. Biological interconnectedness was both a source of inspiration and a source of concern in the postwar era. The same worries that had led Canada and the United States to send scientists to Grosse Île in 1942 led them to send scientists back to the island in 1950, tasked to produce both vaccines and a biological weapon. The story of that effort reminds us that the mid-twentieth-century sense of global interdependence inspired multiple kinds of imaginings and actions. How one viewed it depended upon one's larger mission: building a global community, maintaining an empire, or ensuring national security. The rinderpest vaccines played a role in all of them.

This book argues that the rinderpest vaccines encouraged humans to think differently and to act differently at a global level, and that those thoughts and actions had ramifications that went well beyond the struggle against the virus. The vaccines affected what humans thought they could do, what they wanted to do, and what they tried to do. Following the vaccines off the island and around the globe enhances our understanding of development, internationalism, and national security in the postwar world by revealing the biological component that played a role in all of them. These concepts were not just framed by economic and political concerns; they were also framed by environmental ones. Rinderpest, and the vaccines humans created to fight it, helped to ensure that.[11]

They were not alone. Scholars studying twentieth-century international relations have recently begun talking and writing more about its nonhuman participants.[12] This book is a contribution to that ongoing conversation. It argues that we get a fuller picture of twentieth-century international relations when we bring rinderpest and the vaccines humans

---

[11] In this, this book finds inspiration in John R. McNeill's *Mosquito Empires*, in which he wrote, "The book provides a perspective that takes into account nature – viruses, plasmodia, mosquitoes, monkeys, swamps – as well as humankind in making political history" (McNeill, John R. *Mosquito Empires*. Cambridge: Cambridge University Press, 2010), 2.

[12] Dorsey, Kurk. "Bernath Lecture: Dealing with the Dinosaur (and Its Swamp): Putting the Environment in Diplomatic History." *Diplomatic History* 29:4 (September 2005): 573–587; Hamblin, Jacob Darwin. *Arming Mother Nature: The Birth of Catastrophic Environmentalism*. New York: Oxford, 2013; Dorsey, Kurkpatrick. *Whales and Nations: Environmental Diplomacy on the High Seas*. Seattle: University of Washington Press, 2013; McNeill, J. R. and Corinna R. Unger, ed. *Environmental Histories of the Cold War*. Cambridge: Cambridge University Press, 2010; Mitchell, Timothy. *Rule of Experts: Egypt, Techno-Politics, Modernity*. Berkeley: University of California Press, 2002; Cullather, Nick. *The Hungry World*. Cambridge: Harvard University Press, 2010; Biggs, David. *Quagmire: Nation-Building and Nature in the Mekong Delta*. Seattle: University of Washington Press, 2010; Cueto, Marcos. *Cold War, Deadly Fevers: Malaria Eradication in Mexico, 1955–1975*. Washington, DC: Woodrow Wilson Center Press, 2007; Manela, Erez. "A Pox on Your Narrative: Writing Disease Control into Cold War History." *Diplomatic History* 34:2 (April 2010): 299–323; Manela, Erez. "Globalizing the Great Society," in *Beyond the Cold War: Lyndon Johnson and the New Global Challenges of the 1960s*. ed. Francis J. Gavin, and Mark Atwood Lawrence. New York: Oxford University Press, 2014, 165–181; Reinhardt, Bob H. "The Global Great Society and the US Commitment to Smallpox Eradication." *Endeavour* 34:4 (December 2010): 164–172; Reinhardt, Bob H. *The End of a Global Pox: America and the Eradication of Smallpox in the Cold War Era*. Chapel Hill: University of North Carolina Press, 2015.

created to fight it into the narrative, primarily because doing so helps us to think more about the role of the idea of biological interconnectedness, but not only because of that.

The history of the rinderpest campaigns adds a new dimension to the rapidly expanding historiography of twentieth-century internationalism, or, as some scholars have persuasively argued, *internationalisms*, for internationalism was never just one thing. It took many forms and had many genealogies.[13] There were, as Sunil Amrith recently wrote, many different paths to internationalism. This book helps to recover a largely forgotten one. Amrith insisted that "histories of international institutions, internationalist ambitions and international initiatives all need to be embedded in the broader political debates to which they emerged as a response."[14] This author wholeheartedly agrees, and adds that they need also to be embedded in the physical environment that helped to give them life. The history of the rinderpest campaigns enhances our understanding of both the origins and the outcomes of twentieth-century international imaginings and international actions by explicitly tying them to a virus, to vaccines, and to animals.

Bringing rinderpest into the story also enhances the accepted narrative of twentieth-century development.[15] It does so in three main ways. First, it

---

[13] Sluga, Glenda and Patricia Clavin, ed. *Internationalisms: A Twentieth-Century History.* Cambridge: Cambridge University Press, 2017 and Sluga, Glenda. "Turning International: Foundations of Modern International Thought and New Paradigms for Intellectual History." *History of European Ideas* 41:1 (2015): 103–115. See also, Sluga, Glenda. *Internationalism in the Age of Nationalism.* Philadelphia: University of Pennsylvania Press, 2013; Amrith, Sunil and Glenda Sluga. "New Histories of the United Nations." *Journal of World History* 19:3 (2008): 251–274; Kott, Sandrine. "International Organizations – A Field of Research for a Global History." *Zeithistorische Forschungen* 8 (2011): 446–450; Pedersen, Susan. "Back to the League of Nations." *The American Historical Review* 112:4 (October 2007): 1091–1117.

[14] Amrith, Sunil S. "Internationalising Health in the Twentieth Century," in *Internationalisms*, ed. Sluga and Clavin, 246.

[15] Key texts in this literature (not already cited in Footnote 12) include Gilman, Nils. *Mandarins of the Future: Modernization Theory in Cold War America.* Baltimore: Johns Hopkins Press, 2003; Latham, Michael E. *Modernization as Ideology: American Social Science and "Nation Building" in the Kennedy Era.* Chapel Hill: University of North Carolina Press, 2000; Latham, Michael E. *The Right Kind of Revolution: Modernization, Development, and U.S. Foreign Policy from the Cold War to the Present.* Ithaca: Cornell University Press, 2010; Staples, Amy L. S. *The Birth of Development: How the World Bank, Food and Agriculture Organization, and World Health Organization Changed the World, 1945–1965.* Kent: Kent State Press, 2006; Ekbladh, David. *The Great American Mission: Modernization and the Construction of an American World Order.* Princeton: Princeton University Press, 2010; Immerwahr, Daniel. *Thinking Small: The United States and the Lure of Community Development.*

moves attention away from the American-led and Soviet-led efforts that have dominated the literature to focus on the role of international institutions, who approached development differently than did the two major powers. The history of the rinderpest campaigns broadens our understanding of development as an international, rather than as a Cold War act. Doing so puts development into a different timeline – one that begins at the turn of the twentieth century and that continues today.[16] In highlighting the inter-imperial origins of the vaccines themselves, it also reaffirms the point that we need to think more seriously about the lasting impact of colonial development efforts.[17]

Cambridge: Harvard University Press, 2015; Engerman, David C., Nils Gilman, Mark H. Haefele, and Michael E. Latham, *Staging Growth: Modernization, Development, and the Global Cold War*. Amherst: University of Massachusetts Press, 2003; Adas, Michael. *Machines as the Measure of Men: Science, Technology, and Ideologies of Western Dominance*. Ithaca: Cornell University Press, 1990; Adas, Michael. *Dominance By Design: Technological Imperatives and America's Civilizing Mission*. Cambridge: Harvard University Press, 2006; Simpson, Bradley R. *Economists with Guns: Authoritarian Development and U.S.-Indonesian Relations, 1960–1968*. Stanford: Stanford University Press, 2008; Engerman, David C. *Modernization from the Other Shore: American Intellectuals and the Romance of Russian Development*. Cambridge: Harvard University Press, 2003; Nunan, Timothy. *Humanitarian Invasion: Global Development in Cold War Afghanistan*. Cambridge: Cambridge University Press, 2016; Macekura, Stephen J. *Of Limits and Growth: The Rise of Global Sustainable Development*. Cambridge: Cambridge University Press, 2015; Westad, Odd Arne. *The Global Cold War: Third World Interventions and the Making of Our Times*. Cambridge: Cambridge University Press, 2005; Helleiner, Eric. *Forgotten Foundations of Bretton Woods: International Development and the Making of the Postwar Order*. Ithaca: Cornell University Press, 2014; McVety, Amanda. *Enlightened Aid*. New York: Oxford University Press, 2012; Parmar, Inderjeet. *Foundations in the American Century: The Ford, Carnegie, and Rockefeller Foundations in the Rise of American Power*. New York: Columbia University Press, 2012; Hodge, Joseph Morgan. *Triumph of the Expert: Agrarian Doctrines of Development and the Legacies of British Colonialism*. Athens: Ohio University Press, 2007; Escobar, Arturo. *Encountering Development: The Making and Unmaking of the Third World*. Princeton: Princeton University Press, 1995; Packenham, Robert A. *Liberal America and the Third World: Political Development Ideas in Foreign Aid*. Princeton: Princeton University Press, 1973; Arndt, H. W. *Economic Development*. Chicago: University of Chicago Press, 1987.

[16] For more on this see, Wang, Jessica. "Colonial Crossings: Social Science, Social Knowledge, and American Power from the Nineteenth Century to the Cold War," in *Cold War Science and the Transatlantic Circulation of Knowledge, History of Science and Medicine Library*, v. 51, ed. Jeroen van Dongen. Leiden: Brill, 2015, 184–213.

[17] There is an expansive literature on colonial development. For some examples, see Hodge, *Triumph of the Expert*; Headrick, Daniel R. *Power over Peoples: Technology, Environments, and Western Imperialism, 1400 to the Present*. Princeton: Princeton University Press, 2010; Iyer, Samantha. "Colonial Population and the Idea of Development." *Comparative Studies in Society and History* 55:1 (2013): 65–91; Tilly, Helen. *Africa as a Living Laboratory: Empire, Development, and the Problem of*

The history of the rinderpest campaigns additionally broadens our understanding of development by challenging us to think about development as an environmental, rather than just an economic, act. The rinderpest campaigns were biological campaigns. They revolved around injecting bovines with living, mutated virus that was often transported in the bodies of living carrier animals: rabbits and goats whose infected bodies became the source of hundreds of vaccines. This was a very different kind of development from building dams and schools, and it worked (although not without many twists and turns along the way). And that is the third way that the rinderpest campaigns changed the accepted narrative of development: they took what is most often told as a story of failed dreams and unintended consequences and turned it into a story of achieved dreams and intended consequences. FAO announced rinderpest's eradication in 2011.

Scientific Knowledge, 1870–1950. Chicago: University of Chicago Press, 2011; Cooper, Frederick. "Development, Modernization, and the Social Sciences in the Era of Decolonization: The Examples of British and French Africa." *Revue d'Histoire des Sciences Humaines* 10 (2004): 9–38; Wolton, Suke. *Lord Hailey, the Colonial Office and the Politics of Race and Empire in the Second World War*. New York: St. Martin's Press, 2000; Moon, Suzanne. "Empirical Knowledge, Scientific Authority, and Native Development: The Controversy over Sugar/Rice Ecology in the Netherlands East Indies, 1905–1914." *Environment and History* 10:1 (2004): 59–81; Buesekom, Monica M. van. *Negotiating Development: African Farmers and Colonial Experts at the Office du Niger, 1920–1960*. Oxford: James Currey, 2002; Lewis, Joanna. *Empire State-Building: War & Welfare in Kenya, 1925–52*. Oxford: James Currey, 2000; Havinden, Michael and David Meredith, *Colonialism and Development: Britain and Its Tropical Colonies, 1850–1960*. London: Routledge, 1993; Cooper, Frederick. "Modernizing Bureaucrats, Backward Africans, and the Development Concept," in *International Development and the Social Sciences: Essays on the History and Politics of Knowledge*. ed. Frederick Cooper and Randall Packard. Berkeley: University of California Press, 1997, 64–92; Constantine, Stephen. *The Making of British Colonial Development Policy*. London: Frank Cass, 1984; Gonzalo, Javier. *The Idea of Third World Development: Emerging Perspectives in the United States and Britain 1900–1950*. New York: University Press of America, 1987.

I

Rinderpest and the Origins of International Cooperation for Disease Control

In May of 1921, representatives of forty-three countries met in Paris for the first International Conference on Epizootic Diseases of Domestic Animals.[1] Rinderpest brought them there. The infamous cattle plague had been of limited concern in Europe for decades, following a successful continental effort to drive it out of circulation via strict quarantines and cull and kill policies, but recent events had put the disease back in the spotlight. The previous summer, several zebu cattle on their way from India to South America died in holding in the quarantine stables in Antwerp. No one suspected rinderpest, so the stables were not disinfected before being filled with a shipment of North American cattle who were promptly sold to an abattoir in Ghent where they came into contact with a number of German cows on their way, not to butchering, but to farms throughout Belgium. Rinderpest traveled with them, eventually spreading to 222 farms, many close to the French border. The French government quickly closed its border to all animals and animal products and it did successfully keep rinderpest out, but it was an unnerving episode. It was also compounded by the news that Bolshevik troops had carried

---

[1] The official account says that "43 nations, États ou Dominions" were represented at the conference and lists delegates from the following places: Germany, the United States, Argentina, Austria, Belgium, Brazil, Bulgaria, Chile, Denmark, Ecuador, Spain, Finland, France, Great Britain, Ireland, the Union of South Africa, Australia, Canada, New Zealand, Greece, Haiti, Hungary, Italy, Japan, Morocco, Monaco, Norway, Paraguay, the Netherlands, Peru, Poland, Portugal, Romania, Yugoslavia, Sweden, Switzerland, Czechoslovakia, and Tunisia (Ministère de L'Agriculture, Républic Francaise, *Conférence Internationale pour L'Étude des Épizooties, Paris, 25–28 Mai 1921* [Paris: Imprimerie Nationale, 1921], 6–10).

rinderpest into eastern Poland with them when they invaded in July of 1920. The virus infected over 6,500 cattle, lingering longer than the troops, who were soon driven out. European governments moved quickly to help: Denmark sent veterinarians and money; France sent veterinarians and anti-rinderpest serum. Others offered aid as needed. Conference attendees in Paris the following May declared that they "can not do otherwise than note with satisfaction this spontaneous manifestation of solidarity which presents a timely demonstration of interest of a joint action of the civilized nations against the dangers that threaten the animal populations."[2]

Interest in "joint action" was a hallmark of the period. The post–World War I era saw the creation of a new international system that gave rise to an astonishing number of both governmental and nongovernmental international agencies.[3] The new League of Nations rested at the center of it. The League, as its first undersecretary-general wrote, "provides not only for the centralization and coordination of international machinery but for its orderly and systematic development."[4] The League created an "orbit" that drew nations, individuals, foundations, institutions, organizations, empires, and more into closer conversation and cooperation with each other than anything previously seen in the world. The Paris meeting was part of that new system and it resulted in the development of new international machinery for the control of animal diseases.[5]

The delegates at the 1921 Paris meeting called for "an international bureau" for "the campaign against infectious animal diseases." The campaign to which they referred was the general one of humans against the pathogens that attacked their livestock. They were trying to make it more organized and more successful. They wanted a bureau charged with collecting and publicizing information about epizootic disease and how

[2] "International Conference on Epizootic Diseases in Domestic Animals." *Journal of the American Veterinary Medical Association* 60, 13:1 (October 1921): 124–138. For the most detailed account of the meeting, see Ministère de L'Agriculture, Républic Francaise, *Conférence Internationale pour L'Étude des Épizooties, Paris, 25–28 Mai 1921*. Paris: Imprimerie Nationale, 1921.

[3] Sluga, Glenda. *Internationalism in the Age of Nationalism*. Philadelphia: University of Pennsylvania Press, 2013, 63; Iriye, Akira. *Global Community*. Berkeley: University of California Press, 2002: 9–36.

[4] Raymond Fosdick quoted in Mazower, Mark. *Governing the World: The History of an Idea*. New York: Penguin Press, 2012, 148.

[5] Clavin, Patricia and Jens-Wilhelm Wessels, "Transnationalism and the League of Nations: Understanding the Work of Its Economic and Financial Organization." *Contemporary European History* 14:4 (November 2005): 465–492.

to control them, promoting experiments and investigations "relative to the pathology or prophylaxis" of such diseases "wherever they may be occasion to resort to international cooperation," and studying and sharing "international plans and agreements" about sanitary regulations. They were particularly concerned about rinderpest, which is the disease that had brought them all there in the first place.[6]

Participants particularly urged support for "everything relating to the experimental study of rinderpest," because "prophylaxis of rinderpest is rendered difficult and uncertain by our lack of knowledge on a number of questions." That lack had not seemed so important in the past, but now the disease that had been "believed relegated to the Asiatic and African continents, has reappeared in these latter times in Europe and has reached South America." These new outbreaks had reaffirmed what European governments had discovered at the end of the nineteenth century: eliminating rinderpest required "an organized veterinary service and an administrative organization enabling rigorous application of prophylaxis measures." That was theoretically possible everywhere, but it was certainly not the reality everywhere. By the 1920s, with the notable exceptions of the Soviet Union and China, rinderpest dwelled primarily in imperial spaces and primarily traveled along imperially created trade networks. But those networks connected more than just empires – they connected the globe. And the "requirements of quarantine," which rinderpest's existence anywhere along those networks demanded, "constitute a heavy charge on commerce." Prophylaxis measures in both sanitary and biological form promised a tantalizingly cheaper option, provided the virus would cooperate with their efforts. There was reason to think that it might.[7]

***

By 1921, humans had a long, intimate history with the disease rinderpest. Their struggles with it had played a critical role in shaping both how humans understood disease in general and their expectations of what their governments should do about it. Michel Foucault famously argued that the idea that the end of government should be "to improve the condition of the population, to increase its wealth, its longevity, and its health" arose in Europe in the eighteenth century.[8] It was not a coincidence that

[6] "International Conference on Epizootic Diseases in Domestic Animals," 124–138.

[7] Ibid., 127–136.

[8] Foucault, Michel. *Security, Territory, Population: Lectures at the Collège de France, 1977–1978*, ed. Michel Senellart, trans. Graham Burchell. New York: Picador, 2007, 105.

that idea emerged at the same time as a transformation in thinking about disease and contagion. Rinderpest played a critical role in that development. It would also not be a coincidence that the human quest to control disease took a decidedly international turn during the 1920s – one that questioned the utility of the essential nationalism of the existing paradigm. There, again, rinderpest helped to drive the change, primarily by being so troublesome.

Exactly how long it had been so troublesome has been a subject of some debate for a while. Ancient records from around the globe are full of reports of horrific plagues of cattle bearing descriptions that fit with modern descriptions of rinderpest outbreaks, but it is impossible to make a definitive diagnosis centuries later. They could have been outbreaks of rinderpest or the handiwork of other pathogens that may or may not still exist today.[9] Recent molecular clock analyses of RPV (the virus that causes rinderpest) and its relationship to its closest relative, MeV (the virus that causes measles), have brought more clarity to the discussion. Researchers have learned that MeV and RPV – some strains of which are up to 70.50 percent similar[10] – likely arose in an environment where humans and cattle lived in very close proximity to one another: "probably the cattle herds of Central or Southern Asia some 10,000 years ago at the time of domestication of wild aurochs." In the beginning, the two viruses were perhaps one that could infect both species, because they seem to have diverged from each other sometime around 1000 CE, mutating into MeV and RPV.[11] The science is speculative, but intriguing, and it has not gone unnoticed by historians.

Based on extensive research into livestock and human disease in medieval Europe and the recent findings about the divergence between MeV

---

[9] For the most complete overall history of rinderpest, see Spinage, C.A. *Cattle Plague: A History.* New York: Kluwer Academic/Plenum Publishers, 2003. This 700-page wonder is, as Gordon Scott dubbed it in his introduction, "A Veritable Rinderpest Encyclopaedia."

[10] Barrett, Thomas, Ashley C. Banyard, and Adama Diallo, "Molecular Biology of the Morbilliviruses," in *Rinderpest and Peste des Petits Ruminants,* ed. Thomas Barrett, Paul-Pierre Pastoret, and William P. Taylor. Amsterdam: Elsevier, 2006, 31–56.

[11] Furuse, Yuki, Akira Suzuki, and Hitoshi Oshitani. "Origin of Measles Virus: Divergence from Rinderpest Virus between the 11th and 12th Centuries," Short Report, *Virology Journal* 7:52 (2010); Wertheim, Joel O. and Sergei L. Kosakovsky Pond. "Purifying Selection Can Obscure the Age of Viral Lineages." *Molecular Biology and Evolution* 28:12 (December 2011): 3355–3365. See also, Pomeroy, Laura W., Ottar N. Bjørnstad, and Edward C. Holmes. "The Evolutionary and Epidemiological Dynamics of the Paramyxoviridae." *Journal of Molecular Evolution* 66 (2008): 98–106; Roeder, Peter, Jeffrey Mariner, and Richard Kock. "Rinderpest: The Veterinary Perspective on Eradication." *Philosophical Transactions of the Royal Society* 368:20120139 (2013): 1.

and RPV, the historian Timothy P. Newfield has recently proposed that the plagues of 569–570 CE and 986–988 CE "testify to the outbreak of an MV-RPV ancestor that caused mass mortality in cattle and people." Centuries later, Newfield sees evidence of RPV existing on its own, wreaking havoc in the early fourteenth century via a panzootic that stretched across Asia and Europe.[12] The historian Phil Slavin agrees, arguing that the suspected RPV panzootic is "the missing link between the two other great ecological crisis of the fourteenth century, the Great Famine [1314–1322] and the Black Death [1348–1351]." It had devastating consequences. The cattle plague "seems to have created a prolonged 'protein famine' among humans, lasting for about a dozen years." If it did so, he asked, "is it possible that it also ... made them easily susceptible to the plague some thirty years later?"[13] RPV and *Yersinia pestis* may have unwittingly worked together to make the fourteenth century one of the most infamous in history.

Newfield's and Slavin's suspicion that RPV was responsible for the fourteenth-century outbreak rests primarily on the "striking similarity" between its epizootiology and those of "what are commonly thought to be outbreaks of rinderpest in the eighteenth and nineteenth centuries."[14] Scholars know much more about those outbreaks, which were diligently recorded, particularly in Europe, by scientists and by government officials who had been tasked to do something about the devastation.

The effort kept them busy; rinderpest was a challenging opponent. Between 1709 and 1800, the virus claimed an estimated 200 million of

---

[12] Newfield, Timothy. "Human-Bovine Plagues in Early Middle Ages." *Journal of Interdisciplinary History* 45:1 (Summer 2015): 1–38; Newfield, Timothy P. "A Cattle Panzootic in Early Fourteenth-Century Europe." *Agricultural History Review* 57 (2009): 155–190. See also Newfield, Timothy P. "Early Medieval Epizootics and Landscapes of Disease: The Origins and Triggers of European Livestock Pestilences, 400–1000 CE," in *Landscapes and Societies in Medieval Europe East of the Elbe*, Sunhild Kleingärtner, Sébastian Rossignol and Donat Wehner, Papers in Mediaeval Studies 23. Toronto: Pontifical Institute of Mediaeval Studies, 2013, 73–113 and Newfield, Timothy. "A Great Carolingian Panzootic: The Probable Extent, Diagnosis and Impact of an Early Ninth-Century Cattle Pestilence." *Argos* 46 (2012): 200–210.

[13] Slavin, Philip. "The Great Bovine Pestilence and Its Economic and Environmental Consequences in England and Wales, 1318–50." *The Economic History Review* 65:4 (2012): 1240, 1263. See also, DeWitte, Sharon and Philip Slavin. "Between Famine and Death: England on the Eve of the Black Death – Evidence from Paleoepidemiology and Manorial Accounts." *Journal of Interdisciplinary History* 44:1 (Summer 2013): 37–60 and Campbell, Bruce M. S. "Nature as Historical Protagonist: Environment and Society in Pre-Industrial Britain." *The Economic History Review* 63:2 (2010): 281–314.

[14] Newfield. "A Cattle Panzootic," 188.

Europe's cows.[15] Observers noted its arrival with despair: some infected cattle "bellowed, took flight and were extremely restless"; others "died as if struck by lightning." The majority "looked so pitiful, held their heads low, their languishing eyes were filled with tears, and from their nostrils and mouth came mucus and saliva." They "were often attacked by diarrhoea with foetid matter of various colours and usually died during the first week, racked with coughing."[16] Shared knowledge about its movement via an expanded print culture encouraged the development of new theories of contagion in response. A doctor in Augsburg wrote in 1713 that the disease "had proceeded by degrees from Hungary toward the Danube, attacked our territory and produced great destruction to beasts." This was not, he insisted, "caused by any foulness in the atmosphere, but the contagion of oxen brought from infected countries." This was "patent, because it first attacked those pastures adjoining the foreigners, and altogether spared those cattle to which no infected animals had approached, and which had been immediately separated from any in the same herd that were infected."[17] Such observations encouraged official action to control its spread.

Pope Clement XI, frustrated about extensive losses in his own herds, asked his personal physician, Giovanni Maria Lancisi, for a solution. Lancisi recommended that all sick animals be immediately killed and their bodies buried in quicklime. To prevent future outbreaks, he called for strict quarantines: "All roads and by-paths should be carefully guarded, so that no ox or dog be allowed to enter the country. Any animal so entering should be forthwith destroyed and buried."[18] Lancisi published his response in 1715. It included an account of the disease in Latin and instructions for controlling it in Italian, so that more people could read it. Despite disagreement with some of his cardinals, the Pope eventually acted on Lancisi's recommendations, turning them into official edicts. The disease disappeared in the Papal States within nine months, an outcome that did not go unnoticed by other European governments.

---

[15] Scott, G. R. "Rinderpest Virus," in *Virus Infections of Ruminants*. Z. Dinter and B. Morein. Amsterdam: Elsevier, 1990, 343.

[16] Lancisi quoted in Jean Blancou, *History of the Surveillance and Control of Transmissible Animal Diseases*. Paris: OIE, 2003, 162.

[17] Schröck, Lucas (he also went by Schröckius) quoted in Spinage. *Cattle Plague*, 105–108. Spinage spells his name as "Schroëckius." Contagion was by no means a universally accepted idea at the time. For more on this, see Romano, Terrie M. "The Cattle Plague of 1865 and the Reception of 'The Germ Theory' in Mid-Victorian Britain," *Journal of the History of Medicine* 52 (January 1997): 51–80.

[18] Lancisi quoted in Spinage. *Cattle Plague*, 110.

FIGURE 1.1 "God strikes the Netherlands with rinderpest."
Print by Jan Smit, published by Steven van Esveldt and Jacob Maagh, Amsterdam,
1745. *Courtesy of the Rijks Museum.*

Laws, edicts, rulings, and ordinances that restricted cattle sales across
borders, enforced quarantines, required the compulsory slaughter of
infected animals, their burial in lime (with their hides still attached), the
disinfection of their living quarters, and more became the norm throughout Europe. Those who failed to comply risked fines, physical punish-
ment, and sometimes even death.[19]

Rinderpest's enormous cost encouraged the creation of much stronger
state intervention – at the highest levels of government – in the maintenance, movement, and trade of cattle. It also made the pursuit of medical
knowledge about disease a high priority for the state. In 1762, in the

---

[19] Spinage. *Cattle Plague,* 241–262; Wilkinson, Lise. "Rinderpest and Mainstream
Infectious Disease Concepts in the Eighteenth Century." *Medical History* 28 (1984):
130–131; Brantz, Dorothee. "'Risky Business': Disease, Disaster and the Unintended
Consequences of Epizootics in Eighteenth- and Nineteenth-Century France and
Germany." *Environment and History* 17 (2011): 35–51; Koolmees, Peter A. "Epizootic
Diseases in the Netherlands, 1713–2002," in Brown, Karen and Daniel Gilfoyle *Healing
the Herds.* Athens: Ohio University Press, 2010, 25; Hünniger, Dominik. "Policing
Epizootics," in *Healing the Herds,* 76–87. For more information on treatments during
the 1745 epidemic, see Smith, Major-General Sir Frederick. *The Early History of
Veterinary Literature and Its British Development,* vol. II. London: Ballière, Tindall
and Cox, 1924, 47–62.

aftermath of the particularly terrible epidemic of 1742–1760, the French government provided funding to set up the first college of veterinary medicine in Europe. Over the following twenty years, many other governments did the same, as rinderpest returned again to ravage the continent between 1768 and 1786, eager to find ways to better defend and protect their borders.[20] They found philosophical as well as economic reasons to do so. Johann Peter Frank insisted in his 1786 work, *A System of Complete Medical Police*, that "It is one of the foremost tasks of the state to prevent persons or animals, goods, and all objects to which or whom contagions cling, from entering the country."[21] That national effort had clear international implications, because, in Europe, the disease always came in from somewhere else, specifically Russia.

Rinderpest was rampant in China and Russia and usually entered Europe via the massive cattle trade that passed through the Austrian-Hungarian Empire. Europeans were eating more beef and they wanted the cows, but not the rinderpest that regularly came with them.[22] The advent of the railroad in the middle of the nineteenth century exacerbated the problem, promising to directly link the cattle of the east with the markets of the west. The looming threat led to the first international (though still European) conference of veterinary surgeons in Hamburg in 1863. Participants primarily debated quarantine lengths. It was a technical debate waged by technical men: they were not diplomats and they were not making international law. They were simply trying to

---

[20] Youde, Jeremy. "Cattle Scourge No More: The Eradication of Rinderpest and Its Lessons for Global Health Campaigns." *Politics and the Life Sciences* 32:1 (Spring 2013): 45; Spinage, *Cattle Plague*, 133–150; van Veen, Tjaart W. Schillhorn. "One Medicine: The Dynamic Relationship between Animal and Human Medicine in History and at Present." *Agriculture and Human Values* 15 (1998): 115–120.

[21] Frank quoted in Harrison, Mark. "Disease, Diplomacy and International Commerce: The Origins of International Sanitary Regulation in the Nineteenth Century." *Journal of Global History* 1:2 (2006): 201–202.

[22] Appuhn, Karl. "Ecologies of Beef: Eighteenth-Century Epizootics and the Environmental History of Early Modern Europe." *Environmental History* 15:2 (April 2010): 268–287. For more information on rinderpest in Asia during this period, see Spinage. *Cattle Plague*, 447–449, 488–492; Barwegen, Martine. "For Better or Worse?" in *Healing the Herds*, 97; Kishi, Hiroshi. "A Historical Study on Outbreaks of Rinderpest during the Yedo Era in Japan." *The Yamaguchi Journal of Veterinary Medicine* 3 (1976): 33–40; Pastoret, Paul-Pierre, Kazuya Yamanouchi, Uwe Mueller-Doblies, Mark M. Rweyemamu, Marian Horzinek, and Thomas Barrett, "Rinderpest – An Old and Worldwide Story: History to c. 1902," in *Rinderpest and Peste des Petits Ruminants*. ed. Thomas Barrett, Paul-Pierre Pastoret and William P. Taylor. Amsterdam: Elsevier, 2006, 95–100; Yamanouchi, Kazuya. *Sizyo Saidai no Densenbyo, Gyueki*. Tokyo: Iwanami Shoten, 2009.

decide on the best measures to recommend to their national governments.[23]

Europe's veterinary surgeons were far from alone in their effort to find agreement across national lines. The massive expansion of trade and travel in the nineteenth century benefited a number of diseases, including ones that attacked humans. Concerns about the spread of plague and cholera inspired the first International Sanitary Conference in Paris in 1851. Each of the twelve attending nations (all European except for the Ottoman Empire) sent a physician and a diplomat.[24] The 1851 sanitary conference and the ones that followed it, like the 1863 veterinary conference, framed their concerns as the defense of Europe against dangerous Asian diseases. The threat was deemed real enough to inspire some cooperation across national lines, but of a decidedly limited nature. Participant nations saw each other as commercial competitors and they resisted signing treaties that they perceived harmful to their trade strategies, but they did work together more often to try to halt the spread of disease in Europe, and those efforts bore fruit.[25]

In 1871, the Austrian government hosted an international conference in Vienna on rinderpest control. Delegates came from Belgium, Great Britain, Germany, France, Hungary, Italy, Romania, Russia, Switzerland, Serbia, and Turkey. Their final report, "Principles for an International Regulation for the Extinction of the Cattle Plague," was widely circulated following the conference. Extinction, they argued, demanded organized action: communication "by telegraph direct" of every outbreak "as quickly as possible" to "neighboring countries and also to those countries which have expressed a wish for it"; an "arrangement of veterinary matters" in each country to support a "speedy extinction of the cattle plague" should it break out; government compensation

[23] Harrison, Mark. *Contagion: How Commerce Has Spread Disease.* New Haven: Yale University Press, 2013, 215.

[24] Bynum, W. F. "Policing the Heart of Darkness: Aspects of the International Sanitary Conferences." *History and Philosophy of the Life Sciences* 15:3 (1993): 425–427; Harrison, Mark. "Disease, Diplomacy and International Commerce: The Origins of International Sanitary Regulation in the Nineteenth Century." *Journal of Global History* 1:2 (2006):197–217.

[25] Huber, Valeska. "The Unification of the Globe by Disease? The International Sanitary Conferences on Cholera, 1851–1894." *The Historical Journal* 49:2 (June 2006): 453–476. See also, Nagata, Naomi. "International Control of Epidemic Diseases from a Historical and Cultural Perspective," in *Networking the International System.* M. Herren. Switzerland: Springer International Publishing, 2014, 73–88; Howard-Jones, Norman. "Origins of International Health Work." *British Medical Journal* 1 (May 6, 1950): 1032–1046.

to livestock owners whose animals are killed in the act of suppressing an outbreak; and the immediate disinfection of "all objects whatsoever" that were involved with the transportation or care of suspected infected animals. These first four principles were followed by lengthy fifth and sixth ones that prescribed very specific sanitary requirements necessary for ending an outbreak. The participants, the English delegate explained, had to balance the desire "to protect and facilitate commerce as much as possible" with the necessity of repressing rinderpest, "a scourge so powerful and accompanied by results so terrible." It was not an easy task, but it was one that ultimately worked.[26]

The last rinderpest outbreak in Britain, in 1877, was in a shipment of cattle imported from Hamburg. Two hours after it landed, British veterinary authorities were informed by German authorities that rinderpest had just been identified in the stables in Hamburg where the cattle had stayed. Veterinarians quickly examined and isolated the new stock. Meanwhile, Belgium, France, Denmark, and Holland placed an immediate ban on the importation of cattle, hides, and beef from Germany.[27] The increased veterinary surveillance, communication, and quick action in case of an outbreak demonstrated in 1877 would force rinderpest out of circulation in Europe over the course of the next two decades. At the exact same time, however, European imperial expansion dramatically expanded rinderpest's global presence. While driving rinderpest out of their own fields and stables, they spread it to other places.

***

The Philippine archipelago experienced its first serious rinderpest outbreak in late 1886, likely in cattle shipped in to feed Spanish authorities.

---

[26] "Principles for an International Regulation for the Extinction of the Cattle Plague" and "Translation of the Appendix to the Debates of the International Conference at Vienna, Exposé by the English Delegate," in *The Report of the Veterinary Department for 1872*. London: Prince George Edward Eyre and William Spottiswoode, 1873, Appendix, 26–29 in *Parliamentary Papers*, The House of Commons and Command, 26; Spinage. *Cattle Plague*, 299–300.

[27] Spinage. *Cattle Plague*, 203–209. The emphasis on meat is important. Advances in processing and shipping had spurred the growth of the trade in frozen and canned meat, which were easier to control than the trade in live animals, but there were still strong fears of rinderpest spreading through infected meat and hides (Koolmees, "Epizootic Diseases in the Netherlands," 33). For more on these reforms in Britain, see Perren, Richard. "Filth and Profit, Disease and Health: Public and Private Impediments to Slaughterhouse Reform in Victorian Britain," in *Meat, Modernity, and the Rise of the Slaughter House*, Paula Young Lee. Durham, NH: University of New Hampshire Press, 2008, 127–150.

The disease spread rapidly. A witness reported that "at the highest point of these problems ... the cadavers infested the air and the rivers."[28] It was, the historian Ken De Bevoise wrote, "arguably the single greatest catastrophe in the nineteenth-century Philippines." The mortality rate was afterward commonly cited to have been 90 percent. Rivers became clogged with dead animal bodies to the point that they were no longer navigable. Carabao (domestic water buffalo) were the hardest hit, being particularly vulnerable to the disease. Deprived of their draft animals, farmers were forced to reduce the size of their fields. They ate less and repaid fewer of their debts. Meanwhile, deprived of their preferred meal of cattle, mosquitoes bit more humans, spreading malaria as they went.[29]

Tragically, the Philippine case proved not to be an exception, but the rule. The same outbreak that terrorized the archipelago would also terrorize most of Africa. In the final decades of the nineteenth century, rinderpest added pain and suffering to people who were already shouldering a disproportionate burden of it, victims of disasters that were, like the rinderpest panzootic, both "natural" and man-made.[30] They were not, in that sense, just the victims of rinderpest or of famine but victims of imperialism itself. The connections were not unnoticed at the time: "In some respects," Frederick Lugard cynically noted, after observing rinderpest's passage through East Africa, the disease "has favored our enterprise. Powerful and warlike as the pastoral tribes are, their pride has been humbled and our progress facilitated by this awful visitation. The advent of the white man had else not been so peaceful."[31] Rinderpest, like imperialism, was, of course, anything but peaceful. It was devastating.

---

[28] Doeppers, Daniel F. "Fighting Rinderpest in the Philippines, 1886–1941," in *Healing the Herds*, 110–111, 113. Ken De Bevoise argued that it arrived in 1887, perhaps in the bodies of carabao imported from Indochina "for breeding purposes" (De Bevoise, Ken. *Agents of Apocalypse*. Princeton: Princeton University Press, 1995, 159).

[29] De Bevoise, *Agents of Apocalypse*, 158–160.

[30] Mike Davies has dubbed these combinations of events the "Late Victorian Holocausts." These were, he argues, the result of "natural" disasters (most notably droughts caused by "an exceptional intensification of El Niño activity") combined with imperial machinations that massively increased human vulnerability. "The total human toll of these waves of drought, famine and disease could not have been less than 30 million victims," he argues. "Fifty million might not be unrealistic." They died "at the precise moment (1870–1914) when [their] labor and products were being dynamically conscripted into a London-centered world economy." That process of conscription played a central role in the death toll (Davis, Mike. *Late Victorian Holocausts*. London: Verso, 2002, 6–11, 14).

[31] Lugard quoted in Kjekshus, Helge. *Ecology Control and Economic Development in East African History*. London: James Currey, 1977, 1996 impression, 131.

Rinderpest appears to have first arrived in Africa via a shipment of Russian cattle to Alexandria in 1841. Over subsequent years, the disease claimed 90 percent of Egypt's cattle, along with large numbers of buffalo, sheep, and goats. It remained enzootic in Egypt, but did not cross its borders to infect the rest of the continent, in part, it seems, because Ethiopians imposed quarantines against northern cattle when they knew of sickness.[32] These sanitation efforts could work against the endemic, milder forms of rinderpest that existed in Egypt, but were powerless in the face of a strain of RPV that Italian forces brought from India to Massawa, in present-day Eritrea, in 1887.[33]

The Ethiopian historian Richard Pankhurst interviewed survivors decades later who were still shaken by the horror. Alaqa Lamma Haylu ("an old man in his nineties whose remarkable memory helped to establish the chronology of the outbreak") recalled that it began in the rains of 1888. He fell ill for three days and when he got up he "found that all of the cattle were dead." His father, who had owned some 300 cattle, was left with one heifer. A French traveler reported in early 1889 that "a terrible epizootic murrain has carried off all the cattle." An Italian observer estimated that rinderpest killed 90 percent of Ethiopia's cattle and later reports corroborated the claim. Rinderpest infected buffalo, sheep, goats, and numerous wild ungulates – any animal in which it could gain footing. The Emperor Menelik alone lost 250,000 head. He worried, "if the oxen disappear there will be no more grain, and if there is no more grain there will be no men." He ordered everyone to pray, but to no avail. Famine came. "The chief cause of the famine," Alaqa Lamma Haylu remembered, "was the death of cattle." Plagues of locusts and caterpillars also descended on the country, gobbling up what could be harvested. Little was left to go around. A missionary reported in 1890: "Everywhere I meet walking skeletons and even horrible corpses, half eaten by hyenas, of starvelings who had collapsed from exhaustion." Millions died. An Italian diplomat wrote in December of 1890 of the land between Harar and Addis Ababa: "Previously, the country was inhabited; there were very beautiful fields of durra and barley, numerous herds of cattle, sheep and goats, and whole area had an atmosphere of abundance and prosperity. At present, it is one

---

[32] Spinage. *Cattle Plague*, 497–498; Pankhurst, Richard. "The Great Ethiopian Famine of 1888–1892: A New Assessment." *The Journal of the History of Medicine and Applied Science* (April 1966): 99–100.

[33] There may have been a separate outbreak in Egypt at the same time, which also played a role in the origins of the great pandemic of the late 1880s and 1890s (Spinage. *Cattle Plague*, 521).

of continuous desolation." Decades later, these were still remembered as the *Kefu Qan* or Evil Days.[34] And they were not confined to Ethiopia.

Rinderpest moved south and west, consuming East Africa. A European stationed in Tanganyika wrote in 1892, "all trade in cattle has come to a standstill . . . and will most likely not reach pre-plague dimensions within the next 10–15 years. There are now hardly 100 head of cattle left in all of Unyanyembe from the former herds of 30–40,000." The Masai were particularly devastated. "Abandoned villages were, almost without exception, the only trace I found of the Masai people," a German traveler observed the same year. "All of their cattle have been wiped out."[35] And it was not just cattle that were dying. A British officer visiting Lake Mweru (on the border between present-day Zambia and the Democratic Republic of the Congo) in the fall of 1892 reported, "Here enormous quantities of game have died." The banks were covered with the bodies of buffalo as well as pookoo and lechwe, varieties of antelope. "Dead and dying beasts were all around."[36] Wild animals helped RPV move along different paths than just the human-controlled market routes, furthering its expansion. Frederick Lugard, who followed the disease through Uganda in the early 1890s, reported in 1893, "The enormous extent of the devastation it has caused in Africa can hardly be exaggerated. Most of these tribes possessed vast herds of thousands on thousands of cattle, and of these, in some localities, hardly one is left; in others, the deaths have been limited to perhaps 90 per cent." Others agreed, tending to estimate a death toll of 95 percent.[37]

The damage was catastrophic. Cattle provided African families with milk, meat, and labor in the fields. Their hides provided clothing, mats, and shoes. Their dung provided the main source of domestic cooking fuel. It could not be replaced, because colonial officials made wood-gathering illegal, hoarding it for their building programs. But cattle provided more than just subsistence to African pastoralists, they "stored the entire wealth of the family and the society in moveable property which increased over

---

[34] Pankhurst. "The Great Ethiopian Famine of 1888–1892," 100–115; Pankhurst, Richard. *Economic History of Ethiopia, 1800–1935*. Addis Ababa: Haile Sellassie I University Press, 196, 216–220; Marcus, Harold G. *The Life and Times of Menelik II* . Oxford: Clarendon Press, 1975, 135–137; Spinage. *Cattle Plague*, 500–506. The American visitor Robert P. Skinner wrote of the great pandemic in his 1906 account of a trip to Ethiopia, "It was estimated that not more than 7 or 8 percent of the animals of the country were saved" (Skinner, Robert P. *Abyssinia of Today*. New York: Longmans, Green & Co., 1906, 195–196).

[35] Kjekshus. *Ecology*, 128–129.   [36] Spinage. *Cattle Plague*, 518.

[37] Kjekshus. *Ecology*, 130.

the years."[38] Cattle were capital. Rinderpest destroyed "a complex and inter-woven social, political and economic system" when it destroyed herds.[39]

The Zambezi River confined the virus for several years, probably because of an absence of a lively cattle trade across the river and because the area was thick with tsetse flies (which also helps explain the minimal trade). Rinderpest finally made it across in early 1896 and spread swiftly southward from there, following the main trade routes, which quickly became clogged with dead bodies. "Hundreds of carcasses lay here and there, on the roadside or piled up in the fields," a French missionary wrote in 1897. "More than nine hundred wagons loaded with merchandise, without teams or drivers, stood abandoned along the Bulawayo road. . . . Never within the memory of men had such a thing been seen." Another traveler despaired that "the whole country is but a charnel-house of dead rinderpest oxen." As they had in the north, mortality rates climbed over 90 percent. Panicked farmers pressured government officials to stop the spread, but that was, of course, far easier asked for than accomplished.[40]

Local officials turned first to quarantines and then added slaughter, costly moves that angered many and heightened tensions between white and black farmers, who blamed each other for the disease.[41] The Cape Colony government, for example, declared a ban on the entry of cattle from any place north of the Molopo River (posting police along the river to enforce it) and announced that all infected cattle inside the Protectorate would be immediately shot. Not surprisingly, "stamping out" policies started with African, not European-owned herds. Fences were built, cattle

---

[38] Phoofolo, Pule. "Face to Face with Famine: The BaSotho and the Rinderpest, 1897–1899." *Journal of Southern African Studies* 29:2 (June 2003): 503; Phoofolo, Pule. "Epidemics and Revolutions: The Rinderpest Epidemic in Late Nineteenth-Century Southern Africa," *Past & Present* 138 (February 1993): 116.

[39] van Onselen, C. "Reactions to Rinderpest in Southern Africa 1896–97." *The Journal of African History* 13:3 (1972): 484. For more on this, see Homewood, Katherine. *Ecology of African Pastoralist Societies*. Oxford: James Currey, 2008.

[40] Phoofolo, "Face to Face," 504, 511; Spinage, *Cattle Plague*, 525.

[41] Racism dictated the way whites in southern Africa understood the progress of the disease. One of the Cape Colony veterinarians wrote in 1902 of the epizootic: "There is abundant evidence to show that the natives were the main cause of transporting the infection" (Hutcheon, Duncan. "Rinderpest in South Africa." *Journal of Comparative Pathology and Therapeutics* 15:4 [December 31, 1902]: 307). For more on this, see Phoofolo, "Epidemics" and Ballard, Charles. "The Repercussions of Rinderpest: Cattle Plague and Peasant Decline in Colonial Natal." *The International Journal of African Historical Studies* 19:3 (1986): 421–450.

were shot, carcasses were burned, but rinderpest still came, prompting governments to alter their strategy. Prophylaxis policies that had worked in Europe were not working in Africa, but there was another option: biological prophylaxis.[42]

By the late 1890s, the "germ theory" revolution pioneered by Louis Pasteur and Robert Koch was well underway.[43] Eager for new answers, the Cape Colony government asked Koch, then the head of the Institute for Infectious Diseases in Berlin, to come to Africa to help them find a prophylactic inoculation for rinderpest. No one knew what kind of microbe caused the disease, but Koch was undaunted. After all, he wrote, "we know the microbe of neither small-pox nor of rabies, and yet we have succeeded in devising prophylactic inoculations against both diseases dependent upon the fact that infective material can be weakened and converted into a so-called vaccine in the one case by passage through the animal body, in the other by drying."

Koch turned to these techniques first, passaging rinderpest in antelopes, donkeys, mules, dogs, eagles, pigeons, chickens, rabbits, mice, guinea pigs, dogs, and a secretary bird to no avail. He had some luck with goats and pigs, but the process was slow and expensive as he needed fresh animals for every passage. He tried drying infected blood, reliquefying it with water, and then injecting it into animals to prevent the disease. Drying the blood did seem to kill the pathogen, but injections of it did not protect cattle from subsequent infection. Passaging and drying were not, however, the only possibilities. Other germs, he noted, "can be

---

[42] For more on the tension between veterinary "stamping out" methods and the use of vaccines, see Sunseri, Thaddeus. "The Entangled History of Sadoka (Rinderpest) and Veterinary Science in Tanzania and the Wider World, 1891–1901." *Bulletin of the History of Medicine* 89 (2015): 92–121. See also, Gilfoyle, Daniel. *The Many Plagues of Beasts.* Saarbrücken, Germany: VDM Verlag Dr. Müller, 2009, 112–138; Mack, Roy. "The Great African Cattle Plague Epidemic of the 1890's." *Tropical Animal Health and Production* 2:4 (December 1970): 210–219; Phoofolo, "Epidemics," 141; Spinage, *Cattle Plague,* 525–570. For a detailed history of the effort to battle rinderpest in Namibia, see Miescher, Giorgio. *Namibia's Red Line: The History of a Veterinary and Settlement Border.* New York: Palgrave Macmillan, 2012.

[43] See Paul de Kruif's 1926 classic, *The Microbe Hunters.* For criticisms of that narrative, see Latour, Bruno. *The Pasteurization of France.* trans. Alan Sheridan and John Law, Cambridge: Harvard University Press, 1988; Cunningham, Andrew and Perry Williams, *The Laboratory Revolution in Medicine.* Cambridge: Cambridge University Press, 1992; Worboys, Michael. "Was there a Bacteriological Revolution in Late Nineteenth-Century Medicine?" *Studies in History and Philosophy of Biological and Biomedical Sciences* 38:1 (2007): 20–42; Cohen, Ed. "The Paradoxical Politics of Viral Containment; or, How Scale Undoes Us One and All." *Social Text* 106 29:1 (Spring 2011): 15–35.

weakened by treatment with chemicals" or via "the use of serum obtained
from animals which have acquired immunity by having the disease." He
tried those options as well.[44] In the end, after two months of testing, and
the valuable assistance of other scientists, Koch ended up recommending
that the Cape Colony inject healthy cattle in noninfected areas with gall
(more commonly called bile) from an infected ox.[45]

Koch was forced to cut his efforts short, ordered by the German
government to join an expedition to study bubonic plague in India (he
would also look at rinderpest there, bringing his bile technique with him),
but work continued in southern Africa, in part because of the limited
success of his findings.[46] Experiments in the field with bile inoculation in
the spring of 1897 yielded disappointing results: rinderpest regularly
broke out following inoculation and about 80 percent of the cattle slaugh-
tered for bile failed to produce any. Local scientists ended up recommending the "serum-simultaneous method" instead, which involved injecting
infected cattle blood on one side of the animal and serum (blood plasma
with the clotting factors removed) on the other side.[47] The blood–serum

[44] Koch, Robert. "Prof. Robert Koch's Berichte über seine in Kimberley ausgeführten
Experimentalstudien zur Bekämpfung der Rinderpes." Special reprint of *Deutschen
Medicinischen Wochenschrift*, numbers 15 and 16 (1897): 7–13; Koch, Robert.
"Special Report to the 'British Medical Journal' by Professor R. Koch on His Research
into the Cause of Cattle Plague." *The British Medical Journal* 1:1898 (May 15, 1897):
1245–1246.

[45] Koch, Robert. "Prof. Robert Koch's Berichte über seine in Kimberley ausgeführten
Experimentalstudien zur Bekämpfung der Rinderpest." Special reprint of *Deutschen
Medicinischen Wochenschrift* numbers 15 and 16 (1897): 1-15; Miescher, *Namibia's
Red Line*, 26–30; Spinage, *Cattle Plague*, 426–427. "Koch's bile method," Spinage
noted, "was really simply an improvement of a method already developed by some
farmers in the Transvaal known as the Waterberg or Grobler method" (429). The bile
method distracted him from his work on combining serum with blood: "This mixture
acted much more satisfactorily, but just at that time Koch had discovered the efficacy of
bile obtained from sick animals, which he considered both safer and more effective than
serum" (Hutcheon, "Rinderpest in South Africa," 315). For more on this research, see
Gilfoyle, *The Many Plagues of Beasts*. Saarbrüken, Germany: VDM Verlag Dr. Müller,
2009, 119–129.

[46] Lingard, Alfred. "Preliminary Notes on Rinderpest: Experiments Commenced at Imperial
Bacteriological Laboratory by Professor Robert R. Koch, 4th to 15th June 1897,"
(August 5, 1897); sent to the author by Kazuya Yamanouchi. See also, Pastoret et al.,
"Rinderpest – an Old and Worldwide Story: History to c. 1902," 95–97; Spinage, *Cattle
Plague*, 437.

[47] Gilfoyle, Daniel. "Veterinary Research and the African Rinderpest Epizootic: The Cape
Colony, 1896–1898." *Journal of Southern African Studies* 29:1 (March 2003): 143–144;
Mack, "The Great African Cattle Plague," 215–216; [Anonymous.] "An Immunising
Serum against Rinderpest," *The British Medical Journal* 2:1925 (20 November
1897):1517; Spinage, Cattle Plague, 433.

mixture worked in the sense that it appeared to provide immunity, but it was a far-from-ideal method. The mixture was laborious to make, had a short shelf-life ("as brief as nine days in warm weather"), created a live infection that could provoke a rinderpest outbreak in a virgin population, and, most importantly, allowed for the transference of other kinds of microbial infections. It lacked purity. Rinderpest was far from the only creature that liked to make its home in cattle blood, and inoculators sometimes unwittingly spread other pathogens.[48]

Neither the blood and bile method nor the serum-simultaneous method were ideal, but they did help. Daniel Gilfoyle's research has demonstrated that "mortality rates among cattle in individual districts and areas roughly correlated with the degree of veterinary activity and inoculation."[49] Unfortunately, that activity correlated very closely with the skin color of the humans who owned the cattle. White farmers tended both to have greater access to inoculation and to be more willing to subject their uninfected cattle to it, as they did not have the same reasons to distrust government-sponsored veterinarians as their black counterparts did. "In consequence black-owned stock in Natal fell by 77 percent in 1897, compared to a decrease for white-owned stock of 48 percent."[50] Such discrepancies were also no doubt due in part to African farmers' tendency to communally graze livestock and European farmers' tendencies to fence their herds, but the evidence, Gilfoyle concluded, "suggests that inoculation, however contested in its development and practice, modified the course of the rinderpest epizootic at the Cape."[51] Prophylaxis inoculation was certainly not perfect, as the 48 percent mortality rate testified, but it did help in the fight against the virus.

It also helped encourage enhanced inter-imperial coordination against rinderpest. At the first Pan-African Veterinary Conference in 1903, officials recommended the liberal use of the serum-simultaneous method, with a back-up reliance on Koch's bile method if the serum was not readily

[48] Taylor, William P., Peter L. Roeder, and Mark M. Rweyemamu, "History of Vaccines and Vaccination," in *Rinderpest and Peste des Petits Ruminants*, ed. Thomas Barrett, Paul-Pierre Pastoret, and William P. Taylor. Amsterdam: Elsevier, 2006, 223–225; Hutcheon, "*Rinderpest in South Africa*," 322–323.

[49] Gilfoyle, "Veterinary Research," 151. "Veterinary services were, however, largely allocated to European pastoralists, while the African-occupied areas of the colony were assigned a low priority" (Gilfoyle, 153).

[50] Campbell, Gwyn. "Disease, Cattle, and Slaves: The Development of Trade between Natal and Madagascar, 1875–1904." *African Economic History* 19 (1990–1991): 113.

[51] Gilfoyle, "Veterinary Research," 151.

available.[52] These efforts – combined with strict legal prophylaxis measures – eliminated rinderpest from South Africa by 1905, but the virus remained endemic in East Africa, transforming much of what had long been prime pastoral grazing land into "tsetse-infested bush and woodland inhabited only by wild animals." The veterinary framework was far weaker in East Africa than it was in South Africa: too weak to get rid of rinderpest.[53] It was a common problem in the imperial periphery.

By 1900, it was understood that successful animal disease control depended on a strong, active national veterinary service able to rigorously apply both sanitary and biological prophylactic measures combined with regular communication about outbreaks with trading partners. The United States created the Bureau of Animal Industry within the Department of Agriculture in 1884 to manage its livestock disease control. The bureau immediately built "the first significant microbiological laboratory in the United States." By 1905, it also had the power to impose quarantines on cattle moving between states and in and out of the country.[54] This kind of control, coupled with the fact that the United States (along with Canada) was a net exporter rather than importer of cattle, helped to ensure that rinderpest never broke out in North America. It was, however, a serious problem in its Philippine colony. A similar situation existed throughout the globe: the metropole no longer suffered from regular outbreaks, but its colonial peoples did. Governments readily turned to science at home with expanding expectations about what the state *should* do and expanding confidence in what it *could* do, given recent technological advances in a wide number of fields. Increasingly, however, key figures in imperial administration began calling for similar action throughout their empires, convinced that they could be turned into more economically profitable holdings. Such efforts focused specifically on the expansion of export agriculture, which made rinderpest a key target for attack in imperial holdings around the globe.[55]

---

[52] Mack, "The Great African Cattle Plague," 216.

[53] Reader, John. *Africa: The Biography of a Continent*. New York: Vintage, 1997, 592; Brown, Karen. "Tropical Medicine and Animal Diseases: Onderstepoort and the Development of Veterinary Science in South Africa 1908–1950." *Journal of South African Studies* 31:3 (September 2005): 513–529.

[54] Olmstead, Alan L. "The First Line of Defense: Inventing the Infrastructure to Combat Animal Diseases." *The Journal of Economic History* 69:2 (June 2009): 334. See also Olmstead, Alan L. and Paul W. Rhode. *Arresting Contagion: Science, Policy, and Conflicts over Animal Disease Control*. Cambridge, MA: Harvard University Press, 2015.

[55] Hodge, Joseph. *Triumph of the Expert: Agrarian Doctrines of Development and the Legacies of British Colonialism*. Athens: Ohio University Press, 2007, 56.

When Japan annexed Korea in 1910, it began immediately working on the construction of "an immune belt" along the border of China to keep rinderpest out. Officials used the serum-simultaneous method to do it, constructing a massive serum manufacturing institute in Busan. The institute opened in 1911 under the direction of Chiharu Kakizaki, who was already working on trying to find a better method of biological prophylaxis.[56] Researchers in Turkey had discovered in 1902 that the pathogen responsible for rinderpest was a virus, which was an important step forward in knowledge about the microbe, but what governments really wanted was vaccines.[57] Kakizaki turned to chemicals to make one, inactivating a mixture of infected cattle blood and spleen with glycerin. Korean cattle injected twice with the mixture were immune (for a while) to subsequent injections of infected blood.[58] Kakizaki had created the first rinderpest vaccine.

Japanese officials saw the potential right away. At the time, its Manchuria colony was the third largest producer of cattle in the world, behind only the United States and Australia, but it ranked "first in the number of cases of livestock diseases." Rinderpest was one of the most troublesome. Although focused on the industrial development of the region, Japanese authorities had also been trying to modernize farming in Manchuria for years, importing new species of plants and breeds of livestock, creating experimental farms, and building agricultural research stations, laboratories, and scientific institutions. In the aftermath of Kakizaki's discovery, they added another. In 1925, the South

[56] Knight, R. F. and C. G. Thomson, "Brief Report on the Veterinary Institutes of Japan," *Philippine Agricultural Review*.4 (March 1911): 111–118; Hunter, James A. "The Rinderpest Epidemic of 1949–50 Taiwan (Formosa)." *Chinese-American Joint Commission on Rural Reconstruction, Animal Industries Series*, No. 1. Taipei: February 1951; sent to the author by Kazuya Yamanouchi; Yamanouchi, Kazuya. "Scientific Background to the Global Eradication of Rinderpest." *Veterinary Immunology and Immunopathology* 148 (July 15, 2012): 2; Kishi, "A Historical Study on Outbreaks of Rinderpest During the Yedo Era in Japan," 33–34; Kakizaki, Chiharu, Shunzo Nakanishi, and Takashi Oizumi, "Experimental Studies on Prophylactic Inoculation Against Rinderpest, Report III." *Journal of the Japanese Society of Veterinary Science* 5:4 (1926): 221–280; Kakizaki, Chiharu, Shunzo Nakanishi, and Junji Nakamura, "Experimental Studies on the Economical Rinderpest Vaccine." *Journal of the Japanese Society of Veterinary Science* 6:2 (1927): 107–120.
[57] Nicolle, M. and Adil-Bey, "Etudes sur la peste bovine. Troisième mémoire. Expériences sur la filtration du virus." *Annales de l'Institut Pasteur* 16 (1902): 56–64; Özkul, Türel and R. Tamay Başagac Gül, "The Collaboration of Maurice Nicolle et Adil Mustafa: The Discovery of Rinderpest agent." *Revue de Médecine Vétérinaire* 159 (2008): 243–246.
[58] Kakizaki, Chiharu. "Study on the Glycerinated Rinderpest Vaccine." *Kitasato Archives of Experimental Medicine* 2 (1918): 59–66.

Manchuria Railway Company built the Cattle Disease Research Institute in Mukden. The research facility soon began producing hundreds of thousands of units of inactive vaccine for widespread use around the region. Subsequent mandatory vaccinations under the direction of a mobile veterinary corps reduced the rate of infection.[59]

Japan profited first from Kakizaki's vaccine, but he published his research in English in 1918, which meant that other nations could profit as well. That act spoke to the growing scientific internationalism of the day – a spirit of cooperation that helped to shape the international society of the 1920s and 1930s. New possibilities of biological prophylaxis, however, would likely not have resulted in greater cooperation in disease control if the diseases themselves had not continued to prove a problem. Influenza was by far the deadliest, but it was two simultaneous outbreaks of different diseases in Poland – one that affected humans and one that affected cattle – that helped inspire the creation of new international bureaucracy for disease control.[60]

The effort to better regulate international quarantine agreements about human diseases had resulted in 1907 in the creation of the Office International d'Hygiène Public in Paris. The Office "provided a modest but consistent forum for discussion of contagious disease."[61] In 1920, however, "typhus and relapsing fever epidemics originating in Russia spread into Poland, and threatened western countries." The Council of the League of Nations ordered a medical commission immediately to Poland to investigate. Its members promptly reported back that "the eradication of typhus in the areas described appears to us ... to be a question of international importance" and quarantines would not be enough. "Measures taken at the frontiers of other countries can diminish the danger," they argued, "but they cannot obviate it altogether." Poland needed immediate sanitary action, but it did not have the bureaucracy to do it. The League responded with assistance and with the creation of a Health Organization in July 1921.[62]

---

[59] Perrins, Robert John. "Holding Water in Bamboo Buckets," in *Healing the Herds: Disease, Livestock Economies, and the Globalization of Veterinary Medicine.* ed. Karen Brown and Daniel Gilfoyle. Athens: Ohio University Press, 2010, 195–208.

[60] On influenza, see Barry, John M. *The Great Influenza: The Epic Story of the Deadliest Plague in History.* New York: Penguin Books, 2004.

[61] Sealey, Anne. "Globalizing the 1926 International Sanitary Convention." *Journal of Global History* 6 (2011): 433.

[62] Fitzgerald, J.G. "An International Health Organization and the League of Nations." *The Canadian Medical Association Journal* 14:6 (June 1924): 532; Pottevin, Dr., TH. Madsen, and R. Norman White, "Typhus and Cholera in Poland: The Action of the

This sense of the outdatedness of relying on quarantines in the age of sanitary and biological prophylaxis also led to the Paris conference on animal diseases that May. Participants had even been disturbed by the threat of disease coming from the same place: Russia via Poland. The "rinderpest invasion of Russia in Europe and its spread to regions invaded by the Bolshevist armies could be foreseen as an inevitable consequence of the disorganization of administrative services and particularly the sanitary police," participants acknowledged. "On the other hand, the introduction of the contagion by the commercial route had seemed unlikely." And, yet, rinderpest had also broken out in Antwerp. The combination of the two outbreaks "demonstrate[d] the necessity of a more complete study of epizootic diseases."[63] Many of the nations participating in the 1921 meeting had been at the Vienna meeting of 1871, for this was still a European-centered internationalism, but there were critical differences in the tenor of the conversation. In 1871, discussions about animal disease control centered on quarantines, sanitation, and cull-and-kill policies. In 1921, "the campaign against infectious animal diseases" focused on prophylaxis methods and encouraged more study of the diseases themselves. Participants also stressed the necessity of creating better control all over the world as opposed to just in Europe. Internationalism was becoming more international *and* more inter-imperial.[64]

In 1924, twenty-eight nations signed the agreement creating the Office International des Epizooties or OIE: the Argentine Republic, Belgium, Brazil, Bulgaria, Denmark, Egypt, Spain, Finland, France, Great Britain, Greece, Guatemala, Hungary, Italy, Luxemburg, Morocco, Mexico, the Principality of Monaco, the Netherlands, Peru, Poland, Portugal, Rumania, Siam, Sweden, Switzerland, the Czechoslovakia Republic, and Tunisia.[65] Some were imperial powers; most were not. That fit with what was happening throughout the League of Nation's growing orbit. The League fostered the creation of European-dominated international

League of Nations," *The Lancet* (December 4, 1920): 1159–1160; Boroway, Iris. *Coming to Terms with World Health: The League of Nations Health Organization 1921–1946.* Frankfurt am Main: Peter Lang, 2009, 41–76.

[63] "International Conference on Epizootic Diseases in Domestic Animals." *Journal of the American Veterinary Medical Association* 60, 13:1 (October 1921): 127–128.

[64] Knab, Cornelia. "Infectious Rats and Dangerous Cows: Transnational Perspectives on Animal Disease in the First Half of the Twentieth Century." *Contemporary European History* 20:3 (August 2011): 292–295.

[65] International Agreement for the Creation of an Office International Des Epizooties in Paris (January 25, 1924); available at http://www.oie.int/en/about-us/key-texts/basic-texts/inter national-agreement-for-the-creation-of-an-office-international-des-epizooties/.

machinery, but it also opened the door to more truly global cooperation. As Susan Pedersen has persuasively demonstrated, the League was both "a key agent in the transition from a world of formal empires to a world of formally sovereign states" and a "harbinger of global governance."[66] OIE membership would grow throughout the 1920s and 1930s to include up to forty-four member states, a list that included colonies and mandates, and, notably throughout the 1930s, the Soviet Union, Japan, and Germany. The organization brought veterinary experts together regularly for meetings and published a monthly *Bulletin* of meeting notes and information about outbreaks, sanitary measures, and disease research. The OIE also helped facilitate the transfer of limited amounts of financial assistance to laboratories in countries that needed support. The organization could not make its member nations follow up on its recommendations, but it did put pressure on them to do so by making their efforts or lack thereof international public knowledge. In this effort, it would be assisted somewhat by the creation of a series of veterinary subcommittees under the umbrella of the League's Economic and Financial Organization that operated between 1928 and 1939.[67]

Above all, the OIE and the League's New Health Organization (LNHO) together helped to persuade more countries to invest more time and money on disease control at the national, imperial, regional, and international levels. They received some essential help in that effort from scientists, because new discoveries in biological prophylaxis options throughout the 1920s and 1930s encouraged government investment in the quest for disease control through technical advances as opposed to just through continued quarantines and regulations. This was particularly true for the two diseases that had most directly inspired the creation of the LNHO and OIE in the first place: typhus and rinderpest.

The League sent aid to Poland to help it keep typhus from spreading. It also sent Harvard bacteriologist Hans Zinsser to Russia to study the epidemic at the source. Both efforts bore fruit: typhus did not keep moving

---

[66] Pedersen, Susan. "Empires, States, and the League of Nations," in *Internationalisms: A Twentieth Century History*, ed. Glenda Sluga and Patricia Clavin. Cambridge: Cambridge University Press, 2017, 116 and "Back to the League of Nations." *The American Historical Review* 112:4 (October 2007): 1092. See also, Pedersen, Susan. *The Guardians: The League of Nations and the Crisis of Empire*. New York: Oxford University Press, 2015; Gorman, Daniel. *The Emergence of International Society in the 1920s*. Cambridge: Cambridge University Press, 2012; Mazower, Mark. *No Enchanted Palace: The End of Empire and the Ideological Origins of the United Nations*. Princeton: Oxford University Press, 2009.

[67] Knab, "Infectious Rats and Dangerous Cows," 295–305.

west and Zinsser helped to discover a great deal more information about typhus and to develop a vaccine against one form of it. His ability to do so, as he readily acknowledged at the time, was because he was in constant contact with other scientists around the world who were working on the same problems. He traveled to Mexico to study an outbreak with colleagues. He carried a strain of Rickettsiae (the bacterium that causes typhus) across the Atlantic in guinea pigs that he smuggled aboard a ship so that he could get that strain to the Pasteur Institute. In his fascinating 1935 book, *Rats, Lice and History*, he celebrated the scientific internationalism that had played such a decisive role in his own career. He wrote,

> while the world is an armed camp of suspicion and hatred, and nations are doing their best, by hook and crook, to push each other out of the world markets, to foment revolutions and steal each other's political and military secrets, – organized government agencies are exchanging information concerning epidemic diseases; sanitarians, bacteriologists, epidemiologists, and health administrators are cooperating, consulting each other, and freely interchanging views, materials and methods, from Russia to South America, from Scandinavia to the tropics.[68]

It was not quite that clear-cut, of course, but the international cooperation that the League encouraged throughout the 1920s and 1930s did lead to real results in the study of diseases, both human and animal. It proved true with typhus and it proved true with rinderpest.[69]

By 1921, rinderpest remained a devastating problem throughout much of Asia, the Middle East, and Africa. Officials fighting the disease relied upon the serum-simultaneous method as their main form of biological prophylaxis: French authorities sent some to Poland in 1920 and the Soviet Union would soon begin a campaign using the serum-simultaneous method that would, with the assistance of strict sanitary prophylaxis

[68] Zinsser, Hans. *Rats, Lice and History*. 1935. New Brunswick, NJ: Transaction Publishers, 2008, 293.

[69] Ibid. See also, Zinsser, Hans. *As I Remember Him: The Biography of R.S.* Boston: Little, Brown and Company, 1940; Olitsky, Peter K. "Hans Zinsser and His Studies of Typhus Fever." *The Journal of the American Medical Association* 115:10 (8 March 1941): 907–912; Lindenmann, Jean. "Typhus Vaccine Developments from the First to the Second World War (On Paul Weindling's 'Between Bacteriology and Virology ...')," *History and Philosophy of the Life Sciences* 24:3/4 (2002): 467–485; Weindling, Paul. "Between Bacteriology and Virology: The Development of Typhus Vaccines between the First and Second World Wars." *History and Philosophy of the Life Sciences* 17:1 (1995): 81–90. For more on the typhus outbreak, see Patterson, K. David. "Typhus and Its Control in Russia, 1870–1940." *Medical History* 37 (1993): 361–381 and Irwin, Julia F. "The Great White Train: Typhus, Sanitation, and U.S. International Development during the Russian Civil War." *Endeavor* 36:3 (2012): 89–96; Allen, Arthur. *The Fantastic Laboratory of Dr. Weigl*. New York: Norton, 2014.

efforts, eliminate rinderpest from Russia by 1928.[70] Kakizaki's successful creation of a vaccine, however, encouraged bacteriologists around the world to continue the effort to find better methods of control. Kakizaki's vaccine was far from perfect; it was the starting point, not the end.

In the Philippines, Raymond Alexander Kelser, a member of the Army Medical Department Research Board, created a viable inactive wet tissue vaccine by subjecting a suspension of infected cattle spleen, liver, and lymph nodes to 0.75 percent chloroform. The final product could fool the immune system into producing antibodies without infection and without risking starting an outbreak where there was not one already. Kelser's vaccine required three separate injections over several weeks and only conferred short-term immunity, but it was far better than the serum-simultaneous method. The Philippine Bureau of Agriculture utilized it in a massive campaign that vaccinated between 200,000 and 300,000 cattle and carabao each year between 1924 and 1931.[71]

Kakizaki's and Kelser's vaccines, and others like them created in the 1920s, worked by using chemicals to destroy the virus's pathogenicity, but not its immunogenicity, which meant that they would still provoke an immune response in the host.[72] The virus was dead in their inactivated vaccine, but its shell (so to speak) remained, still containing the antigens that stimulated the immune system. But, and this was an important "but," inactive vaccines stimulate the immune system at a far lower degree than an actual infection would, which means that they do not provide lasting immunity. They work for a while – long enough to potentially

---

[70] Taylor et al., "History of Vaccines and Vaccination," in *Rinderpest and Peste des Petits Ruminants*, ed. Thomas Barrett et al., 225.

[71] Kelser, R. A., S. Youngberg, and T. Topacio, "An Improved Vaccine for Immunization Against Rinderpest." *Journal of the American Veterinary Medicine Association* 74 (1929): 28–41; Shope, Richard E. *Raymond Alexander Kelser: A Biographical Memoir.* Washington, DC: The National Academy of Sciences, 1954, 205–206. Kelser was not the first American to create a rinderpest vaccine in the Philippines. William H. Boyton did so before him, but Kelser's vaccine proved better and more useful. For information about Boyton's vaccine, see Boynton, William H. "Rinderpest, with Special Reference to Its Control by a New Method of Prophylactic Treatment." *The Philippine Journal of Science* 36 (May 1928): 1–35.

[72] At the same time as Kelser was doing his work with chloroform, other scientists were discovering that rinderpest could also be inactivated by exposure to toluol, eucalyptol, and formalin (Shope, "Inactivated Virus Vaccine," in *Rinderpest Vaccines: Their Production and Use in the Field*, 1st ed. ed. K.L. Kesteven. Washington, DC: FAO, 1949, 23).

drive the virus out of circulation in a closed community, like an island – but long-term immunity requires repeated injections, which are both costly and time-consuming.

There was, however, another possibility (one that Koch had also tried in Cape Colony): attenuation by the passage of the virus through foreign animal bodies. Attenuated vaccines are mutated forms of the original virus. They were created at that time by injecting an animal not usually infected by the virus with infected blood and seeing if the virus would "take" to the new host; this effort requires multiple passages through different individual animals. If the virus takes, pathogenicity (danger to the original host, in this case cattle and buffalo) is sometimes weakened. The resulting mutated virus is still very much alive and provokes an immune response in the original host, but it ideally either produces a milder sickness or no sickness at all in the process. Attenuation via passaging in living animals is tricky and does not often result in effective vaccines. It ultimately worked in a few select cases with rinderpest because RPV mutates quickly and because it exists as a single immunologic type or serotype.[73] What this means is that antigens do not change as the virus changes, probably because the part of the virus that "the immune system recognizes as an antigen plays an integral role in the function of the virus itself. If it changes shape, the virus cannot survive."[74] Mutations that include mutations of the antigens die off, but mutations that do not touch the antigens can possibly survive. Because the antigens remain unchanged, new mutations will still trigger the same immune response.

---

[73] RNA viruses are beset by extraordinarily high rates of mutation, the vast majority of which are deleterious. At a basic level, the problem is that RNA replication is far more error-prone than DNA replication, because DNA polymerase proofread for errors and fix them during replication. RNA polymerase has no such ability. It is too small to have it, and it is too small to have it because RNA is too error-prone to support longer genomes (a catch-22 known as Eigen's paradox). In general, RNA viruses have a mutation rate of about one in every thousand base pairs each generation. Most of these are not advantageous for the virus, but some are, and these give the virus strength, helping ensure its survival even in changing circumstances. Mutations can help a virus move into different kinds of cells, even different kinds of species. They can also help a virus evade established immunity in a host, which explains why there are new influenza vaccines every year (Holmes, Edward C. *The Evolution and Emergence of RNA Viruses*. Oxford: Oxford University Press, 2009, 37–46). For more on the genetic makeup of morbilliviruses, see Barrett, Banyard, and Diallo, "Molecular Biology of the Morbilliviruses," 31–56. For a basic introduction to virology, see Crawford, Dorothy H. *Viruses: A Very Short Introduction*. Oxford: Oxford University Press, 2011.

[74] Barry, *The Great Influenza*, 99–100. See also, Pomeroy et al., "The Evolutionary and Epidemiological Dynamics of the Paramyxoviridae," 98–106.

The great benefit (for humans and for cattle) of this particular fact about morbilliviruses goes beyond the ability to create attenuated vaccines for them; immunity to one strain of the virus protects against all strains of the virus, because the antigens are not different. This meant that research- ers could share more than just their published research findings with one another. They could also share actual strains of living vaccine, sending them thousands of miles from where they had been created. This mattered a great deal, because creating successful attenuated vaccines is a difficult and time-consuming business. Most attempts to create a vaccine via passaging end in failure, even when the virus is antigen-stable, so being able to share strands that worked was of significant benefit to the effort to combat the virus. Not every vaccine would work as well as another in every case; the pathogenicity of strains of rinderpest, especially attenuated ones, varied widely, provoking different pathogenic responses in different breeds of cattle and buffalo. However, provided an animal survived vaccination with one of the attenuated vaccines (sometimes the virus was still too strong for some breeds), they would be immune to any strain of rinderpest found anywhere on earth for far longer than was possible with an inactive vaccine.

RPV's antigen stability had more than just scientific consequences; it also had political ones. The biology of the virus encouraged international cooperation in the effort to drive it out of circulation along the globe's economic networks. The fact that newly discovered vaccines *could* be shared helped to ensure that they *were* shared. The disease rinderpest had long encouraged international cooperation in terms of sharing infor- mation about outbreaks and enforcing quarantines. During the 1920s and 1930s, RPV's mutability – its "extraordinary biological plasticity," to steal a phrase from Hans Zinsser – encouraged a whole new level of sharing, one that involved not just publishing and communication (although those things remained critically important) but also the actual movement of people and viral strains between laboratories.[75] As the typhus example shows, this type of scientific cooperation was far from limited to rinderpest. What was special about RPV, however, was that it specifically encouraged *inter-imperial* cooperation for largely economic reasons. When Kelser described the importance of his vaccine, it was that "in the Philippine Islands agriculture constitutes the very backbone of the country's resources. Thus, any factor which impedes or in any way

---

[75] Zinsser used the phrase to describe "filterable viruses" in general (Zinsser, *Rats, Lice and History*, 64).

interferes with agricultural development, to an equivalent extent hinders the economic development of the Islands generally."[76] Sharing information about rinderpest was sometimes framed as a sanitary issue, but it was primarily framed as an economic one, because its victims were valuable livestock.

In the early 1920s, at the Imperial Bacteriological Laboratory in Mukteshwar (then called Muktesar), which the British had created in 1893 to produce anti-rinderpest serum, James Thomas Edwards was trying to solve the problem of the "almost inevitable" simultaneous infection of other pathogens that lurked in the injected blood when using the serum-simultaneous method. Such pathogens created complicating infections that made vaccination even less desirable to cattle owners than it already was. This was a particularly troublesome problem in India, where Hindu owners regularly expressed "antipathy" to the "use of ox blood." They did not want to kill some cattle to protect others. Trying to find a way around both problems, Edwards seized on the idea of an inactivated vaccine "as soon as Kakizaki's researches became known," but failed to find one that worked on local hill bulls.[77]

Frustrated, but not daunted, Edwards went in the other direction and started trying to create an attenuated vaccine by passaging rinderpest through rabbits. He had some success, maintaining direct passage for fourteen months and discovering that the virus "seemed to have suffered a certain degree of degradation of virulence toward cattle" along the way. Unfortunately, before he could get the virus to mutate into a significantly

[76] Kelser, R. A., S. Youngberg, and T. Topacio, "An Improved Vaccine for Immunization Against Rinderpest." *Journal of the American Veterinary Medicine Association* 74 (1929): 29. For more on inter-imperial sharing of medical knowledge in general, see Neill, Deborah J. *Networks in Tropical Medicine: Internationalism, Colonialism, and the Rise of a Medical Specialty, 1890–1930.* Stanford: Stanford University Press, 2012.

[77] Edwards, J. T. "Rinderpest: Some Properties of the Virus and Further Indications for Its Employment in the Serum-Simultaneous Method of Protective Inoculation." *Transactions of the Congress – Far Eastern Association of Tropical Medicine* 3 (1927): 705; Edwards, J. T. "Rinderpest: Active Immunization by Means of the Serum Simultaneous Method: Goat Virus." *Agricultural Journal of India* 23 (1928): 185; Edwards, J. T. *The Problem of Rinderpest in India.* Bulletin No. 199, Imperial Institute of Agricultural Research. Pusa: Calcutta Government of India, 1930: 12; Edwards, J. T. "The Uses and Limitations of the Caprinized Virus in the Control of Rinderpest (Cattle Plague) Among British and Near-Eastern Cattle." *The British Veterinary Journal* 105:7 (July 1949): 211. For more on the creation of the Mukteshwar laboratory, see Mishra, Saurabh. "Beasts, Murrains, and the British Raj: Reassessing Colonial Medicine in India from the Veterinary Perspective, 1860–1900." *Bulletin of the History of Medicine* 85 (2011): 587–619.

lower pathogenicity for cattle, he lost all of his subjects in an epizootic of pasteurellosis. The dead rabbits took the rinderpest strain that had been amenable to passaging in the rabbits with them and his subsequent efforts to repeat the adaptation in new rabbits failed. Edwards was not the first researcher, nor would he be the last, to have difficulty getting RPV to passage consistently in rabbits. The virus rarely mutated into survival in that particular host, but there were other options. He and his team next tried sheep, but quickly moved on to goats and found success. Edwards promptly began inoculating cattle with goat-passaged blood and serum and soon discovered that passage through the goats had attenuated the virus so much that they did not even need the serum to go with it.[78] Additional research by others, who built on what Edwards readily shared in publications and in public lectures, convinced colonial authorities to give up serum altogether and just use the passaged goat blood.[79] Delighted

---

[78] Edwards, "Rinderpest: Some Properties of the Virus and Further Indications for Its Employment in the Serum-Simultaneous Method of Protective Inoculation," 702–705; Edwards, "Rinderpest: Some Properties of the Virus and Further Indications for its Employment in the Serum-Simultaneous Method of Protective Inoculation," 699–706; Brotherston, J. G. "Rinderpest: Some Notes on Control by Modified Virus Vaccines, II," *Veterinary Reviews and Annotations* 3 (1957): 45–56; Scott, Gordon R. "Rinderpest," in *Advances in Veterinary Science*, v. 9, ed. C. A. Bradley and E. L. Jungherr. New York: Academic Press, 1964, 133–134. Edwards actually discovered the extent of the attenuation through an accident. Injections of the goat-passaged strains were always accompanied by injections of serum, just as injections of ox blood were in the field. During one test, however, after the cattle had been injected with the goat-passaged virus, the serum vial was broken, and the subjects did not get the second injection. The injected cattle suffered fevers, mouth lesions, and diarrhea, but they did not develop full-blown rinderpest. The accident did not convince the team to drop serum injections altogether, but did persuade them to use a much smaller dose of serum in the simultaneous injection, a move which further reduced the cost of inoculation (told to the author by Kazuya Youmanouchi, who was told it by Gordon Scott).

[79] In 1930, Major R. F. Stirling, the director of veterinary services in the Central Provinces, began researching the possibility of inoculating cattle with Edwards's goat-passaged virus without serum to great success. The continued use of serum was frustrating, since it cost "over two shillings a head." Stirling had 12 million cattle in his province, so that even inexpensive serum was still costly. He reported later that he got the idea of inoculating without serum when he came across an address Edwards had made at the Conference of the Central Provinces Veterinary Association in November of 1928, which was subsequently published in 1930. Stirling promptly published the results of his own experiments in the field so that others could benefit from them (Stirling, R. F. "Some Experiments in Rinderpest Vaccination: Active Immunisation of Indian Plains Cattle by Inoculation with Goat-Adapted Virus Alone in Field Conditions." *The Veterinary Journal* 88 (1932): 192–204). In 1932, in Madras, P.T. Saunders and Rao Sahib K. Kylasam Ayyar created their own goat-virus strain, later known as Madras No. 1 (Saunders, P. T. and Rao Sahib K. Kylasam Ayyar, "An Experimental Study of Rinderpest Virus in Goats in a Series of 150

with his findings, officials soon began shipping the goat-passaged virus (in the bodies of living goats) to other imperial laboratories in the hope that it would prove as useful against rinderpest in other places.

The Mukteshwar strain arrived at the new Central Veterinary Research Institute in Kabete, Kenya, in 1936. There, R. Daubney, the Director of Veterinary Services in Kenya Colony, was trying to find a vaccine for what he insisted was "economically... undoubtedly the most important disease in tropical Africa and Asia." Daubney, like Edwards before him, had initially tried to create an inactive vaccine along the lines of Kakizaki's and Kelser's, but he was intrigued by the goat-passaging results in India and was eager to try Edwards's strain in the field. The results were disappointing. The strain was too strong for Kenya's indigenous zebus and the colonists' imported European breeds.[80]

Daubney could not use the viral strain, but he could use the research that had created it. He and fellow researcher J. R. Hudson decided to create their own goat-passaged virus. Their final product – Kabete O – came from a highly lethal strain of rinderpest isolated in 1911 in Kenya that researchers had kept alive ever since via passage in cattle in the laboratory.[81] Over several years of extensive testing, they showed that Kabete O proved the best option for caprinization (goat-passaging), even more so than its Indian competitors. Kabete Attenuated Goat vaccine (or KAG, as it came to be known) still caused a fever in almost all vaccinated animals, but the vast majority of domestic African cattle survived. For European breeds, the survival rate was closer to 50 percent, which meant that veterinarians still needed to use an inactive vaccine on them.[82]

Campaigns against rinderpest now employed multiple kinds of vaccines. During an expansive outbreak in East Africa, which was complicated by the presence of immense numbers of wild animals that "were infected throughout wide areas of cattle country," veterinary officials used four treatments: two kinds of chemically inactivated vaccines, the serum-simultaneous method, and KAG. The latter traveled around East Africa in the body of its living goat hosts. One goat could yield "500 to 800 doses of virus infected

Direct Passages." *The Indian Journal of Veterinary Science and Animal Husbandry* 6 (1936):4).

[80] Daubney, R. "Rinderpest: A Résumé of Recent Progress in Africa." *The Journal of Comparative Pathology & Therapeutics* 50 (1937): 405–409.

[81] Ibid., 408; MacOwen, K.D.S. "Virulent Rinderpest Virus (R.B.K.)." *Annual Report Veterinary Service Department Kenya* (1955), 29; Daubney, R. "Goat-Adapted Virus," in *Rinderpest Vaccines*, ed. Kesteven, 8.

[82] Daubney, "Goat-Adapted Virus," in *Rinderpest Vaccines*, 8–9.

blood"; it was an incredibly efficient means of transportation.[83] Defeating rinderpest in this late-1930s epizootic required the vaccination of both indigenous African-owned cattle and imported settler-owned cattle, an act that East African authorities had previously avoided, preferring instead to limit vaccination to settler-owned cattle alone because "the vets believed that rinderpest was useful in limiting the number of cattle owned by Africans and thus limiting what they regarded as overstocking and overgrazing – and hence soil erosion." In 1934, the governor of the Tanganyika Territory used a similar argument when asking the Colonial Office "for permission to abandon rinderpest eradication." London refused, but it did not repair recent cutbacks to government veterinary staff in the region.[84] The change in vaccination policy at the end of the 1930s was not only a reaction to the outbreak; it was also a reflection of a transformation in British imperial policy that increasingly chose action over inaction.[85]

In 1940, London passed the Colonial Development and Welfare Act, which committed the British government to improving not just infrastructure and agriculture in the colonies, as previous acts had done, but to actually improve the living standards of its colonial subjects. The transformation came from the ground up. "Officials and experts in London," Joseph Hodge has argued, "were confronted in the late 1930s by an empire that appeared to be in the grips of a series of dramatic social, economic, and ecological crisis" that were challenging "the very legitimacy of the colonial project."[86] Colonial poverty and exploitation had

---

[83] Hornby, H.E. *Report on the Special Anti-Rinderpest Campaign in Tanganyika Territory, 1938* (November 28, 1938); CO 691/173/13; NAUK; Branagan, D. and J. A. Hammond, "Rinderpest in Tanganyika: A Review," *Bulletin of Epizootic Diseases of Africa* 13:3 (September 1965): 225–245; Spinage, *Cattle Plague*, 594.

[84] Gilfoyle, Daniel. "South Africans Abroad: The Origins of the International Control of Rinderpest in East Africa, 1917–1951," presented at the BIHG Conference (5–7 September 2013), unpublished paper. David Anderson also writes about this colonial attitude in Anderson, David. "Kenya's Cattle Trade and the Economics of Empire, 1918–48," in *Healing the Herds*, ed. Brown and Gilfoyle, 250–268. See also, Waller, Richard "'Clean' and 'Dirty': Cattle Disease and Control Policy in Colonial Kenya, 1900–40." *The Journal of African History* 45:1 (2004): 45–80.

[85] There was a 62 percent rate of increase in animal ("nearly all" cattle) protective inoculations in Britain's African colonies between 1931 and 1937. Inoculations in India almost tripled. During those years, colonial officials recorded giving a total of 17,191,967 rinderpest inoculations to cattle and buffalo in India (over 10 million there alone), Nigeria, Kenya, Uganda, Sudan, Gold Coast, and Tanganyika (Rogers, Leonard. "Prophylactic Inoculations against Animal Diseases in the British Empire." *The British Medical Journal* 1:4080 (March 18, 1939): 565).

[86] Hodge, Joseph Morgan. *The Triumph of the Expert: Agrarian Doctrines of Development and the Legacies of British Colonialism.* Athens: Ohio University Press, 2007, 179.

become too politically embarrassing for the government to ignore, precisely because the interwar international community had decided to start drawing attention to them and had started to act to change them. That community, centered in the League of Nations, had by the late 1930s created an international society "characterized by the close intersection of international welfare issues, the establishment of systematic information exchange and the pursuit of economic interest" that challenged imperialism like nothing before.[87]

That society had not started out that way. The League became "more holistic," embracing "social health and welfare," only in the aftermath of the Great Depression and the breakdown of its efforts to secure stronger inter-governmental cooperation. Determined to stay relevant, the League shifted focus. The transition was most evident in the creation in 1935 of the "Mixed Committee on the Problem of Nutrition," which brought together people from the League of Nations Health Organization (LNHO), the International Labor Organization (ILO), the International Institute for Agriculture (IIA), and the Economic and Financial Organization (EFO) to analyze the global food situation.[88] The resulting report on the committee's work published statistics on food production and consumption and recommendations of minimum caloric requirements. By doing so, the report highlighted global inequality in access to food and "in the process created a new language for discussing poverty and consumption on a global scale." By insisting that a laboring adult anywhere in the world needed 2,500 calories a day, the report established a benchmark that could be used to pressure government authorities.[89] That work complemented the LNHO's *International*

France would take similar action in 1946. See Cooper, Frederick. "Reconstructing Empire in British and French Africa." *Past and Present*, Supplement 6 (2011): 196–210 and Cooper, Frederick. "Development, Modernization, and the Social Sciences in the Era of Decolonization: The Examples of British and French Africa." *Revue d'Histoire des Sciences Humaines* 10 (2004): 9–38.

[87] Knab, Cornelia and Amalia Ribi Forclaz, "Transnational Co-Operation in Food, Agriculture, Environment and Health in Historical Perspective: Introduction." *Contemporary European History* 20:3 (August 2011): 249. See also, Pedersen, "Empires, States and the League of Nations," in *Internationalisms: A Twentieth-Century History*. ed. Glenda Sluga, and Patricia Clavin. Cambridge: Cambridge University Press, 2017, 113–138 and Gorman, *The Emergence of International Society in the 1920s*. Cambridge: Cambridge University Press, 2012.

[88] Clavin, Patricia. *Securing the World Economy: The Reinvention of the League of Nations, 1920–1946*. Oxford: Oxford University Press, 2013, 8–9, 161.

[89] Amrith, Sunil and Glenda Sluga. "New Histories of the United Nations." *Journal of World History* 19:3 (2008), 251-274; Boroway, *Coming to Terms with World Health*,

Health Yearbook, which provided "a survey of the progress made by the various countries in the domain of public health."[90] The League also supported ILO and EFO efforts to collect and publish comparative data on standards of living – a concept "that became a lens through which to see the world as an object to be improved by liberal capitalism, western science, and international organization."[91] League efforts were aided by the efforts of individual economists to compile and publish national income estimates.[92] A 1939 League report explained that the work was "making men and women all over the world more keenly aware of the wide gap between the actual and potential conditions of their lives," which, in turn, was making them "impatient to hear that some real and concerted effort is being made to raise the standard of their lives nearer to what it might become."[93]

Knowledge of global inequality changed what people demanded from their governments, be they national or imperial. It also changed ideas about the purpose of international cooperation. By the end of the 1930s, the League was insisting that the "really vital problems" facing the world did "not lend themselves to settlement by formal conferences and treaties" and, therefore, "the primary object of international co-operation should be rather mutual help than reciprocal contract – above all, the exchange of knowledge and the fruits of experience."[94] There was, of course, more than a little bit of self-interest in the shift. The League had clearly failed in its original purpose of preventing another global war and it was looking for a way to remain relevant. But the decision to emphasize the exchange of knowledge and technical skill was more than just a defensive maneuver; it was the outcome of a decade-long shift in priorities at Geneva toward "positive security" that sought peace through

---

379–394; Cullather, Nick. *The Hungry World: America's Cold War Battle against Poverty in Asia.* Cambridge, MA: Harvard University Press, 2010, 38, 32–33.

[90] A.G.N., "The International Health Year-Book of the League of Nations, 1928," *Canadian Medical Association Journal* 22:1 (January 1930): 93; Boroway, *Coming to Terms with World Health,* 177–183.

[91] Clavin, *Securing the World Economy,* 172–179.

[92] This will be discussed in greater detail in Chapter 3. See, Schmelzer, Matthias. *The Hegemony of Growth: The OECD and the Making of the Economic Growth Paradigm.* Cambridge: Cambridge University Press, 2016; Mitchell, Timothy. *Rule of Experts.* Berkeley: University of California Press, 2002; Arndt, H. W. *Economic Development.* Chicago: University of Chicago Press, 1987.

[93] Quoted in Amrith and Clavin, "Feeding the World," 35.

[94] League of Nations, *The Development of International Co-Operation in Economic and Social Affairs,* quoted in Clavin, *Securing the World Economy,* 231.

higher living standards, and higher living standards through technical and scientific cooperation for the expansion of human welfare.[95]

The League had been engaged to a limited degree in exactly that kind of exchange from the beginning via the work of some of its agencies and affiliates, but now it argued that such efforts could become the *purpose* of international cooperation. It was driven to do so by the need to stay relevant; it was able to do so by the beneficial results that had come out of those exchanges.[96] Notably, those exchanges had not been limited to League efforts. As Helen Tilly and others have shown, "national, imperial, and international scientific infrastructures were constituted simultaneously" during the interwar period.[97] The OIE, which was not a League institution, aided in the construction of all of them. So, too, did rinderpest, as the effort to control the virus engaged the attention of

[95] Clavin, *Securing the World Economy*, 231–240. See also, Packard, Randall M. *A History of Global Health: Interventions in the Lives of Other Peoples*. Baltimore: The Johns Hopkins University Press, 2016, 51–90.

[96] On League technical assistance, see Madsen, Thorvald. "The Scientific Work of the Health Organization of the League of Nations." *Bulletin of the New York Academy of Medicine* 13:8 (August 1937): 439–465; Akami, Tomoko. "A Quest to Be Global: The League of Nations Health Organization and Inter-Colonial Regional Governing Agendas of the Far Eastern Association of Tropical Medicine 1910–25." *The International History Review* 38:1 (2016): 1–23; Amrith and Clavin, "Feeding the World"; Amrith, Sunil. *Decolonizing International Health: India and Southeast Asia, 1930–1965*. London: Palgrave MacMillan, 2006; Pemberton, JoAnne. "New Worlds for Old: The League of Nations in the Age of Electricity," *Review of International Studies* 28 (2002) 311–336; Hell, Stefan. "The Role of European Technology, Expertise and Early Development Aid in the Modernization of Thailand before the Second World War." *Journal of the Asia Pacific Economy* 6:2 (2001): 158–178; Gorman, Daniel. *The Emergence of International Society in the 1920s*. Cambridge: Cambridge University Press, 2012, 52–108; Zanasi, Margherita. "Exporting Development: The League of Nations and Republican China." *Comparative Studies in Society and History* 49:1 (2007): 149; Osterhammel, Jürgen. "'Technical Co-Operation' between the League of Nations and China." *Modern Asian Studies* 13:4 (1979): 661–680; Farley, John. *To Cast Out Disease: A History of the International Health Division of the Rockefeller Foundation (1913–1951)*. Oxford: Oxford University Press, 2002; Weindling, Paul. "Philanthropy and World Health: The Rockefeller Foundation and the League of Nations Health Organization." *Minerva* 35 (1997): 269–281.

[97] Tilly, Helen. *Africa as a Living Laboratory*. Chicago: The University of Chicago Press, 2011), 7. See also Neill, Deborah. *Networks in Tropical Medicine: Internationalism, Colonialism, and the Rise of a Medical Specialty, 1890–1930*. Stanford: Stanford University Press, 2012; Bennett, Brett M. and Joseph M. Hodge, eds. *Science and Empire: Knowledge and Networks of Science across the British Empire, 1800–1970*. New York: Palgrave MacMillan, 2011) and Akami, Tomoko. "Beyond Empires' Science: Inter-Imperial Pacific Science Networks in the 1920s," in *Networking the International System: Global Histories of International Organizations*, ed. Madeleine Herren. Switzerland: Springer International Publishing, 2014, 107–132.

nations, empires, international bureaucracies, and the scientists that they employed. The successes that had come from technical cooperation in the interwar period encouraged calls for a continued reliance on such cooperation in the future. The international machinery that operated throughout the 1920s and 1930s had not stopped the world from descending into war, but it had convinced more people than ever before that international collaboration was necessary to fix the problems that many argued lay at the heart of the war: economic insecurity and hunger.

The widespread adoption of this recognition would ultimately transform imperialism, nationalism, internationalism, and international society in general. It would also transform the human struggle against rinderpest, turning it from a primarily imperial economic concern into a global economic and welfare concern. The new vocabulary of "standards of living" would play a central role in that shift, but so, too, would the war itself, which made food not only an international humanitarian concern but an essential tool for victory.

2

GIR-1: Rinderpest in World War II

In 1942, Junji Nakamura, a scientist at Japan's Busan Institute in Korea, published a report in *The Japanese Journal of Veterinary Science* titled "Rinderpest: On the Virulence of the Attenuated Rabbit Virus for Cattle." The report detailed how he and his team had inadvertently created a new rinderpest vaccine while trying to research RPV's pathogenic mechanism – the way in which it causes an infective illness. Nakamura decided to use rabbits as his test subjects instead of cattle, because they were cheaper, more plentiful, and far easier to house and work with than cattle. He knew of both successful and failed attempts to establish the virus in rabbits from articles that had been published in scientific journals and from his own visit to Mukteshwar in 1935, where he saw the passaging of rinderpest through goats and undoubtedly also heard about Edwards's initial, thwarted attempt to passage the virus in rabbits.[1] To his delight, the viral strain Nakamura had in Busan took to his rabbits. In fact, it took so well that it began mutating to flourish better in its new home, becoming more skilled at taking over rabbit cells, which, correspondingly, made it less adept at taking over bovine ones. One strain in particular, strain III

---

[1] Hatsukuma Tokishige, a scientist at Tokyo Imperial University passaged rinderpest through rabbits three times, but then, to his dismay, found that the infection (which could only be confirmed by then injecting healthy cattle with blood from the injected rabbits) disappeared within a week (Tokishige, H. The Fourth Report of Epizootics in Japan by the Agriculture Bureau. [1910]: 207–210); available in Japanese at http://kindai .ndl.go.jp/info:ndljp/pid/841866/1. Narrative of Nakamura's visit to Mukteshwar during a 1934–1935 trip to laboratories around the world related to the author by Kazuya Yamanouchi based on Nakamura's memoir, his official CV from the Nippon Institute for Biological Sciences, and Yamanouchi's conversations with Nakamura's former collaborators there.

(also known as the L strain), proved, after over 100 passages, "very highly infective for rabbits" and "very poorly potent for cattle." Nakamura had created a new vaccine.[2]

Nakamura's success had depended in no small part on the interwar period's spirit of scientific internationalism. He had access to other scientists' rinderpest research through articles and firsthand visits. He even published his findings in a journal that often concluded Japanese-language articles with summaries in English or German. However, scientific internationalism looked very different in 1942 than it had back in 1935 when Zinsser had celebrated it and Nakamura had visited Mukteshwar, along with laboratories in Paris, London, and the United States.

The Japanese government quickly seized on Nakamura's new vaccine, not as an imperial aid but as a military one. In 1942, they renamed the Rinderpest Serum Manufacture Institute in Busan the Animal Health Research Institute and opened a second branch in Anyang, appointing Nakamura director.[3] Massive production of his vaccine commenced immediately to help secure food supply lines critical to their war effort. Defense against rinderpest was not the only thing on their minds when they thought about the virus, however, and here they would turn to Nakamura as well. He was Japan's leading expert on the subject, a status that would demand of him significant involvement in his nation's imperial eradication efforts

---

[2] Nakamura, Junji, Shosaburo Wagatuma, and Kanemato Fukusho. "On the Experimental Infection with Rinderpest Virus in the Rabbits." *The Japanese Journal of Veterinary Science* 17 (1938): 185–204; Nakamura, Junji and Sadao Kuroda. "Rinderpest: On the Virulence of the Attenuated Rabbit Virus for Cattle." *The Japanese Journal of Veterinary Science* 4:2 (1942): 75–102 and Isogai, S. "On the Rabbit Virus Inoculation as an Active Immunization Method Against Rinderpest for Mongolian Cattle." *The Japanese Journal of Veterinary Science* 6:5 (1944) 371–390. Additional information from Nakamura's memoir: *Ichi-Zyueki Kenkyusha no Syuki (Memoirs of a Veterinary Scientist)* (Iwanami-shoten, 1975), told to author by Kazuya Yamonouchi. See also, Jacotot, H. "Sur la sensibilité du lapin au virus de la peste bovine." *Bulletin de la Société de Pathologie Exotique* 23 (November 12, 1930): 904–909 Through additional research, Nakamura and his team discovered that rinderpest could be attenuated further in rabbits than was possible in goats, producing a milder vaccine. They began field trials in Mongolia in 1941 with strain III that had now been passaged through rabbits over 370 times. The test went very well, with the cattle showing no marked reaction to the new vaccine (Yamanouchi, "Scientific Background to the Global Eradication of Rinderpest," 2. Additional information from author correspondence with Yamanouchi).

[3] Animal and Plant Quarantine Agency, Korea, "Our History," available at http://www.qia.go.kr/english/html/About_QIA/01About_QIA_004-05.jsp. See also (in Japanese) http://www.digital.archives.go.jp/DAS/meta/Detail_F0000000000000040687. Additional information from author correspondence with Yamanouchi.

FIGURE 2.1 Junji Nakamura and Keikichi Yamawaki, who established what is now Japan's National Institute of Animal Health, in front of the statue of Jean-Baptiste Jupille at the Pasteur Institute in Paris. The photograph was taken in 1935 while Nakamura was working in William Elford's laboratory in London. Elford had recently discovered how to use gradocol membranes to filter viruses and Nakamura learned the technique under him.
Photograph courtesy of Kazuya Yamanouchi.

and, though he likely did not know it yet, additional involvement in its biological weapons research.

At the same time that Nakamura was moving to Anyang to continue his vaccine research, across the Pacific – on a continent that had never experienced an outbreak of rinderpest – the American scientist Richard

Shope received notice from the United States government that he had been
selected to run a top-secret joint US–Canadian rinderpest vaccine production project. North American authorities were worried that the Axis
powers were plotting to unleash the virus upon them and they wanted to
be prepared. They were right to worry.

***

World War II was, in part, about food from the beginning: securing it for
one's self and denying it to one's enemies. Japan's empire ran on food
imported from its imperial expansions. Hitler insisted that Germany,
which had lost its empire after World War I, needed *Lebensraum*, or
living space. "It is a *battle for food*," he explained, "a battle for the
basis of life, for *the raw materials* the earth offers, *the natural resources*
that lie under the soil and *the fruits* that it offers to the one who cultivates
it." He looked to the east to recreate what the United States had done in its
west.[4] Recognizing that food was critical to Germany's success, Churchill
dubbed it contraband and made it part of his total blockade in August of
1940. Recognizing that food was critical to Britain's success, the United
States made it a part of its lend-lease aid. It also created the Anglo-
American Food Committee in the spring of 1941 to coordinate the movement of American food commodities. This was replaced in 1942 by the
Combined Food Board "to coordinate further the prosecution of the war
effort by obtaining a planned and expeditious utilization of the food
resources of the United Nations."[5] Half of the world's population was
now connected into the Allied food network.[6]

The war meant big business for American farmers. They produced
50 percent more food per person during the war than they had before it
began. Unable to keep up with the ever-rising demand, they lobbied
Congress for draft exemptions and got them by the millions. Although

[4] Collingham, Lizzie. *The Taste of War: World War II and the Battle for Food.* New York:
Penguin, 2013, 30, 42. For additional information on this, see Mazower, Mark. *Hitler's
Empire: How the Nazis Ruled Europe.* New York: Penguin, 2009 and
Baranowski, Shelley. *Nazi Empire: German Colonialism and Imperialism from
Bismarck to Hitler.* Cambridge: Cambridge University Press, 2010. For a look at the
way the United States thought about the concept at the time, see Kruszweski, Charles.
"International Affairs: Germany's Lebensraum." *The American Political Science Review*
34:5 (October 1940): 964–975.
[5] Rosen, S. McKee. *The Combined Boards of the Second World War.* New York: Columbia
University Press, 1951, 191–206.
[6] For a fascinating argument about how "nature" helped draw America into war, see
Russell, Edmund. *War and Nature.* Cambridge: Cambridge University Press, 2001.

reserving most of its production for the explosives industry, the United States government channeled enough out of its synthetic-nitrogen-processing plants to triple the amount of fertilizer farmers used in their fields. It was all part of the war effort. Food, the United States War Food Administration explained in 1945, has been a "weapon of war" that "ranks with ships, airplanes, tanks, and guns." "American food," it insisted "has been essential to the fighting efficiency of our allies as well as our own military forces, and has been required to maintain colossal industrial productivity here and in other allied countries." In consequence, American farm incomes rose 156 percent over the course of the war.[7]

The Americans did not, of course, do it alone. Iceland provided fish, Australia provided mutton, and New Zealand provided butter, cheese, and meat. The most important other producer was Canada. Canadian farmers played a major role in the food effort, notably shifting from grain to livestock production. In response to the expansion in North America and the heightened demands on shipping, Britain halted its own livestock production in favor of grain production, dropped all animal feed imports, and prioritized the import of "condensed high energy foods such as meat and dairy products."[8] Walt Disney produced the cartoon short "Food Will Win the War" for the United States Department of Agriculture in 1942 to emphasize the importance of the "food for freedom" that was daily leaving American shores in "victory ships."[9]

This massive expansion in food production was most important for the calories that it provided soldiers, sailors, and civilians around the world, but it also had a psychological importance. In the 1941 Atlantic Charter, Churchill and FDR declared that "after the final destruction of the Nazi tyranny, they hope to see established a peace which will afford to all nations the means of dwelling in safety within their own boundaries, and which will afford assurance that all the men in all lands may live out their lives in freedom from fear and want."[10] Both fighting and food production were key to the fulfillment of that vision, which was taken at its word by people around the globe.[11] Anglo-American soldiers were by far the best fed in the world, but, even more than that, in both the United

[7] Collingham, *The Taste of War*, 78–82, 88.   [8] Ibid., 90, 96–101, 420.
[9] Luske, Hamilton. *Food Will Win the War.* Walt Disney Productions (July 21, 1942).
[10] *The Atlantic Charter (August 14, 1941)*, available at http://avalon.law.yale.edu/wwii/atlantic.asp
[11] Borgwardt, Elizabeth. *A New Deal for the World.* Cambridge: Harvard University Press, 2005), 14–45.

States and Great Britain the diets of the working classes actually improved during the war, primarily because they gained access to more meat and dairy products. American aid also fed civilians in the Soviet Union, in China, and across liberated Europe.[12] Food was a particularly valuable "weapon of war" – one that needed special safeguards because it was vulnerable to special kinds of attack.

On December 19, 1941, just a few days after the United States officially entered the war, the newly formed Canadian biodefense committee known as M-1000 met in Ottawa to discuss concern of a biological attack on North American soil.[13] They had reason for concern. Biological weapons had been used in the previous war. In 1915, an American physician of German descent was recruited by Germany to infect horses and mules, which were on their way to Europe, with anthrax and glanders. He was aided in his efforts by a German steamship captain who had been trapped in Norfolk, Virginia, by the British blockade. The physician made the material. The captain ensured its distribution, recruiting a number of stevedores (eager for extra money) to foment labor unrest, randomly start fires, and infect horses and mules. The effort lasted until the fall of 1916 and allegedly encompassed Norfolk, Newport News, Baltimore, and New York. It appears, however, to have been largely ineffective and the historian who has studied it notes that "the absence of reports by any of the numerous participants of human illness associated with the operation suggests that the cultures may have been non-viable or avirulent."[14] Germany also attempted biological attacks on animals in Romania, Spain, Norway, and Argentina. Agents trying to infect horses along the Western Front were periodically captured and admitted their assignment. The efforts seem more extensive in hindsight than they were at the time; none had any real success and most were unknown for some time. German efforts in the United States did not become public until the 1930s, but that was particularly damning timing, as they added fuel to what was by that time a growing flame of concern.[15]

---

[12] Collingham, *The Taste of War*, 339–341, 416–420.

[13] Avery, Donald. *Pathogens for War*. Toronto: University of Toronto Press, 2013), 23–24.

[14] Wheelis, Mark. "Biological Sabotage in World War I," in *Biological and Toxin Weapons*, ed. Erhard Geissler and John Ellis van Courtland Moon. Oxford: Oxford University Press, 1999, 39–46.

[15] Ibid., 46–62; Historical Report of the War Research Service (November 1944-Final), Box 180, Material from Colonel Van Ormer's File to GWM Barcelona File, Security-Classified Correspondence File of Dr. G. W. Merck, NM84 488, RG 165: Records of the War Department General and Special Staffs, NARA. See also, Bacteriological Sabotage: Rinderpest (1939–1940), MAF 35/558, NAUK.

The Geneva Protocol of 1925 banned "the use of bacteriological methods of warfare." Neither Japan nor the United States ratified the Protocol, but, regardless, the language only prohibited "use," not research, and most of the countries that did sign it added formal reservations explaining that they would not be bound to it if they were attacked first with such weapons.[16] Indeed, French research into biological weapons was already well under way by 1925, in direct response to growing knowledge of past German activities. After signing the treaty, France did dramatically reduce its research efforts in 1926, but ramped them back up in 1935 in response to political events in Germany.[17] In 1934, British journalist Henry Wickham Steed published several incendiary articles, quoting alleged German documents discussing the possibility of biological weapon attacks on the Paris Metro and the London Underground.[18] Both the French and British governments cast public doubt on the authenticity of these claims, but the French were already reopening their own research and the British were hearing from spies in Berlin that a course "for military specialists in gas weapons included material on bacteriological warfare." In 1936, Britain's Imperial Defense Ministry created the Bacteriological Warfare Subcommittee, though most of its members remained doubtful that Germany was making any serious progress along those lines.[19]

Canadian doctor and Nobel Laureate, Frederick Banting of the University of Toronto, disagreed, insisting that the German government, which recognized that in modern warfare nations were fighting *nations*, as opposed to simply soldiers, had to be looking at ways to knock out its enemies' infrastructure. In September of 1939, following the invasion of Poland, Canada's National Research Council agreed to send Banting to London to try to persuade the leadership there of the necessity of

[16] Guillemin, Jeanne. *Biological Weapons*. New York: Columbia University Press. 2005, 4–5.

[17] Lepick, Olivier. "French Activities Related to Biological Warfare, 1919–45," in *Biological and Toxin Weapons*, ed. Geissler and Moon, 70–90.

[18] Steed, Wickham. "Ariel Warfare: Secret German Plans." *The Nineteenth Century and After* 116 (July 1934): 1–14; Steed, Wickham. *"The Future of Warfare." The Nineteenth Century and After* 116 (August 1934): 129–140. The editor of the journal followed Steed's first article with a note explaining that since Steed could not disclose his source, "it is impossible to guarantee ... the authenticity of the information it contains." He decided to "take the risk of publication anyway," for "even if the whole tale is a forgery, the fact that anyone should deem it worth while to perpetrate such an ingenious fraud is not without a sinister significance, which makes it all the more essential that ordinary citizens should realize the kind of horrors with which modern war is threatening them" (Editorial Noted, *The Nineteenth Century and After* 116 [July 1934]:14–15).

[19] Guilleman, Biological Weapons, 40–46.

biological weapons research, so that the United Kingdom and its dominions could retaliate in kind against a German biological weapons attack. The British government seemed to politely deny Banting's request and he left London in frustration. War Cabinet member Maurice Hankey, however, began secretly pushing for more research on the possibility of Banting's proposal following the Canadian's departure. After Winston Churchill became prime minister in May of 1940, Hankey was forced out of the War Cabinet but he was not deterred. He persuaded the minister of supply, who maintained authority over the Chemical Defense Experimental Station at Porton Down in Wiltshire, to open a biology department at the facility in August of 1940. Hankey belatedly secured Churchill's approval that October.[20]

British officials were worried about several possible pathogens, including rinderpest, that could be used for attack. A government report noted that "the malicious distribution of the virus of rinderpest (cattle plague) is a possibility." The disease, it continued, "exists in Abyssinia and virus could conceivably be procured from there through the intermediary of the Italian Government" though it would not, admittedly, be an easy task. "The virus . . . is artificially non-cultivable and can only be obtained from an infected animal," it acknowledged. "To obtain virus for sabotage would almost certainly involve the setting up of the disease in animals in a laboratory in enemy country." Rinderpest needed living hosts. This was exactly the problem vaccine researchers had struggled with for decades and why they had tried to establish the disease in goats and rabbits for research purposes before they realized that they could use them to create attenuated vaccines. Living hosts were expensive and, more importantly, dangerous to have around. "No disease is so feared in Europe," the report noted, which meant that the Germans "might hesitate to set up the necessary virus factory owing to the danger of escape of infection." Likewise, the Italians "might refuse to supply virus" for the same reason.[21] But what if the Italians did not refuse and what if the Germans did not hesitate?

It seemed unlikely that the Germans would unleash the virus on the continent, but on Britain was another story, and the report recommended defensive preparations. "Localized infections could be overcome by means of the slaughter policy, but extensive outbreaks would require

[20] Balmer, Brian. *Britain and Biological Warfare*. London: Palgrave, 2001: 29–41; Avery, *Pathogens for War*, 22.
[21] Bacteriological Sabotage: Rinderpest (1939–1940), MAF 35/558, NAUK.

other measures. To be adequately armed would mean a large provision of serum." In addition, the report recommended exploration of the preparation of tissue vaccines. "It is true," it concluded, "that the world has no experience of the successful establishment of disease on a wide scale by artificial means but it would not be reasonable to assume that this is impracticable."[22] The Germans had attacked horses in the first war; it did not seem irrational to assume that they might attack cattle in this one. With that in mind, the United Kingdom's Chief Veterinary Officer, Sir Daniel Cabot, called J. T. Edwards (formerly of Mukteshwar, India) in for consultation about a "possible invasion" of rinderpest. Edwards insisted he could research the virus "as an appendage" to work already in progress at the Foot-and-Mouth Disease Research Station, Pirbright, insisting it would not cost much. Cabot promptly set up Edwards, along with an assistant and a large number of young Devon steers, at the station. Researchers at Kabete, Kenya, sent them KAG. Experimentation commenced.[23]

In 1940, the North Americans had even more reason than the British to be worried about a rinderpest attack. It was far less likely that an outbreak unleashed in Canada or the United States would reach Germany or Italy. Back in Ottawa, unaware that his efforts in London had born fruit, Banting convinced the National Research Council to begin supporting biological weapons research in Canada.[24] The Americans were also concerned. A Japanese Army doctor tried to get a yellow fever strain from the Rockefeller Institute for Medical Research in New York in February of 1939. When turned down, he tried to bribe a laboratory technician to provide it. Meanwhile, reports of Japanese bacterial warfare in China began arriving in Washington in 1940 and continued into 1941. The United States was not yet at war, but was preparing for it, and the reports sparked concern about the possible future use of such agents against American soldiers and American food supplies.[25]

---

[22] Ibid.

[23] Edwards, "The Uses and Limitations of the Caprinized Virus in the Control of Rinderpest (Cattle Plague) Among British and Near-Eastern Cattle," 224–233.

[24] Wheelis. "Biological Sabotage in World War I," in *Biological and Toxin Weapons,* ed. Geissler and Moon, 46–51; Avery, *Pathogens for War,* 16–23; Carter, Grandon B. and Graham S. Pearson, "British Biological Warfare and Biological Defense, 1925–45," in *Biological and Toxin Weapons,* ed. Geissler and Moon, 168–175; Guillemin, *Biological Weapons,* 53.

[25] Historical Report of the War Research Service (November 1944-Final), Box 180, Material from Colonel Van Ormer's File to GWM Barcelona File, Security-Classified

In response, during the summer of 1941, "because of the evidence of increasing enemy preparation in b.w.," the Army's Surgeon General sent a request to the National Research Council "for the formation of a special committee of civilian scientists to survey all phases of this matter and to advise [him] on the entire subject." The Chemical Warfare Service (CWS), noting that "it has become increasingly apparent that there are certain possibilities inherent in biological weapons," also joined the conversation. All involved believed that further research would be wise, but along separate lines. The concerned parties agreed at an August meeting that the Surgeon General's office would "be responsible only for the defensive aspects of the problem" and the CWS for the "offensive aspects." They asked Secretary of War Henry Stimson to gather additional outside assistance from the scientific community. Officials knew that they would need help from people who knew the pathogens.[26]

The result, via consultation with the National Academy of Sciences, was the War Bureau of Consultants Committee (WBC), whose purpose was "conceived to be a consideration of those biological agents that may be purposefully used to produce harmful effect on man, animals, plants or food supplies, and methods of control of these agents." The committee was charged to produce "a study of both offensive and defense methods," so that its work could assist both the Surgeon General's office and the CWS. The WBC Committee sought advice from "experts in agriculture, plant disease, entomology, veterinary medicine, and sanitary engineers." They also established liaison officers, one of whom was Colonel R. A. Kelser, the Chief of the Surgeon General's Veterinary Division, and the man who had created the most effective inactive rinderpest vaccine back in the 1920s in the Philippines. The WBC also reached out to their Canadian counterpart, the M-1000 biodefense committee, inviting its representative, Prof. Everitt G. D. Murray, to its second meeting that December.[27] Murray came fresh out of an M-1000 meeting that discussed a possible rinderpest attack on Canada's eight million cattle, noting that, if it were unleashed, the virus would no doubt spread "rapidly, extensively, and disastrously" in that virgin population. Participants had recommended that Canada develop its own vaccine supply. They hoped to have US assistance to do it, which is partially why Murray readily agreed to attend the WBC meeting.[28] He quickly

Correspondence File of Dr. G. W. Merck, NM84 488, RG 165: Records of the War Department General and Special Staffs, NARA.
[26] Ibid.   [27] Ibid.   [28] Avery, *Pathogens for War*, 23–24.

discovered that the Americans were just as concerned. They, after all, had about seventy-four million cows of their own to worry about.[29]

At the WBC meeting, there "was a long discussion of the question of Rinderpest and the potential danger of its introduction into North America." Participants agreed that the threat was severe enough to warrant action.[30] At the conclusion of the meeting, the WBC recommended "that immediate steps should be taken to establish a laboratory as soon as possible where facilities will be provided" for the production of a known vaccine "on a large scale" and "for the study of the virus ... with particular reference to the possibility of its propagation by the newer methods of virus culture to the end that a more economical vaccine be produced than now available."[31] The committee noted Kelser's inactivated vaccine, but explained that it was "expensive to prepare in that the tissues of one animal will produce sufficient vaccine for only one hundred animals." Also, it did not last very long, even refrigerated, so that a large stock could not be made and stored. The United States needed something better, but the task was not without risk. "In order to make the vaccine the importation of virus would be necessary," the committee warned, and "a properly selected and isolated spot, preferably an island, would be essential in order to obviate the danger of the escape of the virus to our livestock."[32] No one had ever studied rinderpest on the North American continent – it had always been deemed too dangerous. Now, however, it was deemed too dangerous not to study it.

Everitt Murray wrote to Otto Maass, the Director of Chemical Warfare in Canada, that spring about the December meetings in Canada and the United States, explaining, "Quite independently the two bacteriological committees (M. 1000 and W.B.C.) came to identical conclusions and placed Rinderpest first on the list of dangers with recommendations for active measures for protection. The importance of the cattle on this continent in the war is self evident," he wrote, "and for their destruction Rinderpest is the most likely choice by the enemy."[33] Both committees concluded that they should work together to produce "active measures for

---

[29] Historical Report of the War Research Service (November 1944-Final), Box 180, Material from Colonel Van Ormer's File to GWM Barcelona File, Security-Classified Correspondence File of Dr. G. W. Merck, NM84 488, RG 165: Records of the War Department General and Special Staffs, NARA.

[30] Ibid.

[31] Everitt Murray to Otto Maass (27 April 1942), Department of National Defense, Chemical Warfare Files, 4354-1-23, Reel C-5019; LAC.

[32] Ibid.   [33] Ibid.

protection." Murray continued to attend WBC meetings and, in turn, he invited American specialists up to Canada.[34] The political structures necessary for cooperation were already in place. Roosevelt had initiated the US-Canada Joint Defense Board in 1940 to "work out defense plans for the northern half of the Western Hemisphere."[35] Rinderpest was a potentially grave threat for which there was a known defensive maneuver available. Vaccines already existed and it seemed clear that better vaccines could be created, given time and money. The WBC and M-1000 committees began working together to convince the Joint Defense Board to establish a joint rinderpest vaccine project.

Before they could propose the project, however, they had to propose a location in which to house it. That, indeed, was the trickiest part, because the virus was so dangerous. A possible solution had been raised at the December M-1000 meeting: a small, recently uninhabited island in the Saint Lawrence River called Grosse Île. The government had shut down the quarantine station in 1937 after 105 years of operations, turning the island over to the Department of Public Works who had not yet done anything with it.[36]

The geography that had made Grosse Île an ideal quarantine station location seemed to make it an ideal biological research station. It was "far enough removed from coastal waters to make it difficult to attack by casual coastal raids, and sufficiently close to be within reach of the mainland for the obtaining of supplies and equipment."[37] But the island's advantages were not limited to geography alone. As Murray would later explain, the island is "isolated, it is large enough, it is uninhabited, it has on it numerous buildings easily adaptable for our purpose. It has machinery and plant [sic] for disinfection, for electric and water service and it has residences for the staff." Even more, "the distribution and types of buildings would permit the development of the station to deal with the investigation and large scale production of whatever needs arise in the

---

[34] Avery, *Pathogens for War*, 23–24.

[35] Oral History Interview with John D. Hickerson by Richard D. McKinzie (10 November 1972), Harry S. Truman Presidential Library, available online at http://www.trumanlibrary.org/oralhist/hickrson.htm.

[36] O'Gallagher, Marianna. *Grosse Ile: Gateway to Canada, 1832–1937*. Quebec City: Livres Carraig Books, 1984. For more information about the island's rich history, see Thematic Guide to Sources Relating to the Grosse Île Quarantine Station, in Lower Canada, the Province of Canada and Canada, 1832–1937, LAC; available at http://www.collectionscanada.gc.ca/the-public/005-1142.08-e.html.

[37] Memo by J.V. Young (August 6, 1942), Department of National Defense, Chemical Warfare Files, 4354-1-23, Reel C-5019, LAC.

possibilities of Bacterial Warfare ... for both offense and defense."[38] To top it off, the island and the buildings on it were all "the property of the Crown," and ready for use "if required."[39]

Although a couple of members expressed concern that the island was not isolated enough, M- 1000 sent the proposal to the WBC, which concluded that Grosse Île "could be used with reasonable safety to the cattle population of the continent," providing certain precautions were observed. The recommended precautions were stringent: they limited where experimental animals could be kept ("well-constructed buildings" with electrified screens) and how their remains and their waste products would be disposed of ("sterilized by incineration or otherwise"). They also insisted on the need for "adequate policing ... including double fencing, electrified, of the portion of the island on which the laboratory buildings are located to minimize the possibility of the escape of animals or other infectious material in the event that the buildings were bombed or shelled."[40] All of this would be possible on Grosse Île. Everything on it had, after all, been constructed with the purpose of keeping the sick away from the healthy. The committees had found their location. Now they needed to get their project.

Both M-1000 and the WBC were advisory bodies, composed of scientists (some of whom were military personnel). They had no authority and no funding to act on their own, so they lobbied the people who could. The WBC Committee wrote up a report to be passed along to Secretary of War Stimson in February of 1942.[41] In April of that year, Stimson wrote about the report in a letter to Roosevelt, stating that it was full of "disturbing warnings." The committee, he continued, found "that real danger from biological warfare exists both for human beings and for plant and animal life." It "recommends prompt action along a number of lines," but "the matter which the committee considered as requiring the most immediate attention is the great danger of attacks on our cattle with the

---

[38] Everitt Murray to Otto Maass (April 27, 1942), Department of National Defense, Chemical Warfare Files, 4354-1-23, Reel C-5019, LAC.

[39] Memo by J.V. Young (August 6, 1942), Department of National Defense, Chemical Warfare Files, 4354-1-23, Reel C-5019, LAC.

[40] Everitt Murray to Otto Maass (April 27, 1942), Department of National Defense, Chemical Warfare Files, 4354-1-23, Reel C-5019, LAC.

[41] Report of the W.B.C. Committee (February 19, 1942), in Historical Report of the War Research Service (November 1944-Final), Box 180, Material from Colonel Van Ormer's File to GWM Barcelona File, Security-Classified Correspondence File of Dr. G. W. Merck, NM84 488, RG 165: Records of the War Department General and Special Staffs, NARA.

disease of 'Rinderpest.'" Stimson seconded the committee's recommendation for action, requesting that it be of a civilian, not a military kind.
"Entrusting the matter to a civilian agency would help in preventing the
public from being unduly exercised over any ideas that the War
Department might be contemplating the use of this weapon offensively,"
he wrote. Of course, he quickly added, "a knowledge of offensive possibilities will necessarily be developed because no proper defense can be
prepared without a thorough study of means of offense." And any such
knowledge procured "should be known to the War Department."[42]
The Secretary of War was hardly inclined to shut off his department
from potentially useful new weapons.

Roosevelt "gave verbal approval" to the letter and the loosely outlined
plan in May. Even though Stimson had explained the benefit of using
a civilian agency within the federal government to house the program
would be to prevent public concern, the process was, from that moment
on, conducted "under the highest secrecy categories." The orders for what
would become the War Research Service (under the Federal Security
Agency) that summer "were never written but were given in verbal directives of the President and the Secretary of War." Biological warfare, as
Stimson readily acknowledged, was "dirty business," and the less
known – at home and abroad – about US activities within the field, the
better.[43]

The WBC Committee would meet again in June and submit another
formal report, but it had already fulfilled its main purpose with
its February report, answering Stimson's questions about the threat biological weapons posed to the United States and what America's response
should be.[44] The committee had convinced him to urge action. Its
Canadian counterpart, M-1000, did the same in Ottawa.

The M-1000 committee had sent a recommendation to Canada's
Department of Defense at the end of December urging action in the fields
of rinderpest, plague, and "enemy agents." Murray followed up on the

[42] Historical Report of the War Research Service (November 1944-Final), Box 180,
Material from Colonel Van Ormer's File to GWM Barcelona File, Security-Classified
Correspondence File of Dr. G. W. Merck, NM84 488. RG 165: Records of the War
Department General and Special Staffs, NARA.

[43] Ibid.

[44] Report of the W.B.C. Committee (June 18, 1942) in Historical Report of the War
Research Service (November 1944-Final), Box 180, Material from Colonel Van
Ormer's File to GWM Barcelona File, Security-Classified Correspondence File of
Dr. G. W. Merck, NM84 488, RG 165: Records of the War Department General and
Special Staffs, NARA.

matter in a letter that April (two days before Stimson wrote his letter to Roosevelt) to Otto Maass, the Director of Chemical Warfare.[45] Maass agreed to bring M-1000 under Chemical Warfare's wing and to lend his support to the pursuit of a joint rinderpest vaccine project with the United States. Maass quickly appointed a special committee, dubbed the C.1. Commission, to organize the effort.[46]

C.1. held its first meeting on June 22 in Ottawa. It opened with reports on British biological warfare activities, special note being made of Edwards's experiments with rinderpest at Pirbright, which were, the commission noted, "going on in the middle of Surrey in a good agricultural district." Grosse Île would be far safer and potentially all rinderpest research could be transferred there. There was a lot that needed to be investigated, including, according to the commission, a comparison of Kelser's vaccine with the inactive one being used as Pirbright and the "propagation of the virus by the egg technique and the development of any advantage this method allows." The commission also recommended that researchers study the "preservation of infective virus" and "methods of transmission of Rinderpest and routes of infection. The infectivity of virus on Alum Hydroxide, peat, and dried dung."[47] The last recommendations hint at the dual nature of the research the scientists were hoping to conduct on their island laboratory. Learning about methods of preservation and transmission could be useful for defensive purposes, but they could also be useful for offensive purposes. There was to be more going on at Grosse Île than vaccine manufacturing.

Henry Stimson appointed four Americans to the C.1. Commission in early July. Kelser, who had by far the most experience with rinderpest,

---

[45] Everitt Murray to Otto Maass (April 27, 1942), Department of National Defense, Chemical Warfare Files, 4354-1-23, Reel C-5019, LAC; recommendations also available in Avery, Donald H. and Eaton, Mark, ed., The Meaning of Life: The Scientific and Social Experiences of Everitt and Robert Murray, 1930–1964. Toronto: The Champlain Society, 2008, 104–106.

[46] The commission was made up of: Murray; G. B. Reed, a professor of bacteriology at Queens University; C. A. Mitchell, a pathologist at the Animal Disease Research Institute in Hull; and James Craigie, an assistant professor of hygiene at Connaught Laboratories in Toronto. Also present at the first meeting was Captain Luman F. Ney of the US Army who came as a liaison for the US Chemical Warfare Service. Everitt Murray to Dean E.B. Fred (May 30, 1942), Department of National Defense, Chemical Warfare Files, 4354-1-23, Reel C-5019, LAC; C.1. Minutes – Meeting No. 1. (June 22, 1942) submitted by Murray on 10 July 1942, Department of National Defense, Chemical Warfare Files, 4354-1-23, Reel C-5019, LAC.

[47] Ibid.; Thistle, Mel, ed. The Mackenzie-McNaughton Wartime Letters. Toronto: University of Toronto Press, 1975, 122–123.

was appointed chairman.[48] The new, fully formed commission met for the first time in Quebec City in late July. They toured Grosse Île and decided it "eminently suited" for the project that they now, for reasons of security, referred to as GIR-1. All agreed that their governments should fund the project, deciding that "in view of the considerable contribution of the Canadian Government in land, buildings and facilities," the cost of the project henceforth should be divided on a twenty-five/seventy-five split.[49]

The United States government agreed, on the condition that Canada provide an additional $300,000 up front. A letter confirming US assent arrived at the Canadian Department of National Defense on August 11, explaining that funding had been allotted because the US War Department "was most anxious to start working on the undertaking."[50] It was also anxious to keep the project a secret. In an August twenty-first memo, following a recent visit to Washington, a high-ranking Canadian official reported that there had been "some talk of clearing [the funding] through diplomatic channels, but that this had been rejected as increasing the number of people 'in the know.'"[51] The funding problem "was circumvented," a researcher later explained, "by the Department of War purchasing monthly reports of the work, the charge being made for these reports on the basis of 75% of the cost of the project."[52] The Canadian government did not care how the funding arrived, just that it arrived.

---

[48] Members included: Kelser; E. B. Fred, a professor of bacteriology at the University of Wisconsin; H. W. Schoening, Assistant Chief in the Bureau of Animal Industry in the Department of Agriculture; and Rolla E. Dyer, the Director of the National Institute of Health .Stimson originally appointed Lewis H. Weed, Chairman of Medical Sciences at the National Research Council, but he was too busy to do it and was quickly replaced with Dyer.

[49] GIR 1 Meeting (July 25, 1942), Department of National Defense, Chemical Warfare Files, 4354-1-23, Reel C-5019, LAC.

[50] Francis J. Graling to Colonel Flood (August 12, 1942), Department of National Defense, Chemical Warfare Files, 4354-1-23, Reel C-5019, LAC; Historical Report of the War Research Service (November 1944-Final), Box 180, Material from Colonel Van Ormer's File to GWM Barcelona File, Security-Classified Correspondence File of Dr. G. W. Merck, NM84 488;RG 165: Records of the War Department General and Special Staffs, NARA.

[51] Memo by J.V. Young (August 6, 1942), Department of National Defense, Chemical Warfare Files, 4354-1-23, Reel C-5019, LAC; Memorandum of conversation with Mr. Norman Robertson, Under-Secretary of State for External Affairs (August 21, 1942) (author's signature undecipherable), Under-Secretary Office Papers, Department of External Affairs; Biological Warfare – Grosse Isle Station (GIR-1), Record Group 25, Series A-2, Volume 829, Page 7, File 3, LAC.

[52] Charles A. Mitchell, Memorandum: Re Project at Grosse Isle (January 19, 1946), Diseases of Animals, Rinderpest Control Vaccination Project, Grosse Isle, QC, Record Group 17, Volume 3029, Page 83, File 37–23, LAC.

The Department of National Defense approved the creation of the Joint Canada-United States Commission to pursue GIR-1 at the Cabinet War Meeting on August 14.[53]

The Joint Commission had its island; it needed its research team. GIR-1 was a unique case, run, as it was, through the Joint Commission, but it still worked closely with the new War Research Service and its director, George W. Merck. In all WRS projects, the program's final report insisted, "in the selection of the project leader, every effort was made to secure an individual with a background of experience, training and accomplishment which would qualify him as one of the outstanding workers in his particular field."[54] GIR-1 was no different. The commission needed an expert virologist. They turned to Richard Shope of the Department of Animal Pathology at the Rockefeller Institute for Medical Research in Princeton, New Jersey.

***

Shope was an inspired choice. He knew nothing about rinderpest, but he knew a great deal about viruses. He had been hunting them, studying them for decades. He was also "an extrovert and unconventional thinker receptive to novel ideas."[55] As a young researcher, he and Paul Lewis had discovered that the cause of swine influenza, *Bacillus influenzae suis*, was virtually identical to *Bacillus influenzae*, the bacteria that had been identified as the source of human influenza back in 1892. Shope and Lewis then identified a virus that seemed also connected to influenza, complicating the accepted narrative of infection. On the heels of this important discovery, Lewis set off for Brazil to study a yellow fever outbreak. Shope had volunteered to go, but his boss told the twenty-eight-year-old with a wife and infant son that it was too dangerous and refused him. Shope stayed behind to continue the swine influenza research. It proved a fortunate decision; Lewis soon died of yellow fever from an unexplained laboratory accident. Shope would later tell his sons that the rumor was

---

[53] "Defense of Canada – Biological Warfare," Under-Secretary Office Papers, Department of External Affairs; Biological Warfare – Grosse Isle Station (GIR-1), Record Group 25, Series A-2, Volume 829, Page 7, File 3, LAC. For more information on the Joint Commission, see Chemical Warfare – Correspondence – United States, RG24-C-1, microfilm reel C-5018, LAC.

[54] Historical Report of the War Research Service (November 1944-Final), Box 180, Material from Colonel Van Ormer's File to GWM Barcelona File, Security-Classified Correspondence File of Dr. G. W. Merck, NM84 488, RG 165: Records of the War Department General and Special Staffs, NARA.

[55] Beveridge, W.I.B. *Influenza: The Last Great Plague*. London: Heinemann, 1977, 5.

that the researcher had infected himself accidently through a contaminated cigarette.[56]

Back in Princeton, Shope struggled to understand why the bacillus he and Lewis had identified in infected pigs did not produce the disease when injected into healthy pigs and the virus produced only a mild form of infection. The answer, he realized, was that the disease required the presence of both. The virus (which was later revealed to have descended directly from the human influenza virus of 1918) had either mutated into a milder form or pig immune systems had gotten better at fighting it. The introduction of *Bacillus influenzae suis* into a pig already infected with the virus, however, turned a mild infection into a lethal one. Shope was quick to note the potential implications of his discovery. He wrote in his first report "that Pfeiffer's bacillus and a filterable virus might act in concert to cause human influenza." He was right. Two years later, researchers inspired by Shope's work demonstrated that human influenza was also caused by a filterable agent, which meant a virus.[57]

During the 1930s, Shope discovered that viruses were responsible for "mad itch" in cattle (a disease he had observed while studying swine flu on farms in Iowa), fibromas in cottontail rabbits in New Jersey (a disease he had noticed while out hunting), and papillomatosis in midwestern cottontail rabbits (which he had also found in Iowa). He also continued work on swine influenza.[58] By the end of the 1930s, he had become a well-known expert virus hunter. That is why they wanted him at Grosse Île.

Shope was eager to go; he had always liked a challenge and this one seemed "importantly significant." He "immediately cleared affairs at Princeton and set to formulating plans for the project as [he] visualized

---

[56] Statement concerning the scientific work of Richard E. Shope, Box 1, Folder 6, Collection RU, RG 450 Sh77, RAC; Peyton Rous, "Presentation of the Kober Medal to Richard Shope," Box 1, Folder 2, Collection RU, RG 450 Sh77, RAC; Biography of Richard E. Shope (December 29, 1964), Box 1, Folder 4, Collection RU, RG 450 Sh77, RAC; Barry, *The Great Influenza*, 434–435. Also based on author conversation with Shope's grandson, Richard Shope, on July 7, 2015.

[57] Shope, Richard E. "Influenza: History, Epidemiology, and Speculation." *Public Health Reports* 73:2 (February 1958): 165–178; Van Epps, Heather L. "Influenza: Exposing the True Killer." *Journal of Experimental Medicine* 4 (April 17, 2006): 803; Peyton Rous, "Presentation of the Kober Medal to Richard Shope," Box 1, Folder 2, Collection RU, RG 450 Sh77, RAC;Barry, *The Great Influenza*, 439–447.

[58] Statement concerning the scientific work of Richard E. Shope, Box 1, Folder 6, Collection RU, RG 450 Sh77, RAC; Peyton Rous, "Presentation of the Kober Medal to Richard Shope," Box 1, Folder 2, Collection RU, RG 450 Sh77, RAC; Biography of Richard E. Shope (December 29, 1964), Box 1, Folder 4, Collection RU, RG 450 Sh77, RAC.

FIGURE 2.2 Richard Shope during World War II.
Photograph courtesy of Richard Shope.

it might best be set up."[59] His research team consisted of six officers of the US Army Veterinary Corps, one from the US Navy Medical Corps, two Canadian scientists, and numerous technicians from both countries. They would not be alone. In November, Canadian researchers started another project, GIN, on the island to produce weaponized anthrax spores.[60] Both sets of researchers would be supported by local civilian staffers cleared through security by members of the Canadian military.

Shope arrived in Canada anxious to begin work, but was stopped short by the situation on the island, where very little progress had been made on updating the facilities. Informed that "the last of December" was the "latest possible time we could have access to the island for the purposes of construction or the taking on of animals" and that "even this was admittedly an optimistically late date," Shope and his team resorted to emergency measures. They began rehabilitating buildings themselves, "planning rather drastic alterations in some of them to adapt them to our uses" and "winterizing a utilities system that was adapted for only mild weather." They did so in four months "at a location which, to state the situation as charitably as possible," Shope wrote, "is difficult to access."[61]

Shope wanted to begin his research immediately, but he was now working for the military (he was put into active service as a Commander in the US Naval Reserve) in a foreign country on a jointly administered project. It was complicated. Otto Maass would tell George Merck, "Shope was a fine man who took off his shirt and pitched in when no one was around to work, etc., but that he didn't understand the administrative angles."[62] Murray would write to Maass that Shope was "unduly impatient and magnifies difficulties out of proportion because of his anxiety to get started," but he wrote it in December, which proves that Shope was right to be frustrated.[63] He was not able to begin his research until the deep freeze of winter, which comes early in Quebec, had settled over the island. The only useful part

[59] Richard Shope to Otto Maass (October 19, 1942), Department of National Defense, Chemical Warfare Files, 4354-1-23, Reel C-5019, LAC.

[60] For more on the anthrax program, see Avery, *Pathogens for War*, 31–32 and Avery and Eaton, ed., *The Meaning of Life*, 87–153.

[61] Richard Shope to Otto Maass (October 19, 1942), Department of National Defense, Chemical Warfare Files, 4354-1-23, Reel C-5019, LAC.

[62] WRS Diary (December 28, 1942), Box 187, WRS Diary, Ruth Hunsberger to WRS, Mary E. Switzer, Security-Classified Correspondence File of Dr. G. W. Merck, NM84 488, RG 165: Records of the War Department General and Special Staffs, NARA.

[63] Quoted in Avery, *Pathogens for War*, 30.

of the delay was that it no doubt let Shope read up on rinderpest, so that he was better prepared to tackle this challenge. He needed to become a rinderpest expert quickly.

The Joint Commission gave him two "immediate objectives" and he wanted to get started on them immediately: "First, we were to produce as rapidly as possible an adequate stock of virulent rinderpest virus, and, secondly, we were to convert this virus into a safe vaccine effective in protecting cattle against rinderpest."[64] Additional tasks, such as those discussed in the C.1. meetings over the summer, would come later. These first ones were daunting enough, considering the situation. Shope and his team had no virus, no animal hosts, and less-than-adequate research facilities. A lot needed to get accomplished in a very short period of time and winter was looming.[65]

The host problem was the simplest to address: they bought calves on the mainland and brought them to the island on boats, putting them up in the facilities that they had until a new stable could be built. They attempted to address the virus problem by ordering Kabete O (the strain that had been needle-passed in cattle since 1911) from Kenya. It arrived in dried pieces of bovine lung tissue, encased in sealed glass ampules. There was concern that it might not have survived the trip, but, to everyone's relief, it proved ready and willing to thrive after introduction to new hosts. "No difficulty was encountered in establishing the virus in calves," Shope reported. "Administered subcutaneously to calves, it induced a disease that was clinically characteristic of rinderpest, as described in textbooks, after an incubation period of from forty-eight to seventy-two hours."[66] No one on Shope's immediate team had witnessed an outbreak in person, so they had to rely on descriptions. They kept the virus alive by needle-passaging it in calves. By the time they ended their research, years later, they had brought to the island over 900 calves – mostly Holstein-Friesians, Jerseys, and Ayrshires, but also a few

---

[64] Shope, Richard E., Henry J. Griffiths, and Dubois L. Jenkins. "I: The Cultivation of Rinderpest Virus in the Developing Hen's Egg." *American Journal of Veterinary Research* 7:23 (April 1946): 135, which includes the notation: "Received by the Commission on July 30, 1943 for publication which had to be delayed on account of security regulations in effect at that time."

[65] The team would be trapped on the island all winter. Shope wrote to Murray in the summer of 1943, "If our Station were in Labrador or up in Hudson Bay ... I could see some sense to being completely cut off for 5 months. But here we are within 35 miles of the city of Quebec where there are facilities ... [and] supplies. This extremely unpleasant experience did little to boost our morale here ... because you get too damned hungry" (Avery, *Pathogens for War*, 30).

[66] Ibid., 135–136.

Herefords, Red Polls, Shorthorns, Guernseys, and Canadians – from the Montreal stockyards and local farmers.[67]

Once they had the virus, the team began working on the vaccine. "There were, at the time," Shope later wrote, "at least three vaccines reputed to be effective against the disease": caprinized, formalin-inactivated, and chloroform-inactivated. Western scientists did not yet know about Nakamura's vaccine and they would not until after the war.[68] The Grosse Île researchers rejected the caprinized vaccine as too dangerous in that it risked causing an outbreak, weighed chloroform inactivation versus formalin inactivation and, not surprisingly, perhaps, considering Kelser was on the Joint Commission, chose Kelser's chloroform-inactivated vaccine. They quickly encountered its limitations. The Commission wanted at least 100,000 doses of vaccine ready "at the earliest possible moment," but production was not uncomplicated. The problem was that inactivated tissue vaccines were made from the tissues of infected cattle and "a good size calf made only about 350 doses of vaccine." It took 270 calves to produce the requested 100,000 doses in 1943.[69]

By February of 1944, Shope and his team could report, "We have on hand a sufficient amount of vaccine to combat initial, scattered outbreaks of the disease." They noted that they had also frozen "a relatively large amount of infected tissue which can, on short notice, be made into vaccine." They did not know how long the vaccines would stay viable, having tested them so far at only fifteen weeks, but Kelser's vaccine maintained its potency up to a year, and since theirs was "such a close replica of his," they believed it would be the same. They also did not know how long the immunity it conferred would last, but reasoned that even it were only two or three months it "should be sufficient to handle outbreaks of the type that are visualized as possibilities in the present emergency."[70]

The successful completion of the first task had highlighted the importance of the second. Bovine tissue vaccines were a stop-gap measure. The cattle population of North America, Shope estimated, was somewhere around sixty million (others put the estimate closer to

---

[67] Ibid., 136.

[68] Shope, Richard E. "Experimental Wartime Studies on Rinderpest." *Journal of the American Veterinary Medical Association* 110 (April 1947): 216–218.

[69] Shope, "Inactivated Virus Vaccine," in *Rinderpest Vaccines,* ed. Kesteven, 23–25.

[70] Walker, R.V.L., Henry J. Griffiths, Richard E. Shope, Fred D. Maurer, and Dubois L. Jenkins. "III: Immunization Experiments with Inactivated Bovine Tissue Vaccines." *American Journal of Veterinary Research* 6:23 (April 1946): 145–151.

eighty million), so the production of enough vaccine for complete vaccination would require about 170,000 calves, not to mention about 22,666
days of labor at Grosse Île. "Mathematical calculation," Shope later
explained, "proved to us that it would be impractical for use in controlling
any widespread outbreak in Canada or the United States."[71] Kelser's
vaccine was fine as an emergency intervention for stopping the expansion
of localized outbreaks, but it could not protect North America from
a significant, multi-location attack.

Assessing the risk, Shope and his team decided to try something new:
growing the virus in eggs. A few years earlier, the Australian researcher
Frank Macfarlane Burnet had discovered that the influenza virus could
grow in embryonated hen's eggs, so there was reason for the Grosse Île
researchers to believe that they could do the same with rinderpest. Burnet
had, not incidentally, also shown that the passaged egg virus could immunize mice and ferrets against non-attenuated strains.[72] Shope and his
team, however, were not after a live vaccine but after a much more
efficient method of growing virus that could be inactivated. Eggs were
easier to get, easier to store, and unquestionably easier to sacrifice than
cattle in the name of science.

To their delight, the team found that Kabete O could live in eggs
(specifically, fertile Barred Plymouth Rock eggs).[73] That was great news,
but it was quickly followed by bad. The team discovered that this egg-
propagated virus could not survive inactivation with either chloroform or
formalin.[74] This turned out, however, to be only a minor setback, for the

---

[71] Shope, "Inactivated Virus Vaccine," in *Rinderpest Vaccines*, ed. Kesteven, 24–25.

[72] Burnet, F. M. "Influenza virus on the developing egg. IV. The pathogenicity and immunizing power of egg virus for ferrets and mice." *British Journal of Experimental
Pathology* 18:1 (1937): 37–43.

[73] It required a two-step process of growing the virus (initially a suspension of virus-infected
cattle spleen in broth) on the chorioallantoic membrane for between eight and twelve
serial passages and then moving it to the yolk sac. At that point, the virus multiplied at
a very rapid rate, filling the embryo and extra-embryonic fluids within twenty-four hours
(Shope, Richard E., Fred D. Maurer, Dubois L. Jenkins, Henry J. Griffiths, and James
A. Baker. "IV: Infection of the Embryos and the Fluids of Developing Hens' Eggs."
American Journal of Veterinary Research 6:23 [April 1946]: 152–163).

[74] Further research revealed that this was likely because egg and embryo proteins "preoc-
cupied" the "antibody mechanism of inoculated calves" so that it did not respond to the
"small amount of virus protein" (Shope, "Inactivated Virus Vaccine," in *Rinderpest
Vaccines*, ed. Kesteven, 25). Maurer, Fred D., R. V. L. Walker, Richard E. Shope,
Henry J. Griffiths, Dubois L. Jenkins. "V. Attempts to Prepare an Effective Rinderpest
Vaccine from Inactivated Egg-Cultured Virus." *American Journal of Veterinary Research*
6:23 (April 1946): 164–169.

team soon realized that the Kabete O strain that they had put into the eggs was attenuating as they kept passaging it. At the forty-first egg passage, the virus still produced a mild illness in a calf; at the sixty-seventh passage, it appeared to produce none at all. "Within ten days of injection," the team reported to the Joint Commission in the summer of 1944, "the attenuated virus solidly immunizes calves against fully virulent bovine spleen virus." Critically, the "mild rinderpest" it causes – "so mild as to practically escape recognition" – "does not appear to be contagious."[75] This last part was particularly wonderful news. As Shope later pointed out, "in Canada, the United States, Australia and New Zealand, where rinderpest is not endemic, we must use either an inactivated virus vaccine, or one whose attenuation is sufficient to allow us to depend upon its pathogenic properties." The "live" vaccine had to be strong enough to work, but weak enough so that it could not cause an outbreak.[76]

The new "avianized" vaccine was deemed a solid success. It was easier and cheaper to produce than the bovine wet tissue vaccine: they could produce the material necessary for about three and a half vaccines from one egg, which meant that 100 eggs could produce as much as one calf.[77] Even more than that, however, it was easier and cheaper to produce than the "inactivated egg-cultivated vaccine" that the researchers had hoped (and failed) to make. Additional research demonstrated that the new vaccine could be safely freeze-dried in order improve its "keeping qualities." Dried vaccine "packed in vacuum" could "maintain its potency for as long as fifteen months" when refrigerated.[78] The researchers also discovered that chicks could "be substituted for calves in tests for the presence and concentration of rinderpest virus in eggs," which saved even more money. Efforts to employ rabbits, guinea pigs, and mice to the same purpose proved unsuccessful.[79] These experiments would all become

---

[75] Jenkins, Dubois and Richard E. Shope. "VII. The Attenuation of Rinderpest Virus for Cattle by Cultivation in Embryonating Eggs." *American Journal of Veterinary Research* 6:23 (April 1946): 174–178.

[76] Shope, "Inactivated Virus Vaccine," in *Rinderpest Vaccines*, ed. Kesteven, 26.

[77] Maurer et al., "V. Attempts to Prepare an Effective Rinderpest Vaccine from Inactivated Egg-Cultured Virus." 164.

[78] Hale, M. W., R. V. L. Walker, Fred D. Maurer, James A. Baker, and Dubois L. Jenkins. "XIV. Immunization Experiments with Attenuated Rinderpest Vaccine Including Some Observations on the Keeping Qualities and Potency Tests." *American Journal of Veterinary Research* 6:23 (April 1946): 212–221.

[79] Baker, James A. and A.S. Greig. "XII. The Successful Use of Young Chicks to Measure the Concentration of Rinderpest Virus Propagated in Eggs." *American Journal of Veterinary Research* 6:23 (April 1946): 196–198. While the researchers focused most of their attention on the avianized vaccine, some also tried to keep Kabete O alive via passage in

public record after the war (more on this later). They were not, however, the only experiments with rinderpest happening on Grosse Île during the war. Only traces exist of the others, which were never published. The history has to be pieced together from a few clues.

On June 24, 1943, George Merck hosted an all-day meeting on GIR-1. The Joint Commission members were in attendance, as was Shope, who presented a report on his team's work. He had, he explained, been able to "propagate the 'R' agent" and was now testing it to ensure that he was "working with the pure virus." Discussion followed about additional issues regarding Grosse Île, including how to get an electric incubator there, along with some personnel issues. The notes about the meeting from the WRS diary also included the aside, "because of the original directive from Mr. Bundy limiting the activity of the Commission to defensive work, it is necessary to have the directive changed if offensive measures are to be considered" (emphasis in the original). "Mr. Bundy" was Harvey Hollister Bundy, special assistant to Secretary of War Henry Stimson. Merck sent Bundy a memo on July 5, requesting that he "give General Kelser clearance for complying with the request of General Porter that GIR-1 be allowed to perform certain 'offensive' experiments not provided for in the original directive." Major General William N. Porter was Chief of the Chemical Warfare Service. That same day, Merck also wrote to Porter, "asking that CWS issue instructions to the Chairman of the US Commission (Kelser) directing concurrence with the Canadian request that offensive aspects be studied." Merck was putting events in motion.[80]

In a later summary by Kelser about the results of GIR-1, after a description of the vaccine research, he wrote that "subsequent instructions directed that the project include studies for possible 'O' activities." The "'O' investigations," Kelser continued, "have resulted in some interesting and important findings." We have learned that "the virus as

---

rabbits. These efforts were directed primarily toward finding alternate means of keeping the virus alive in the laboratory. Hopes of replicating Nakamura's findings, however, were not fulfilled. "Using the strain of virus available in this laboratory," they reported to the Joint Commission, "it was impossible with the methods described by others to transform rinderpest virus in rabbits beyond one passage." Researchers found that they could only passage Kabete O in their New Zealand white rabbits by alternating passages with calves, which was certainly not a better solution than using eggs alone (Baker, James A. "VIII. Rinderpest Infection in Rabbits." *American Journal of Veterinary Research* 6:23 [April 1946]: 179–182).

[80] WRS Diary (June 24, 1943 and July 5, 1942), Box 187, WRS Diary, Ruth Hunsberger to WRS, Mary E. Switzer, Security-Classified Correspondence File of Dr. G. W. Merck, NM84 488, RG 165: Records of the War Department General and Special Staffs, NARA.

contained in the blood and solid tissues of infected animals will not ordinarily produce disease if fed to susceptible animals," but that the "virus as it occurs in saliva and fecal material will readily produce the disease when fed to susceptible animals." These findings, he concluded, "are not only interesting but open up a number of important possibilities in connection with the spread of other virus diseases of man and lower animals. Further studies of 'O' possibilities are in progress."[81] These were relatively basic offensive studies, likely identical to the "attempts to infect cattle with rinderpest by ingestion" research that British scientists conducted during the war.[82] And there were others.

In a February 1944 memorandum to the Canadian Chiefs of Staff about biological weapon (BW) research, Everitt Murray praised a new technique of using "finely ground peat (LP) 'for the distribution of pathogenic bacteria maintaining their numbers, viability and virulence.'" This peat moss "carrier" could be "effectively disseminated as a cloud or as ground contamination." He closed with a note that while the rinderpest research at Grosse Île remained primarily defensive, the "possibilities of using this virus as an offensive weapon" were under investigation.[83] It proved too tempting to ignore. At a January 1945 meeting of the C.1. Committee in Ottawa, "the possibilities of R as an offensive weapon particularly the new observations on its survival in L.P. were discussed."[84]

Starting with the approval obtained in the summer of 1943, researchers at GIR-1 began running two different kinds of experiments on Grosse Île: continued testing of the vaccine for defensive purposes and new testing of the virus for offensive purposes. They, along with their counterparts at Pirbright, were not the only ones doing so: a parallel series of events was unfolding at the same time in Japan. They revolved around a man named

---

[81] Historical Report of the War Research Service (November 1944-Final), Box 180, Material from Colonel Van Ormer's File to GWM Barcelona File, Security-Classified Correspondence File of Dr. G. W. Merck, NM84 488, RG 165: Records of the War Department General and Special Staffs, NARA.

[82] Ibid.

[83] Review of Present Position of B.W. in Canada by the C.1. Committee (15 February 1944), in *The Meaning of Life*. Avery and Eaton, ed., 125–128; Avery, *Pathogens for War,* 37.

[84] Minutes of the Meeting of the C.1. Committee Together with Representatives of the U.K. and Suffield (January 8, 1945), British Documents (originals and/or copies) copied for USBWC Steering Committee, CWS, and USBWC Files, Box 186, USBCW British Material on Crop Destruction Agents to WBC, Recommendations, Security-Classified Correspondence File of Dr. G. W. Merck, NM84 488, RG 165: Records of the War Department General and Special Staffs, NARA.

Noboru Kuba, who was an army veterinarian assigned to the Army Ninth Technical Research Institute, which was commonly called the Noborito Institute, in Kawasaki, Kanagawa Prefecture. The institute researched covert warfare. "Operating under a veil of heavy secrecy, the facility's departments developed and produced miniature cameras, invisible inks and other equipment for intelligence gathering; researched chemical and biological weapons, and developed new and deadly poisons."[85] Kuba had begun working at Noborito in 1943 and immediately focused his attention on the question of how to "infect an acute contagious disease" in cattle in "the enemy nation." The "ultimate goal," he wrote, "was to lead the enemy to discontinue the war by implementing the plan." He seized on rinderpest to do it.[86]

To get the virus, he contacted the Mukden Veterinary Institute in Manchuria, which was currently dealing with an outbreak. Kuba sent a colleague to collect lymph nodes from dead cattle. He placed them in glass bottles filled with a "glycerine preservative solution," sealed them with paraffin, and packed them into a "large vacuum bottle stuffed with snow and ice." He then carried them back to Busan, where Kuba and Junji Nakamura were waiting for them. Nakamura was now the director at Anyang, but the two institutions were tightly linked and it was an easy train trip between there and Busan. Both Nakamura and Ichiro Isayama,

[85] Schreiber, Mark. "Balloon Bombs, Poisons All in a Day's Work at Noborito." *The Japan Times* (October 12, 2010), available at http://www.japantimes.co.jp/life/20 10/10/17/general/balloon-bombs-poisons-all-in-a-days-work-at-noborito/#.VXl_hEZ PobR; Mercado, Stephen C. "The Japanese Army's Noborito Institute." *International Journal of Intelligence and Counterintelligence* 17 (2004): 286–299.

[86] Kuba, Noboru. *The Research Outline of the Seventh Research Team of the Ninth Army Technical Research Laboratory: The Attack on the U.S. with Balloon Bombs Containing Rinderpest Virus* (March 31, 1990). This document was given to me by Kazuya Yamanouchi, who was given it by one of Kuba's friends. George Sweat, a native Japanese speaker, translated it for me. Kuba wrote it forty-six years after the events described within it took place. It is, as far as I know, the only account available, because Japanese researchers burned all of their records at Noborito in August of 1945. There is also no record of this research in the 1945 report of the US Scientific Intelligence Survey (commonly called the Compton Survey), which is discussed in Chapter 6 of this book. Shigeo Ban, who had also worked at Noborito, used Kuba's report in his 2001 book (it was published after his death) *Rikugun Noborito Kenkyujo no shinjitsu* (Tokyo: Fuyo Shobo Shuppan, 2001), but I do not know of any reference to it in any book published in English. For more information about Ban's book, see Mercado, Stephen C. Review of Rikugun Noborito Kenkyujo no shinjitsu [The Truth About the Army Noborito Research Institute]. *Studies in Intelligence Studies* 46:4 (posted April 14, 2007), available at https:// www.cia.gov/library/center-for-the-study-of-intelligence/csi-publications/csi-studies/stu dies/vol46no4/article11.html#author1.

who was the head at the Busan facility, had been officially asked to work "provisionally" for Noborito. Both had agreed. Nakamura tested the samples and found that the virus had survived the trip. Kuba had his virus; he could now, as he recalled decades later in a written account of his experiments, start trying "to make it applicable to actual warfare."[87]

The key, Kuba decided, was that he needed to turn the virus into a powder that could be dispersed through the air. This he did by grinding infected lymph node tissue up with a large porcelain mortar, mixing it with refined flour, drying it, and then grinding it some more. Kuba tested his new product's stability by exposing it to extreme heat and cold. He left the dried virus powder in a petri dish outside in the sunshine for three days and found that it still could cause a fatal infection. He had similar results when he used dry ice and ether to reduce the temperature of the virus powder to -70°C. Testing the stability results was another test in and of itself, because Kuba had to figure out how to infect the cattle with the powder. Several attempts at oral inoculation failed. On the suggestion of a colleague, and with Nakamura's consent, he sprayed the virus powder into the nasal cavity. The animal developed typical rinderpest symptoms and died. The spray proved an effective mode of transmission every time. This was an exciting find, because Kuba already knew how he wanted to disperse his weapon: he was going to blow up a box full of rinderpest powder in the air over a large space where animals could breathe it in.[88]

Kuba scheduled his test in May of 1944 along the shore of the Nakdong River to the west of the Serum Production Station in Busan. A family had a house there, but the researchers "requested" that they move and "they willingly accepted." Kuba came with a large group and all the necessary supplies: ten cows, tents for forty people, and cooking utensils. A regiment of military policeman from Pusan brought blankets, and the Noborito Institute provided "tools for blasting" and an anemometer to measure wind speeds. The group dined on a breakfast of miso soup with shijimi clam and got to work. Everyone had an assigned role: "blasting, transport-arrangement-monitoring of cows, wind direction–wind speed monitoring, picture/movie." The team tested the arrangement of the cows ten times using flour instead of rinderpest to see how it spread in the wind. When they were satisfied with the formation – three cows tied thirty meters downwind from his desired launch site and another seven further downwind – they

[87] Kuba, *The Research Outline of the Seventh Research Team of the Ninth Army Technical Research Laboratory.*
[88] Ibid.

switched to the real thing. Kuba attached fifty grams of the powdered virus in a small cardboard box to a skyrocket that had been built for the team at Noborito. He launched his rocket; the box exploded in the air, releasing the powdered virus. It covered all of the cows "as if they were trapped in a large net." The cows were immediately loaded onto a ship on the river and sent to a "special cowshed." Ten days later, all ten developed symptoms and died. The experiment was a stunning success. Kuba had found a way to release the virus, but he needed a way to get it across the ocean. Other researchers at the Noborito Institute offered a solution: balloon bombs (*fusen bakudan*).[89]

These balloons (about ten meters high when fully inflated with hydrogen) were made of paper and paste and each carried a fifteen-kilogram, high-explosive bomb and two or more thermite incendiary bombs. They were designed to be caught in the jet stream and carried across the Pacific by the wind. Researchers predicted the trip would take about three days, so they designed a control system to keep the balloon aloft for that amount of time. At the end, they hoped that the balloons would land on North America and blow up, causing various kinds of damage, most likely forest fires. The researchers suspected that only about 10 percent of the balloons launched would actually reach North America, and many of those might not detonate on impact, so they anticipated that they would have to build and launch ten thousand of them if they hoped to have any serious impact. They needed massive amounts of paper. When the demand could not be met by commercial paper factories, the army recruited thousands of teenage schoolgirls to hand manufacture the paper-and-paste balloons. The initial tests began in late August 1944 and seemed satisfactory. They also raised new possibilities. Perhaps the balloons could carry something besides incendiaries.[90]

In September 1944, Kuba and Nakamura met with army staff officers, representatives from the ministry of agriculture, the Army Veterinary School, additional Noborito staff, and Manchurian Unit 100 to discuss "a plan to annihilate cattle in the U.S. by producing 20 tons of rinderpest powder that can be attached to balloon bombs." The group concluded that it could be done. "Afterwards," however, Kuba later recalled, one of the staff officers took the idea to Hideki Tojo, who, though no longer

[89] Ibid.

[90] Mercado, "The Japanese Army's Noborito Institute," 292–293; Coen, Ross. *Fu-Go: The Curious History of Japan's Balloon Bomb Attack on America*. Lincoln: University of Nebraska Press, 2014, 16–46.

prime minister, was still a powerful general. Tojo responded, "If we attack the U.S. by using balloon bombs that contain rinderpest, the U.S. will consider incinerating our rice plants in the harvest season. Therefore, the plan about the rinderpest attack is aborted." To Kuba's dismay, the operation was "unfortunately" canceled. None of the many balloons that Japan would subsequently send across the Pacific carried rinderpest powder.[91]

The military launched the first fire balloons on November 3rd. Others quickly followed. The US and Canadian governments kept the landings quiet, anxious not to let the Japanese know that the bombs were making it across the Pacific.[92] They worried what they might contain. A January 1945 report on BW intelligence noted that "several large Japanese balloons capable of carrying loads of 450 pounds have landed in the Western United States. The purpose of these balloons is not yet clear, but one of the more likely ones is to spread animal or human diseases directly or by releasing infected insects or animals." Officials set up two-person biological warfare teams with "kits for collection of samples" that could be quickly sent out to located balloons to test for pathogenic organisms. "Since balloons of this type could be sent over in large numbers," the report warned, "they represent a possible BW menace to the Western United States."[93]

Canadian officials – the same ones in charge of Grosse Île – engaged in similar actions and expressed similar concerns. Charles Mitchell was sent to collect samples from a discovered balloon on January 13 for fear of what its cargo might contain.[94] Everitt Murray advised at a February meeting called by the Minister of Defense "to make recommendations regarding counter-measures against the threat of BW from the Japanese balloons" that "Rinderpest and Foot and Mouth Disease were the two animal diseases most likely to be introduced." He then promptly sent

---

[91] Kuba, *The Research Outline of the Seventh Research Team of the Ninth Army Technical Research Laboratory*; Mercado, "The Japanese Army's Noborito Institute," 293.

[92] This decision would have serious consequences: the deaths of one pregnant woman and five children in Oregon in May of 1945. The group touched a downed balloon bomb, not knowing what it was, and the explosion killed all of them. The sight of the tragic event was, as the memorial there reads today, "the only place on the American continent where death resulted from enemy action during WWII" (Coen, *Fu-Go*, 1-5, 159–182, 206).

[93] Summary of BW Intelligence November 5, 1944 to January 15, 1945, BW Intelligence, Box 181, Report of Scientific Intelligence Survey in Japan, BW, Vol. 5 to Final Board Report, Security-Classified Correspondence File of Dr. G. W. Merck, NM84 488, RG 165: Records of the War Department General and Special Staffs, NARA.

[94] Coen, *Fu-Go*, 87–93.

a letter to Mitchell advising a review "to indicate augmentation and improvement for investigation and control of B.W." in "view of the definite possibility of the first indication of enemy intention being an outbreak."[95] There was a reason the C.1. Committee was discussing "the possibilities of R as an offensive weapon" that January: the threat of a rinderpest attack seemed a more imminent danger than it ever had before.[96]

The balloons were frightening, and more of them kept coming. Japan launched about 9,000 balloons between November 1944 and April 1945, when the effort was canceled for its perceived lack of success and a shortage of materials. An estimated 350 of them made it to North America, but the Japanese did not know it. None of them contained rinderpest. None of them contained any biological weapons, but American and Canadian officials kept checking to make sure.[97] A May 15, 1945, summary of BW intelligence warned that "captured documents and questioning of prisoners continue to reveal Japan's interest in both defensive and offensive BW."[98] How could they not keep checking the balloons? And how could they not engage in their own offensive research in turn?

On May 3, 1945, George Merck and John P. Marquand sent a memo to Henry Stimson explaining that "scientists connected with the Chemical Warfare Service are completing an investigation of the use of an organic chemical substance which has proven highly effective in inhibiting and destroying plant growth." The chemical could be turned into powder form and dissolved in the water surrounding paddy crops such as rice where, "over the course of a few weeks," it gets absorbed by the plants until "the great majority fall over and the growth of those still living is so retarded that they produce no rice." The CWS planned to conduct

---

[95] Minutes of Meeting Held in the MGO Conference Room (February 14, 1945) and Murray to Mitchell (February 27, 1945) in *The Meaning of Life*, ed. Avery and Eaton, 133–135.

[96] Minutes of the Meeting of the C.1. Committee Together with Representatives of the U.K. and Suffield (January 8, 1945), British Documents (originals and/or copies) copied for USBWC Steering Committee, CWS, and USBWC Files, Box 186, USBCW British Material on Crop Destruction Agents to WBC, Recommendations, Security-Classified Correspondence File of Dr. G. W. Merck, NM84 488, RG 165: Records of the War Department General and Special Staffs, NARA.

[97] Coen, *Fu-Go*, 187–195, 213–215.

[98] Summary of BW Intelligence from January 15 to May 12, 1945, BW Intelligence, Box 181, Report of Scientific Intelligence Survey in Japan, BW, Vol. 5 to Final Board Report, Security-Classified Correspondence File of Dr. G. W. Merck, NM84 488, RG 165: Records of the War Department General and Special Staffs, NARA.

extensive field trials in Texas when the rice crop grew in, but, "on the basis of the present results," Merck and Marquand wrote, "it is believed that we are in possession of a weapon which can destroy ... the rice crop of Japan." The chemical "is now in commercial production," they continued, and it can be dropped from planes "in powdered form in bomb clusters of the incendiary type."[99] Tojo was right to have been worried.

Merck and Marquand were quick to insist that such an attack would not be an act of either biological or chemical warfare. The agent involved "was not made from living organisms or by any biological process, and it is not a living organism." Neither is it "a poison harmful to man or animals." Its use would not "violate international law."[100] The fact went unspoken that plenty of other research the US was conducting at the time would, if used, violate international law. That was just research even if, as was true at Grosse Île, it was "made from living organisms." But what if it became something more than research?

That April, Kelser had written to Merck that two Canadian members of the Joint Commission had "discussed informally with several British officials engaged in b.w. activities the possibility of a field test." They had received a positive response, promptly being informed that "proper arrangements could undoubtedly be made to subject the avianized virus vaccine ... to a field test in East Africa." Merck passed the information on to the United States Biological Warfare Committee (USBWC) three days later, stressing the point that the avianized vaccine was "to the best of our knowledge the only vaccine which it would be practical to produce in the large quantities which would undoubtedly be required if the disease should be introduced on this continent through enemy activities." He added that it was "highly desirable to subject it to a field test for the purpose of fully establishing its potency under field conditions." Securing approval with the USBWC, Merck took the matter to Henry Stimson on May 4, insisting that "it is necessary to try it in the field in a region where the disease is endemic." The "most suitable region for such a test," he continued, "is British East Africa." And it just so happened that "British authorities, who will guard the security of the project with care, have expressed a desire to perform the proposed field test."[101]

---

[99] John P. Marquand and George W. Merck to Secretary of War (May 3, 1945), USBWC – Steering Committee, Box 184, United Kingdom to USBWC Steering Committee, Security-Classified Correspondence File of Dr. G. W. Merck, NM84 488, RG 165, Records of the War Department General and Special Staffs, NARA.

[100] Ibid.

[101] Kelser to Merck (April 18, 1945); USBWC – Steering Committee, Box 184, United Kingdom to USBWC Steering Committee, Security-Classified Correspondence File of

The USBWC received a note the following day from a member of the British Army Staff emphasizing the point and explaining that, "in order to avoid undesirable attention being drawn to the trials, they can best be done as a normal test of an advance in veterinary practice, without revealing the exact nature or method of production of the material nor its connection with B.W."[102]

British eagerness to test the vaccine in East Africa was, of course, about more than a willingness to help out its allies. The same qualities that made the vaccine so exciting for North American scientists also made it exciting for British colonial authorities. They wanted to know if it would work in East African cattle. Canada and the United States wanted the vaccine for defense. The United Kingdom wanted it for empire. Field tests performed in Kenya in September and October of 1945 and additional tests in Uganda confirmed everyone's hopes: the vaccine worked.[103] By that point, however, the war was over, and the vaccine's fate remained uncertain.

***

None of this later research was under Shope's direction. Once it became clear that he had fulfilled his initial mission to create a safe, effective, and efficient vaccine, the US government decided that it needed his expertise elsewhere. Shope was transferred to Naval Medical Research Unit No. 2 in April of 1944 and shipped to the Pacific. There, he oversaw the construction of a Navy laboratory unit on Guam. He also headed an

---

Dr. G. W. Merck, NM84 488, RG 165, Records of the War Department General and Special Staffs, NARA; Merck to USBWC (April 21, 1945), USBWC – Steering Committee, Box 184, United Kingdom to USBWC Steering Committee, Security-Classified Correspondence File of Dr. G. W. Merck, NM84 488, RG 165, Records of the War Department General and Special Staffs, NARA; Merck to Stimson (May 4, 1945), USBWC – Steering Committee; Box 184, United Kingdom to USBWC Steering Committee, Security-Classified Correspondence File of Dr. G. W. Merck, NM84 488, RG 165, Records of the War Department General and Special Staffs; NARA.

[102] Colonel Paget to USBWC (May 5, 1945), USBWC – CIR1 – Consultations, Box 185, USBWC BW Press Releases to USBWC, GIR 1, Considerations, Security-Classified Correspondence File of Dr. G. W. Merck, NM84 488, RG 165, Records of the War Department General and Special Staffs, NARA.

[103] Hale et al., "XIV. Immunization Experiments with Attenuated Rinderpest Vaccine Including Some Observations on the Keeping Qualities and Potency Tests." *American Journal of Veterinary Research* 6:23 (April 1946): 215; Hudson, J. R. "Rinderpest Virus Attenuated in Eggs." *The Veterinary Record* 59:25 (July 5, 1947): 331; Kelser to Murray (December 11, 1946) in *The Meaning of Life*, ed. Avery and Eaton, 150–151.

expedition of thirty doctors to Okinawa to "find out what diseases were present"; "learn whether investigation *could be done under combat conditions*" (emphasis original); and "do such investigation if this were possible." It was harrowing work. The team landed twelve days after the first American landing in the bloodiest battle of the Pacific theater. They stayed three months. Shope had always loved field work, but this was of a very different variety. He remained in the Navy until 1946, returning then to Princeton, New Jersey, to his family and to his job at the Rockefeller Institute for Medical Research.[104] Shope no doubt left the military satisfied that he and his team at Grosse Île had created for humanity a powerful new weapon against its old foe, rinderpest, but very few people knew about it. The vaccine had to be known to matter.

The United States had funded GIR-1 by paying for the researchers' reports. Those documents had been kept top secret throughout the war, but the war was over, and they contained valuable information about how to produce avianized vaccine. Discussion commenced almost immediately about what to do with them and the other reports like them on the Allied research that had been done on all kinds of pathogens during the war. At an August 23, 1945, meeting of the USBWC Steering Committee, members discussed a proposal to "declassify all work done except that on offensive military operations without announcing that we have worked on the subject from an offensive point of view. Announce publically that we have worked on all defensive aspects of the problem and tell why we did so" (emphasis original). Merck insisted that "declassification" was necessary "in order to obtain proper recognition for those who have done the job, and to solicit financial support for a continuing program."[105] He would employ a different argument outside the committee: a humanitarian one.

In October 1945, George Merck sent a summary report to Secretary of War Robert P. Patterson, describing "the combined efforts of American scientists and industry working with the armed forces and in cooperation with ... the United Kingdom and Canada to develop defenses to enemy attack by biological warfare." The research that had come of out these

[104] Peyton Rous, "Presentation of the Kober Medal to Richard Shope," Box 1, Folder 2, Collection RU, RG 450 Sh77, RAC; Richard E. Shope (January 11, 1965), Box 1, Folder 2, Collection RU, RG 450 Sh77, RAC.

[105] Report on Meeting of USBWC Steering Committee (August 23, 1945), USBWC – Steering Committee, Box 184, United Kingdom to USBCW Steering Committee, Security-Classified Correspondence File of Dr. G. W. Merck, NM84 488, RG 165, Records of the War Department General and Special Staffs, NARA.

efforts, Merck wrote, "contributed significant knowledge to what was already known concerning the control of diseases affecting humans, animals and plants." This work had significance far beyond the war. "Steps are being taken," Merck continued, "to permit the release of such technical papers and reports by those who have been engaged in this field as may be published without endangering the national security." This needs to be done, he argued, because much of the information we learned during the war "will be of great value to public health, agriculture, industry, and the fundamental sciences."[106] Merck's report was passed along to the Joint Chiefs of Staff shortly thereafter with the message that "arrangements have been made whereby this information of value to humanity as a whole will be made available to the public."[107] And it was.

In the aftermath of Merck's efforts to ensure approval, researchers published over 160 "technical articles" on the work they had undertaken on biological warfare during the war.[108] In April 1946, the *American Journal of Veterinary Research* published an entire issue dedicated to the research completed at Grosse Île. The reports that the United States government had paid thousands for during the war via its below-the-radar accounting deal were now freely shared with anyone who could get a copy of the issue.[109] The journal did not, of course, contain information about all of the research that had been conducted on the island. These were the defensive reports, not the offensive ones. They announced the new vaccine to the world.

Publication was a critical part of the effort to share the vaccine, but it was not, some of the researchers involved argued, enough. They were not certain that their efforts could be duplicated elsewhere. They had tried to repeat their egg-growing experiment with "the caprine strain of the virus," the "Indian strain," and the "North African strain," but had

---

[106] Report to the Secretary of War by Mr. George W. Merck, Special Consultant for Biological Warfare (October 1945), CCS 385.2 (December 17, 1943) Sec. 4, Chemical, Biological & Radiological Warfare (JCS 1822), Box 375, Central Decimal File, 1942–45, RG 218, Records of the US Joint Chiefs of Staff, NARA.

[107] Ibid.

[108] Secretary Forrestal Issues Statement on Biological Warfare Potentialities (13 March 1939), CCS 385.2 (December 17, 1943) Sec. 8, Chemical, Biological & Radiological Warfare (JCS 1822), Central Decimal File 1948–50, Box 206, RG 218, Records of the US Joint Chiefs of Staff, NARA.

[109] "Foreword." *American Journal of Veterinary Research* 6:23 (April 1946): 134. For another example of the publication of wartime biological weapons research, see Brandly, C. A., et al., "Newcastle Disease and Fowl Plague Investigations in the War Research Program." *Journal of the American Veterinary Medical Association* 108:831 (June 1946): 369–371.

been unable to produce another line of avianized vaccine.[110] This was not unusual in vaccine research; sometimes the mutation worked and sometimes it did not, but it was disappointing and concerning. They did not want the vaccine to be lost. And that was not a remote threat, for Grosse Île's days were numbered.

Immediately following the victory over Japan, the United States Department of War announced all war projects would be terminated within a month. "An exception was made to the Grosse Île project because of its biological nature, it being self-evident that time is required to inactivate living material," but researchers there still did not get much time.[111] The United States announced that it would end its relationship with and support for Grosse Île on February 28, 1946. As the deadline approached, Charles Mitchell, who had been involved with the project from the beginning but had become its leader after Shope's departure, wrote to the head of Canada's Department of Agriculture asking that the government keep the station open past February so that it could undertake a new mission: sharing the vaccine.

Mitchell proposed that they needed "approximately six months for the purpose of permitting representatives from other parts of the world, particularly those of the British Commonwealth, to visit the Station and train in the methods of vaccine production, also so that the particular strain which has been propagated will not be lost." That event, Mitchell insisted, "would be a major catastrophe having regard to the future protection of food supplies of the world, as it permits the economical production of rinderpest vaccine on a large scale." Grosse Île had the world's best vaccine. It came from one strain of the virus. Losing it might mean losing the vaccine, perhaps forever. Mitchell pleaded for a stay of execution: "Decontamination will be commencing in the last week in January," he warned.[112] Mitchell's efforts, combined with a note of support from the US Department of Agriculture, which, though it could not offer any funding, insisted on its "interest in having the vaccine material made available" and indicated a willingness "to join in any way they could in making the product available through services they

[110] Hale, M. W., et al., "Rinderpest XIII. The Production of Rinderpest Vaccine from an Attenuated Strain of Virus." *American Journal of Veterinary Research* 6:23 (April 1946): 200.
[111] Charles A. Mitchell, Memorandum: Re Project at Grosse Isle (January 19, 1946), Diseases of Animals, Rinderpest Control Vaccination Project, Grosse Isle, QC, Record Group 17, Volume 3029, Page 83, File 37–23, LAC.
[112] Ibid.

may have in other countries," worked.[113] The Canadian government decided that, "having regard for the need of conserving world food supplies and for the economic stabilization of production in those areas affected, or most likely to be affected, provision should be made to make the vaccine available for use in such areas." Grosse Île got six more months.[114]

The team there used that time to host visiting researchers, training them, as Mitchell had promised, in the production of the vaccine. They tried to ensure the survival of the seed virus by sending some to Kabete, Kenya, where its parent strain had come from in the first place.[115] R. Daubney was there waiting for it. He had been involved with the testing of the avianized vaccine the year before and had been very pleased with the results. His caprinized KAG vaccine worked well for most East African zebus – it was presently being used to vaccinate about fourteen million of them in an ongoing imperial effort in the region – but it was too virulent for cattle from Europe, West African cattle, Ankole longhorns, and even some zebu in Uganda. The avianized vaccine could replace the inactive vaccines that were still required for vaccinating these types of cattle. Such vaccines were not only expensive; they were frustrating to work with in a massive vaccination campaign. The duration of immunity that they conferred was always uncertain, "its gradual disappearance," Daubney lamented, "revealed only by the reappearance of the disease in the vaccinated." The avianized vaccine was the solution authorities had been waiting for, because it could be easily adjusted to the desired degree of virulence according the number of passages. It was simply a question of selecting the appropriate egg passages for the herd in question. This flexible vaccine, he cheered, marked "great progress in the fight against rinderpest," opening the door to a revitalized "eradication campaign."[116]

[113] G. S. H. Barton to A. Ross (January 4, 1946), Diseases of Animals, Rinderpest Control Vaccination Project, Grosse Isle, QC, Record Group 17, Volume 3029, Page 83, File 37–23, LAC; Unknown to G. S. H. Barton (January 19, 1946), Diseases of Animals, Rinderpest Control Vaccination Project, Grosse Isle, QC, Record Group 17, Volume 3029, Page 83, File 37–23, LAC.

[114] Order in Council Providing for Continuance of R. Project Being Conducted at Grosse Isle (January 24, 1946), Diseases of Animals, Rinderpest Control Vaccination Project, Grosse Isle, QC, Record Group 17, Volume 3029, Page 83, File 37–23, LAC.

[115] Cheng, S. C. , T. C. Chow, and H. R. Fischman. "Avianized Rinderpest Vaccine in China," in Rinderpest Vaccines, ed. Kesteven, 35–41.

[116] Daubney, R. "Récentes Acquisitions Dans La Lutte Contre La Peste Bovine." Bulletin Office International Des Epizooties 28 (May 1947): 36–45.

Daubney, a British colonial official in East Africa, was predisposed to be excited about the new vaccine. He was already engaged in a large-scale struggle against the virus. He interpreted the vaccine as a new weapon in what was essentially an imperial war. In his case, the vaccine did not transform his political vision; it simply expanded the scope of what he hoped the technology could accomplish. But Daubney and his fellow colonial administrators were not the only ones paying attention to the new technology.

In the spring of 1946, researchers at Grosse Île sent seed virus to Kabete, but that was not the only package that left the island. That May, they also sent one million doses of avianized vaccine to United Nations Relief and Rehabilitation Administration officials in Shanghai, enabling them to add fighting rinderpest to their list of actions undertaken in the name of "relief and rehabilitation" in liberated China.[117] Chinese researchers were also invited to Grosse Île to study. This eradication campaign, unlike Daubney's, would not be undertaken in the name of imperialism, but in the name of internationalism, through the efforts of the United Nations.

That August, an ad hoc committee formed by the new United Nations Food and Agriculture Organization met to discuss whether or not FAO should be involved in "veterinary problems" or if the issue lay outside its purview. The committee, noting "with the liveliest interest, the cooperative research into this disease carried out by the United States and Canadian Scientists which resulted in the preparation of a vaccine of a new type which is now being tested in the field in Africa and China," "strongly recommended" that FAO make animal health a permanent part of its mission and that it make rinderpest eradication one of its primary goals.[118] The Grosse Île research played a key role in convincing the committee that the organization should do so, because the new avianized vaccine, along with the proven success of the caprinized ones, offered hope that rinderpest eradication was suddenly an achievable goal. The vaccines provided the way. The new United Nations machinery of international assistance provided the will.

***

---

[117] Cheng, S.C., T.C. Chow, and H.R. Fischman. "Avianized Rinderpest Vaccine in China," in *Rinderpest Vaccines,* ed. Kesteven, 35–41.

[118] Notes on FAO Veterinary Meeting held in London (August 13–15, 1946), London 1946 Report, ADHOC Committee on Animal Health, Animal Production and Health Division, 10AGA407, FAO.

In September of 1946, a few days after the termination of the Grosse Île project, an internal memo at Canada's Department of Agriculture tackled the question of what to do with "a number of chickens" still on the island. "The chicken colony is required as a source of food for the caretaker's family and also as a nucleus should another project be undertaken similar to the one which has been abandoned," the memo explained. "None of the chickens, or their progeny, or any eggs should be removed from the island."[119] All had to stay. One never knew when researchers might need them again. Just in case, researchers also left some of the seed virus and the avianized vaccines in an ice chest, quarantined on the island. They might need them again, too.[120] Internationalism was all well and good, but there was a reason that the US and Canadian governments had not published the reports of the offensive research that they had encouraged on the island during the war. There was a reason that they kept samples of the virus and the vaccine stored there, as well. The vaccines had helped to make the concept of internationalism more plausible by offering a tangible way forward for action in its name, but they had also done the same for biological warfare. For the moment, attention focused on the former, but the virus in the ice chest remained a reminder that the war had transformed rinderpest in more ways than one. It could giveth – through virus mutated into vaccine – and it could taketh away – through virus mutated into resistance to that vaccine. It was just one of many ways in which the war had made the world both more connected and more dangerous.

[119] Memo from J.H. Craigie for the Deputy Minister (September 24, 1946), Diseases of Animals, Rinderpest Control Vaccination Project, Grosse Isle, QC, Record Group 17, Volume 3029, Page 83, File 37–23, LAC.
[120] Charles A. Mitchell to Glen Gay (July 15, 1949), Biological Warfare – Rinderpest Virus, RG24, Series F-1, Vol. 4224, LAC.

3

"Freedom from Want": UNRRA's Rinderpest Campaigns

In December 1942, the International Bureau of the Fabian Society held a conference at Oxford to discuss "the first steps in the relief and rehabilitation in Europe." The lectures, which were broadcast in Europe and North America at the time, were published in 1944 with an introduction by Leonard Woolf that warned readers that the Nazis will leave behind them not only "burnt and bombed cities" but "millions of individuals all over Europe weakened by undernourishment and disease and threatened by starvation and disease." The disaster will be "gigantic," he acknowledged, but we can through "forethought and by international planning and cooperative action" successfully address it. The conference was an effort to aid in that planning and encourage that action.[1]

Participants had reason to be hopeful. In August 1941, Roosevelt and Churchill had promised via the Atlantic Charter that they were dedicated to securing a peace "which will afford assurance that all the men in all lands may live out their lives in freedom from fear and want."[2] Everyone involved knew that it was, as Woolf wrote, "a problem first of relief, of immediate relief to millions of people who will be in urgent need of food and medical aid." But Woolf and the participants of the 1942 meeting in Oxford were insistent that just sending food and medicine would not be enough. "Such relief would be a mockery," Woolf wrote, "unless at the

[1] Woolf, Leonard. "Introduction," in *When Hostilities Cease: Papers on Relief and Reconstruction Prepared for the Fabian Society*. Huxley et al. London: Victor Gollancz Ltd., 1944, 11.

[2] Borgwardt, Elizabeth. *A New Deal for the World*. Cambridge, MA: Harvard University Press, 2005), 28–35. See also Iriye, Akira. *Cultural Internationalism and World Order*. Baltimore: The Johns Hopkins University Press, 1997, 131–176.

same time steps are taken to restart them in ordinary productive life." For Woolf, those extra steps were repatriation, reconstruction, and the supplying of "raw materials, machinery, seeds, cattle, fertilizers" and more. "Relief will only be effective in any part of the field," he insisted, "if it is a prelude to reconstruction."[3]

The members of the Fabian Society were adamant that relief and reconstruction needed to be embraced as two parts of one whole, and that the Allied governments needed to create international machinery for carrying them out. The British zoologist Julian Huxley noted in his speech at the 1942 conference that "modern technology has brought the world to a state in which ideas, inventions, economic and social processes and policies are no longer even approximately limited by national or even continental boundaries." The tools of "relief and reconstruction" were inherently international, "hence the urgent need for international machinery" to administer them. To forestall resistance from those who would hold up the League of Nations as an example of the inherent limitations of such machinery, Huxley countered that its "failure ... as a political organization ... must not blind us to the success of the special purpose League agencies, such as the International Labour Organization, the Health Organization, and the Economic and Financial Section."[4] What was still left of the League of Nations heartedly agreed.

In 1942, the League (still "soldier[ing] on in a variety of guises and locations") circulated amongst Allied officials a "manifesto" whose principle purpose, Patricia Clavin has argued, was "to shape policy, notably American, on reconstruction and international relations."[5] *The Transition from War to Peace Economy*, which was officially published in May of 1943, argued that international cooperation was necessary to both winning the war and making a lasting peace. Securing the latter would begin with relief, but it could not stop there, as it had after the last war. "Relief, to be effective," the report's authors insisted, "must not simply fill the human belly for a short period of time, but must enable the individuals who require it to continue that process themselves in the future." Relief must not be thought of as charity, but as "the first step in reconstruction," for they are "a single problem." Avoiding a repeat of the

[3] Woolf, "Introduction," 12–15.
[4] Huxley, Julian. "Relief and Reconstruction," in *When Hostilities Cease*, Huxley et al., 20–21, 25.
[5] Clavin, Patricia. *Securing the World Economy: The Reinvention of the League of Nations*. Oxford: Oxford University Press, 2013, 251, 285–294.

economic anarchy of the 1930s demanded aggressive international action to secure human welfare. It needed to start with a commitment "to ensure that the fullest possible use was made of the resources of production, human and material, of the skill and enterprise of the individual, of available scientific discoveries and inventions so as to attain and maintain in all countries a stable economy and rising standards of living."[6] The message did not go unnoticed in Washington, precisely because the American president had already shown himself willing to make that kind of commitment.

Roosevelt had first made "freedom from want" a goal of the war effort in his 1941 Annual Message to Congress, describing it as securing "to every nation everywhere a healthy peacetime life for its inhabitants – everywhere in the world." Roosevelt had added the Four Freedoms to the speech himself: of speech, of religion, from want, and from fear. He had also explicitly made them international goals. "The world order which we seek," he explained, "is the cooperation of free countries, working together in a friendly, civilized society."[7] Roosevelt was already planning for peace long before the United States entered the war. The question by 1942 was not whether the Allied powers would work to create international machinery in search of "freedom from want," for they had already started to do so in pursuit of "relief." The question was how expansive would be the scope of that machinery's charge. Woolf, Huxley, and the League were insistent that "freedom from want" required not just "relief," but also "reconstruction."

US and British officials were not there yet in 1942, but the machinery that they began putting into place for "relief" paved the way over the course of the 1940s to not just a commitment to "reconstruction" but a commitment to "development," creating spaces for action where humans could assist one another, as Julian Huxley later argued, at growing "better" plants, breeding "finer" animals, and eradicating diseases.[8] The rinderpest vaccines assisted in that transformation. Initially sent abroad as "relief," the vaccines were, by their very nature,

---

[6] League of Nations. Report of the Delegation on Economic Depressions, Part I, in *The Transition from War to Peace Economy*. Geneva: League of Nations, 1943, 14, 75–76.

[7] Franklin D. Roosevelt, Annual Message to Congress, 6 January 1941; available at https://fdrlibrary.org/four-freedoms.

[8] Huxley, Julian. "Introductory Note," in *Reshaping Man's Heritage*. London: George Allen & Unwin Ltd., 1947, 5

a transforming power, not a relieving one. They altered other living things and, in consequence, changed the environment itself. In the process, they also altered human hopes about the possibility, as *The Transition from War to Peace Economy* urged, of using "available scientific discoveries and inventions so as to attain and maintain in all countries a stable economy and rising standards of living."[9] The rinderpest vaccines encouraged bolder imaginings of what international cooperation in the name of "freedom from want" could achieve.

***

The Allied effort to tackle want began in the fall of 1941 when the British government created the Inter-Allied Committee on Post-War Requirements. Churchill had promised in a speech the year before that "we can and we will arrange in advance for the speedy entry of food into any part of the enslaved area, when this part has been wholly cleared of German forces and has genuinely regained its freedom," and he wanted to fulfill it. Europeans need to know, he insisted, that the "shattering of Nazi power will bring them all immediate food, freedom and peace." Representatives of all of the European allies whose governments were living in exile in London and observers from the USSR and the US participated in the meeting that fall which focused on trying to figure out what countries were going to need at the end of the war. The outcome of the war was far from certain at that point, but the meeting helped build confidence and cooperation.[10] That fall, the United States created the Office of Foreign Relief and Rehabilitation Operations (OFFRO) within the Department of State. These efforts were followed, in the summer of 1942, with the creation of the Combined Food Board to sponsor "the interchange of information and the development of international plans to make the best use of the free world's diminishing resources of manpower, machinery, fertilizers, and other material on the food front." The Food Board coordinated the production and distribution of food in over seventy

---

[9] League of Nations, *The Transition from War to Peace Economy*, 14.

[10] *"Foreign Relief and Rehabilitation," The Testimony of Dean Acheson, Assistant Secretary of State, before the Committee on Foreign Affairs, House of Representatives*, 78th Cong. First Session, on A Draft Agreement for a United Nations Relief and Rehabilitation Administration, July 7, 1943 (Washington, DC: United States Government Printing Office, 1943), 1–2; Woodbridge, George. *The History of the United Nations Relief and Rehabilitation Administration, v. I.* New York: Columbia University Press, 1950), 8–9; Shephard, Ben. "'Becoming Planning Minded': The Theory and Practice of Relief 1940–1945." *Journal of Contemporary History* 43:3 (July 2008): 405–419.

countries.[11] All of it, Roosevelt explained was designed "to restore each of the liberated countries to soundness and strength, so that each may make its full contribution to United Nations' victory and to the peace which follows."[12] The "and" was important, reflecting, as it did, the message of *The Transition from War to Peace Economy*: peace was going to require victory over a great deal more than just the Axis powers.

In the fall of 1942, an Australian official gave Roosevelt a memo titled "United Nations Programme for Freedom from Want of Food," which proposed the creation of a permanent international organization to address hunger. Roosevelt was intrigued.[13] He told cabinet officials that he wanted to hold a conference "as soon as the preparatory work could be rushed through" about food and agriculture. When Dean Acheson, then Assistant Secretary of State, asked what Roosevelt "wanted done about food and agriculture," he was told to "work it out." Further questioning, Acheson later recalled, "elicited only that the President regarded food and agriculture as perhaps man's most fundamental concern and good place to begin postwar planning."[14]

That spring, the State Department gave notice to officials in Britain, the USSR, and China that the United States believed it was time for the United Nations and the associated nations "to begin joint consideration of certain fundamental economic questions which will confront them and the world after the attainment of complete military victory." As a "first step," the US planned on inviting those countries "to send a small number of appropriate technical and expert representatives" to a conference to explore "post-war plans" for agriculture and trade "in light of possibilities of progressively improving in each country the levels of consumption within the framework of an expansion of its general economic activity." Such considerations were to be at the conference "entirely divorced from the question of the provision of relief."[15]

---

[11] United States Department of Agriculture War Food Administration, Report of the Combined Food Board (April 1945), General Documentation, Combined Food Board (2CFB17), FAO.

[12] Division of Public Information, OFRRA, The Office of Foreign Relief and Rehabilitation Operations, Department of State (Washington, DC, 1943), 3.

[13] Borgwardt, *A New Deal for the World*, 114–115.

[14] Acheson, Dean. *Present at the Creation*. New York: W.W. Norton & Company, 1969, 73.

[15] Welles, Sumner. The Acting Secretary of State to the Chargé in the United Kingdom (8 March 1943), in *FRUS*, 1943, Volume I. Washington, DC: United States Government Printing Office, 1963), 820–821.

It was not that "relief" was not still a supreme concern. Indeed, that spring, while Acheson worked on organizing the United Nations Conference on Food and Agriculture, he was also working with his British, Soviet, and Chinese counterparts to create what would become the United Nations Relief and Rehabilitation Administration (UNRRA). But the Food and Agriculture Conference emphasized the point that sustained postwar peace was going to require sustained postwar cooperation. The conference opened in Hot Springs, Virginia, on May 18, 1943. Attendees insisted that "Freedom from want of food, suitable and adequate for the health and strength of all peoples, can be achieved."[16] They called for the creation of a permanent international Food and Agriculture Organization (FAO) to help do it. Many of the nations at Hot Springs were already members of the International Institute of Agriculture (IIA), which had been created in 1905 to "collect, study, and publish" information about farming, vegetable diseases, prices, and "commerce in agricultural products" in general. Issues "concerning the economic interests, the legislation, and the administration of a particular nation" were "excluded from the consideration of the institute."[17] Such a limited mandate had made sense in 1905, but it could not secure global "freedom from want." The main question on the table at Hot Springs was what kind of powers FAO needed in order to be able to do so. The debate was contentious, with American and British policymakers proposing that FAO focus its energies on technical advising while the representatives of many other countries calling for it to be a force for action to stimulate both production and distribution. The nutrition and agricultural experts at the conference firmly believed that international cooperation had to *mean* more and *be* more in the postwar world than it had in the prewar one. The question was how much national sovereignty (and funding) needed to be surrendered in the pursuit of international prosperity.[18]

[16] "United Nations Conference on Food and Agriculture: Text of Final Act." *The American Journal of International Law* 37:4, Supplement: Official Documents (October 1943): 163.

[17] Convention of the International Institute of Agriculture (7 June 1905), in *Treaties and Other International Agreements of the United States of America, 1776–1949*. Volume 1, ed. Charles I. Bevans. Washington, DC: United States Government Printing Office, 1968–76, 436–440.

[18] Jachertz, Ruth. "Coping with Hunger? Visions of a Global Food System, 1930–1960." *Journal of Global Health* 6 (2011): 99–119; Staples, Amy L. S. *The Birth of Development.* Kent, OH: Kent State University Press, 2006, 76–77; Vernon, James. *Hunger: A Modern History.* Cambridge: Harvard University Press, 2007, 17–40, 83; Cullather, Nick. *The Hungry World.* Cambridge: Harvard University Press, 2010, 13–25. See also

The question was far from confined to FAO. Immediately following Hot Springs, the State Department sent out a draft agreement to all of the United Nations and associated nations for the "immediate establishment of a central United Nations agency to assume responsibility for the relief and rehabilitation of the victims of the war." The two "Rs" marked a shift in thinking. Acheson later explained that the "bombing and obsolescence had pointed to a larger problem and the word 'rehabilitation' was added to 'relief.'" Participants felt that "the occupied countries must not merely be fed, they must be helped to be self-supporting." His use of the passive voice was revealing: "rehabilitation was added." The US Ambassador to Great Britain cabled Acheson to ask what it meant. The latter recorded in his memoirs, "A good question it was, but never answered. To us the word had no definition; rather it was a propitiation by ignorance of the unknown."[19] The House Committee on Foreign Affairs asked him the same thing during a hearing on the draft in July 1943 and Acheson turned to the draft itself for his answer. "Now relief is defined as the provision of food, fuel, clothing, and other basic necessities, housing facilities, medical and other essential services," he explained. Then the draft "goes on to say, and this is what is called the definition of rehabilitation – 'and to facilitate in areas receiving relief the production and transportation of these articles and the furnishing of these services so far as necessary to adequate provision of relief.'"[20] It was not a particularly helpful answer.

"Rehabilitation," was, in fact, a way of avoiding the use of the more expansive "reconstruction." The State Department had invited some of the League authors of *The Transition from War to Peace Economy* to Washington to advise them on the creation of the new organization that spring and, while they adopted many of the report's ideas, they resisted its language.[21] The United States had already picked "rehabilitation" over "reconstruction," using the former for OFFRO back in 1941. The wording of the UNRRA draft almost exactly matched Roosevelt's charge for OFFRO and it was a deliberate choice – not, as Acheson later

---

Cullather, Nick. "The Foreign Policy of the Calorie." *American Historical Review* 112:2 (April 2007): 336–364.

[19] Acheson, *Present at the Creation*, 68–69.

[20] "Foreign Relief and Rehabilitation," *The Testimony of Dean Acheson, Assistant Secretary of State, before the Committee on Foreign Affairs, House of Representatives, 78th Cong. First Session, on A Draft Agreement for a United Nations Relief and Rehabilitation Administration*, July 7, 1943 (Washington, DC: United States Government Printing Office, 1943), 11.

[21] Clavin, *Securing the World Economy*, 297–304.

tried to make it seem, a passive one.[22] Herbert Lehman, the head of OFFRO and future director of UNRRA, explained in 1943 that relief meant "food to halt starvation, clothing and emergency shelter to stop deaths by exposure and medicines and medical facilities." "Immediately thereafter, or perhaps simultaneously, seeds and some farm implements must be provided to enable liberated peoples to get a crop into the ground." We assume, he continued that the subsequent harvest "will mark the peak of actual relief operations in any given area." At that point, we can provide some "raw materials" and "commercial goods" to get the "economy moving once again." With these combined relief and rehabilitation measures, Lehman concluded, "the liberated nations can initiate their own long range measures of reconstruction."[23]

The final UNRRA agreement followed OFFRO's plan: immediate aid in the form of "food, clothing and shelter, aid in the prevention of pestilence and in the recovery of the health of the people," followed by assistance for "the return of prisoners and exiles to their homes" and for "the resumption of urgently needed agricultural and industrial production

---

[22] Division of Public Information, OFRRA, The Office of Foreign Relief and Rehabilitation Operations, Department of State (Washington, DC, 1943), 4–5. Elizabeth Borgwardt wrote, "Wartime discussions of aid usually divided the topic into three categories: relief, rehabilitation, and reconstruction. The logic underlying these divisions was apparently to separate short-term emergency aid from longer-term assistance, but definitions and boundaries remained somewhat mysterious, even to the planners themselves" (Borgwardt, *A New Deal for the World*, 118). I think the word choice was much more deliberate. In his history of UNRRA, George Woodbridge wrote that both the United States and the United Kingdom had "made it clear that they wished to avoid all responsibility for long-range reconstruction in the relief organization. They therefore adopted the term 'rehabilitation' in the draft Agreement and decided that it should cover only the transitional measures needed to restart industrial and agricultural activities and essential services" (Woodbridge, *UNRRA*, V. I, 31). For more discussion of the terminology, see Shephard, "Becoming Planning Minded," 405–419; Mazower, Mark. "Reconstruction: The Historiographical Issues." *Past and Present*, Supplement 6 (2011): 17–28; Reinisch, Jessica. "'We Shall Rebuild Anew a Powerful Nation': UNRRA, Internationalism and National Reconstruction in Poland." *Journal of Contemporary History* 43:3 (July 2008): 451–476; Reinisch, Jessica. "Internationalism in Relief: The Birth (and Death) of UNRRA." *Past and Present*, Supplement 6 (2011): 258–289; Reinisch, Jessica. "'Auntie UNRRA' at the Crossroads." *Past and Present*, Supplement 8 (2013): 70–97; Ekbladh, David. *The Great American Mission*. Princeton: Princeton University Press, 2010, 87; Williams, Andrew J. "'Reconstruction' before the Marshall Plan." *Review of International Studies* 31:3 (July 2005): 541–558; Brown, William Adams, Jr. and Redvers Opie. *American Foreign Assistance*. Washington, DC: The Brookings Institution, 1953, 111.

[23] Lehman quoted in Division of Public Information, OFRRA, The Office of Foreign Relief and Rehabilitation Operations, Department of State (Washington, DC, 1943), 18.

and the restoration of essential services." The key action verbs were relief, return, resumption, and restoration.[24] Congress accepted them, authorizing the United States to participate in March of 1944.[25] In a message accompanying the first quarterly report to Congress on UNRRA expenditures and operations that December, Roosevelt reported that "the colossal task of relieving the suffering victims of war is under way." UNRAA, he continued, is helping the liberated peoples "so that they can help themselves; they will be helped to gain the strength to repair the destruction and devastation of the war and to meet the tremendous task of reconstruction which lies ahead." Their own governments, the report explained, will have the "major responsibility for seeing to it that the peoples liberated from the enemy will be able to liberate themselves also from the hunger and disease that the enemy left behind among them."[26] UNRRA was just a starting point, but everyone agreed it was a vital one.

UNRRA assistance took many forms. Officials divided its supply program, which consumed the vast majority of its resources, into five major commodity divisions: food, agricultural-rehabilitation supplies, industrial-rehabilitation supplies, clothing, and medical and sanitation supplies. Boats, trains, and planes had carried grain, canned goods, vaccines, pesticides, shoes, blankets, fertilizers, seeds, animals, tractors, trucks, freight cars, and much more into seventeen "invaded" nations by the time UNRRA completed its major operations in June of 1947. By that point, the agency had spent almost four billion dollars total – almost

[24] Agreement for the United Nations Relief and Rehabilitation Administration (November 8, 1943); available online at https://www.loc.gov/law/help/us-treaties/bevans/m-us t000003-0845.pdf.

[25] To get that approval, Roosevelt assured Congress that the organization "will not, of course, be expected to solve the long-range problems of reconstruction. Other machinery and other measures will be necessary for this purpose" (*Participation by the United States in the Work of the United Nations Relief and Rehabilitation Organization. Message from the President of the United States recommending that the Congress enact legislation authorizing the appropriation of funds as Congress may from time to time determine to permit the participation by the United States in the work of the United Nations Relief and Rehabilitation [15 November 1943]*; House of Representatives, 78th Cong. 1st Session, Document No. 355). Ellen S. Woodward, a member of the US delegation, seconded the point, reporting back from UNRRA's first conference that "UNRRA cannot do reconstruction on a vast scale – nor do the nations which have suffered destruction want them to. They want to do the work of reconstruction themselves" (*"UNRRA Spells Hope"* [Address before the Mississippi State Society, March 5, 1944], Ellen Sullivan Woodward Papers, 1927–1954, Speeches 1944. A-54, folder 22. SL).

[26] *First Report to Congress on United States Participation in Operations of UNRRA, as of September 30, 1944*. Washington, DC: US Government Printing Office, 5, 8.

90 percent of which went into the procurement and shipping of those supplies.[27] The plan was to focus on food, medicine, and clothing relief first and then move on to agricultural and industrial rehabilitation: securing short-term "freedom from want" while building a foundation from which nations could help themselves secure it permanently. In the pursuit of that vision, UNRRA both scored victories and admitted defeats, the combination of which had a significant impact on the way both donor and recipient nations thought about the possibilities of international assistance. Its most infamous defeat would come in China – the site of its largest program – but there were also victories won in that effort, including the successful establishment of an anti-rinderpest campaign that would eliminate the virus from the country within ten years. That effort, and others like it, transformed UNRRA's "agricultural rehabilitation" mission into a "new economic development" mission, the ramifications of which stretched far beyond both China and UNRRA.[28]

***

The Chinese government sent a request for UNRRA aid in September of 1944. The scope of its need was overwhelming. Japan had occupied as much total space in China as Germany had in Europe, but it had held much of it longer. By the end of the war, as many as 260 million people (out of China's total population of around 461 million) were living in provinces that had been liberated from Japan. Most were hungry. The first UNRRA contingent arrived in Chunking, the wartime capital, from Washington, DC, in November. They officially opened the China office at the end of the year. The Chinese Nationalist government, in turn, created the China National Relief and Rehabilitation Administration (CNRRA) in January 1945 to coordinate efforts. The communist government created the Communist Liberated Areas Relief Administration (CLARA) to fill the same need in its territory. CLARA representatives were allowed to be stationed at CNRRA headquarters in Shanghai and at Tientsin and Kaifeng, but they would be allotted only a small fraction of aid. UNRRA and CNRRA signed the

---

[27] Recipient nations of the supply program included Albania, Austria, Byelorussian S.S.R., China, Czechoslovakia, Dodecanese Islands, Ethiopia, Finland, Greece, Hungary, Italy, Korea, Philippines, Poland, Republic of San Marino, Ukrainian S.S.R., and Yugoslavia. Operations ended in Europe in June of 1947, but continued in China, Korea, and the Philippines into the fall (*12th Report to Congress on Operations of UNRRA, as of June 30, 1947*. Washington, DC: US Government Printing Office, 1948, 3–15).

[28] Ibid., 9.

Basic Agreement in November 1945. Supplies and staff followed. There was a great deal to do.[29]

In a message to Congress that month asking for an additional appropriation of 1.35 billion dollars for UNRRA activities, President Harry S. Truman noted that "China presents the largest of all the relief responsibilities which UNRRA now faces. With inadequate supplies and resources it has struggled bravely for 8 years to combat the enemy as well as the ravages of famine, disease, and inflation."[30] Humans needed food, clothing, shelter, and medicine, and they were not alone. The Chinese government estimated that during the war it had lost 2.6 million water buffalo, 2.2 million cattle, 5.17 million sheep and goats, and 16.8 million pigs.[31] The ones that had survived were more valuable than ever and needed protection. A May 1945 UNRRA memo warned that "the replacement of livestock, particularly work stock, is going to be one of the most serious problems in agricultural rehabilitation." Replacement would help, but that would not solve the problem of keeping fresh livestock alive. The loss to disease "is very great," the memo continued, and "a vigorous program of livestock disease control needs to be gotten underway at once."[32] UNRRA shipments to recipient countries regularly included sulfa drugs, penicillin, serums, and vaccines to help humans. It made sense to do the same to help keep alive the animals that humans depended upon for food and labor.[33]

UNRRA livestock relief-via-replacement would be limited: 3,257 dairy cows from the United States, New Zealand and Canada; ninety hogs from Canada; 1,000 sheep from New Zealand; twenty chickens from New Zealand; and 792 ex-Army mules and one horse from the United States. Instead, the organization focused on livestock "rehabilitation," which it

[29] Ray, J. Franklin, Jr. "UNRRA in China" (Tenth Conference of the Institute of Pacific Relations, International Secretariat, Institute of Pacific Relations, September 1947); Woodbridge, George. *The History of the United Nations Relief and Rehabilitation Administration*, v. II. New York: Columbia University Press, 1950, 371–376; UNRRA, "UNRRA in China," 1945–1947. *Operational Analysis Papers*, no. 53. Washington, DC: UNRRA, 1948), 2–10.

[30] *Message from the President of the United States Transmitting His Request to Congress to Authorize a New Appropriation for Participation in the Activities of United Nations Relief and Rehabilitation Administration (November 13, 1945)*. House of Representatives, 79th Cong. 1st Session, Document #378.

[31] UNRRA, "UNRRA in China," 15.

[32] W. J. Green to Kiser (May 1945), Agriculture Rehabilitation, 1944–1949, Folder S-1121–0000-0006, Box S-0528–0018, China Office, AG-018–001, UNRRA, UN.

[33] Woodward, Ellen S. "UNRRA and War's Aftermath." *Social Security Bulletin* (November 1945): 10.

primarily interpreted to be putting "China in a position to produce many of her own biologics." This required, to start, "supplies for rehabilitating four livestock disease control stations which will make biologics as well as conduct control programs." Efforts did not stop there. UNRRA shipments included "supplies, both biologics and equipment for laboratories and manufacturing" and "high-grade specialists, to assist in organizing livestock disease control programs in the provinces, . . . planning the equipment and staffing of laboratories and manufacturing facilities, . . . [and] setting up of veterinary training schools." UNRRA would try to meet China's need with technology transfers, not livestock transfers.[34] The new rinderpest vaccines played a key role in making that a viable option.

A 1947 UNRRA report estimated that diseases claimed about five percent of Chinese livestock each year. Rinderpest was "much the most serious, causing an estimated loss of 200,000 to 300,000 draft cattle per year, a single outbreak sometimes causing the loss of eighty to ninety percent of the work animals in a community and rendering the proper cultivation of the land difficult or impossible."[35] Humans were trying to farm without them: "Today in many parts of China men and women are hitched to plows attempting to do the work of animals," but that was a terribly inefficient, and humiliating, option for which there was a clear solution.[36] UNRRA livestock disease control efforts focused on rinderpest, because rinderpest was killing China's agricultural labor force *and* because rinderpest could be controlled. The Chinese government wanted the vaccines. In its original UNRRA program request, it asked for "the establishment of five grade-A and ten grade-B serum laboratories and twenty-seven field units, altogether the furnishing of supplies to thirty institutions in twenty provinces." It proposed a wide-ranging war on the virus, and it got one.[37]

---

[34] Ibid.; Agricultural Rehabilitation in China During the UNRRA Period, Agricultural Rehabilitation Reports, 1944–1949, Folder S-1121-0000-0011, Box S-0528-0019, China Office, AG-018-001, UNRRA, UN.

[35] Agricultural Rehabilitation in China, Operational Analysis Report (December 31, 1947), Agricultural Rehabilitation Reports, 1944–1949, Folder S-1121-0000-0011, Box S-0528-0019, China Office, AG-018-001, UNRRA, UN.

[36] Agricultural Rehabilitation in China during the UNRRA Period; Agricultural Rehabilitation Reports, 1944–1949, Folder S-1121-0000-0011, Box S-0528-0019, China Office, AG-018-001, UNRRA, UN.

[37] Agricultural Rehabilitation in China, Operational Analysis Report (December 31, 1947), Agricultural Rehabilitation Reports, 1944–1949, Folder S-1121-0000-0011, Box S-0528-0019, China Office, AG-018-001, UNRRA, UN; Agricultural Rehabilitation in

FIGURE 3.1 Field veterinarians often carried vaccine into rural areas in steel thermoses packed with ice. Here, one gets ready to go to work in a small village office near Taipei.
Photograph courtesy of National Archives, College Park, MD.

The war began in earnest in the spring of 1946. Between May and December, 440 cases of veterinary supplies arrived in China. Sixty-six of them were US Army surplus stock – primarily refrigerators. They needed them to store what was also arriving in 1946: a large assortment of "veterinary biological," including "3,455,000 cc. of sera, 1,041,500 doses of vaccines, 54,900 tests of diagnostic agents and 2,300 cc. of virus."[38] The sera were for hog cholera; the vaccines (another report puts the number of doses at 1,280,000) were for rinderpest. They came from Canada.[39] Ninety-five lots of avianized rinderpest vaccine arrived in

China during the UNRRA Period and Personnel Assistance Needed from FAO, Annual Report China, 1947–1950, 010V2, FAO.

[38] UNRRA Committee of the Council for the Far East, Brief Review of the Agricultural Program for China – submitted by the MOAF (March 4, 1947), Agriculture Rehabilitation, 1944–1949, Folder S-1121–0000-0006, Box S-0528–0018, China Office, AG-018–001, UNRRA, UN.

[39] Elbridge Burnham, Monthly Report of Agricultural Rehabilitation Division, China Office (November 1946), Agricultural Rehabilitation – Monthly Reports 1944–1949, Folder S-1121–0000-0012, Box S-0528–0019, China Office, AG-018–001, UNRRA, UN.

Shanghai from Grosse Île in May. With them came eleven batches of seed virus. All needed to be tested to see if they had survived the journey. Unfortunately, only about half of the vaccines were determined still potent after the journey. They were prepared for immediate distribution.[40]

Jiangxi province, which had been struggling with a nasty outbreak since 1943, received 130,000 doses. UNRRA officials sent them initially to refrigerators in the capital Nanchang, but technicians then had to take them into the field, which meant up to three and a half days of travel. They packed them in "large thermos jugs (all-steel army type)" filled with ice and packed the jugs in large wooden crates filled with ice and sawdust. When the vaccination team arrived in a hsien (county) they transferred the vaccines to a portable US Army kerosene refrigerator that they carried with them in their jeep or truck (the only vehicles that could handle the roads). The team took as much care in ensuring they reached the vast majority of the cattle and buffalo in the county as they did in making sure the vaccines survived the trip. "The team leader always visited the hsien several days before the vaccinators arrived. The program of vaccination ... was discussed with the magistrate, considering the conditions of communication, customs, and traditions of the farmers of the hsien." Sometimes it took five days, sometimes it took thirty, but they made sure that "about 90% of the animals of each hsien were vaccinated" via injection through UNRRA-supplied syringes. Vaccinators branded an "R" on the hip of each cow after injection to help keep track. UNRRA officials pursued similar tactics in the provinces of Guangdong and Hebei, which each received 100,000 doses of Grosse Île vaccine, and Hainan Island.[41] They also sent some to Hong Kong and to the Philippines.[42]

Meanwhile, UNRRA continued its efforts to get China producing its own rinderpest vaccines, sending supplies to thirty different institutions in twenty provinces. Trying to avoid partisanship, officials proposed allocating a significant portion (enough for eight laboratories and ten field units) to

---

[40] Some of the records claim that only 300,000 doses were still potent and some claim 500,000. The total used in the field implies a number closer to the latter.

[41] Cheng, S.C., T.C. Chow, and H.R. Fischman. "Avianized Rinderpest Vaccine in China," in *Rinderpest Vaccines: Their Production and Use in the Field*, ed. K. V. L. Kesteven, second edition. 1949, Rome: FAO, 1955, 35–39; Agricultural Rehabilitation in China, Operational Analysis Report (December 31, 1947), Agricultural Rehabilitation Reports, 1944–1949, Folder S-1121–0000-0011, Box S-0528–0019, China Office, AG-018–001, UNRRA, UN.

[42] Report on Weekly Staff Conference, Agricultural Rehabilitation Division, China Office (September 11, 1946), Agricultural Rehabilitation – Meetings, 1944–1949, Folder S-1121–0000-0010, Box S-0528–0018, China Office, AG-018–001, UNRRA, UN.

CLARA for operation within communist-held China. Rinderpest did not care about political boundaries and it made both strategic and political sense to try to ignore them. Officials focused most of their attention that year, however, on getting the main biological laboratory at Nanjing (then Nanking) into production "as this one center could produce all the biologics required for continuance of present field operations."[43] They desperately needed it, because in the spring of 1947, "work in rinderpest control was largely brought to a halt by the exhaustion of vaccine supplies."[44] Distributing the Grosse Île vaccine had been a good opening salvo, but truly rehabilitating China's agriculture, as opposed to just providing momentary relief, required enabling it to permanently defend itself against rinderpest. China needed to be able to produce its own vaccines.

Attention turned to Nanjing, where UNRRA officials, delighted with the success of the 1946 vaccination campaigns, during which "not a single death traceable to vaccination occurred," worked to open a second front with the avianized vaccine.[45] They stocked the laboratory at the National Research Bureau of Animal Industry accordingly with incubators, microscopes, 400 fertile eggs, and fifty-nine Leghorn chickens.[46] Grosse Île researchers had sent those eleven batches of seed virus back in May of 1946 to aid in just this purpose, but, to the great frustration of everyone involved, only one batch, which had already been stored for 279 days, had proven still potent after the journey. Additional delays in preparing the laboratory meant that the researchers could not use the seed virus to try to start their own vaccine line for another 126 days. By that point, "no virus at all could be demonstrated."[47] This was a serious setback, for the one strain that the researchers had sent was the only one that they had ever been able to turn

---

[43] Proposed Allocation of Rehabilitation Supplies to Communist Controlled Areas, Agricultural Rehabilitation – Meetings, 1944–1949, Folder S-1121-0000-0010, Box S-0528-0018, China Office, AG-018-001, UNRRA, UN; Agricultural Rehabilitation in China During the UNRRA Period, Agricultural Rehabilitation Reports, 1944–1949, Folder S-1121-0000-0011, Box S-0528-0019, China Office, AG-018-001, UNRRA, UN.

[44] Agricultural Rehabilitation in China, Operational Analysis Report (December 31, 1947) Agricultural Rehabilitation Reports, 1944–1949, Folder S-1121-0000-0011, Box S-0528-0019, China Office, AG-018-001, UNRRA, UN.

[45] Cheng, S.C., T.C. Chow, and H.R. Fischman. "Avianized Rinderpest Vaccine in China," in *Rinderpest Vaccines*, ed. Kesteven, 38.

[46] Agricultural Rehabilitation Monthly Report (April–May 1947), Agricultural Rehabilitation – Monthly Reports 1944–1949, Folder S-1121-0000-0012, Box S-0528-0019, China Office, AG-018-001, UNRRA, UN.

[47] Cheng, S.C., T.C. Chow, and H.R. Fischman. "Avianized Rinderpest Vaccine in China," in *Rinderpest Vaccines*, ed. Kesteven, 31.

into a viable avianized vaccine. Now the researchers in Nanjing would have to do the same thing: find a strain that "took" to the eggs. UNRRA gave them supplies and a veterinarian from the University of California to instruct the Chinese researchers on the Grosse Île technique. It would not be enough.[48]

The Nanjing researchers faced several hurdles, the largest being that they would have to try to grow a completely different strain of the virus – one that came from China instead of Kenya. They had two options: the Lanchow strain ("a highly virulent strain fairly recently isolated") and the Szechwan strain ("used in the laboratory for years, causing a very high percentage of fever in calves and about 40 percent mortality"). They tried both, but could not get them to passage on the eggs.[49] After sixteen failed attempts, they decided that they needed new seed virus and proposed that a representative be sent to Cairo and Nairobi to study the techniques being used there and to obtain some new seed virus. UNRRA sent Keith V.L. Kesteven, the head of the Jiangxi vaccination program, off in August in the hope that "the acquisition of this seed virus would speed up the production of the vaccine by several months and enable the field vaccination programs to be resumed at an earlier date."[50] People needed it. In the interim,

---

[48] Agricultural Rehabilitation in China, Operational Analysis Report (December 31, 1947), Agricultural Rehabilitation Reports, 1944–1949, Folder S-1121–0000-0011, Box S-0528–0019, China Office, AG-018–001, UNRRA, UN.

[49] Cheng, S.C., T.C. Chow, and H.R. Fischman. "Avianized Rinderpest Vaccine in China," in *Rinderpest Vaccines*, ed. Kesteven, 30–31.

[50] Agricultural Rehabilitation Division Monthly Report (August 1947), Agricultural Rehabilitation – Monthly Reports 1944–1949, Folder S-1121–0000-0012, Box S-0528–0019, China Office, AG-018–001, UNRRA, UN. J. T. Edwards had arrived in Egypt in April of 1946, fresh from his war research at Pirbright, to fight an eighteenth-month-old outbreak that had stubbornly resisted control via mass vaccinations with inactive tissue vaccine. Edwards had been sent to change the vaccine and the results. He immediately sent cables to Mukteshwar, Kabete, and Vom in Nigeria, "to forward by air samples of their most highly 'attenuated' strains" of goat-passaged virus. Kabete's arrived first and experimentation began in June (Edwards, "The Uses and Limitations of the Caprinized Virus in the Control of Rinderpest (Cattle Plague) Among British and Near Eastern Cattle," 233–253. For more information on the work at Vom, see Daubney, "Récentes Acquisitions Dans La Lutte Contre La Peste Bovine," 36–45). By the time researchers in Cairo received the request from Nanking, however, they had moved on to work on trying to create an avianized vaccine. They first tried to adapt "the Egyptian strain" of rinderpest to eggs and, when that failed, began trying to the do the same with seed virus sent up from Kabete. They sent some of that seed virus to Nanjing in the summer of 1947, hoping that the researchers there would have better luck. They did not (Cheng, S.C., T.C. Chow, and H.R. Fischman. "Avianized Rinderpest Vaccine in China," in *Rinderpest Vaccines*, ed. Kesteven, 40–41).

researchers at the smaller laboratories in Jiangxi and Hebei had been producing blood serum as a stopgap measure to fight outbreaks in their provinces while they waited for more vaccines. It was a less-than-ideal solution.[51] Luckily, researchers quickly found a better one: Junji Nakamura's vaccine.

The Chinese had acquired Nakamura III – at that point in its six-hundred-thirtieth serial passage through rabbits – when the National Research Bureau of Animal Industry, Beijing branch went back into Chinese hands in September 1945. Japanese researchers had been forced to leave it behind when they were pushed out of the city. Chinese researchers at Nanjing, most notably S.C. Cheng and T.C. Chow, picked up where they had left off. They had already investigated the possibilities of caprinized vaccine, but rejected it as a large-scale option, because it was too strong for most of their animals. The lapinized vaccine looked more promising, for "tests on UNRRA cattle had shown it to have no ill effects on the stock." They focused their attention that direction while they waited for Kesteven to return from Africa. They ordered rabbits and left their fifty-nine Leghorn chickens to keep producing eggs.[52]

The Nanjing researchers were easily able to keep Nakamura III going via passages through new rabbits. The benefit of the vaccine for field use was readily apparent: technicians could travel with live rabbits instead of kerosene refrigerators and chilled thermoses. Lapinized vaccine could be made on the spot by killing an infected rabbit, grinding up its spleen and intestinal lymph glands in a sterile mortar, adding a little infected blood and saline, and then straining the mixture through a layer of sterile gauze. Vaccine team members would then inject cattle and buffalo with the resulting suspension. Each rabbit yielded between 300 and 600 doses. No fancy equipment was required. Out in the hsiens, researchers later explained, "vaccine preparation takes place early in the morning in order to deliver the vaccine on time. Usually two rabbits

[51] Agricultural Rehabilitation Division Monthly Report (July 1947) Agricultural Rehabilitation – Monthly Reports 1944–1949, Folder S-1121–0000-0012, Box S-0528–0019, China Office, AG-018–001, UNRRA, UN.
[52] Agricultural Rehabilitation Division Monthly Report (July 1947), Agricultural Rehabilitation – Monthly Reports 1944–1949, Folder S-1121–0000-0012, Box S-0528–0019, China Office, AG-018–001, UNRRA, UN; Agricultural Rehabilitation in China, Operational Analysis Report (December 31, 1947), Agricultural Rehabilitation Reports, 1944–1949, Folder S-1121–0000-0011, Box S-0528–0019, China Office, AG-018–001, UNRRA, UN; Cheng, S. C. and H. R. Fischman. "Lapinized Rinderpest Virus," in *Rinderpest Vaccines: Their Production and Use in the Field,* ed. K. V. L. Kesteven, second edition. 1949, Rome: FAO, 1955, 47–63.

FIGURE 3.2 Four photographs demonstrating the manufacture of lapinized rinderpest vaccine taken in a district of Chongqing, Sichuan province, China in 1949. They were probably created as teaching tools by JCRR staff. The back of the first explains, "Rabbit injected with Rinderpest virus is killed between three to four days after injection."
Photograph courtesy of National Archives, College Park, MD.

are killed every day, and at least two new rabbits are injected intravenously with the infected blood for making a further batch of vaccine."[53] It was an extremely efficient system, provided one had continuous access to rabbits.

Officials quickly began sending the lapinized vaccine to the provinces via infected rabbits, resulting in tens of thousands of successful vaccinations. The plan had great potential, but soon encountered a vital snag: rabbits were scarce in many parts of China, so field operatives had difficulty keeping the infected rabbit lines going. Researchers at Nanjing began investigating the possibility of injecting goats with the lapinized virus to see if they could be used as carriers out to the field instead, since they were more plentiful and could, ideally, produce far more doses per

[53] Cheng, S. C. and H.R. Fischman. "Lapinized Rinderpest Virus," in *Rinderpest Vaccines*, ed. Kesteven, 57–63.

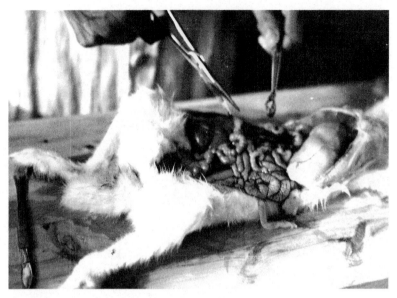

FIGURE 3.3 "The lymph gland is removed."
Photograph courtesy of National Archives, College Park, MD.

FIGURE 3.4 "Grinding the lymph gland from the rabbit."
Photograph courtesy of National Archives, College Park, MD.

FIGURE 3.5 The vaccine was finished by diluting one gram of concentrate with 400 cc of distilled water.
Photograph courtesy of National Archives, College Park, MD.

animal. They also studied the possibilities of producing a dried lapinized vaccine that would require refrigeration, but fewer rabbits.[54]

Meanwhile, Kesteven journeyed to northeastern Africa in search of a seed virus that would grow on eggs. He was anxious to get to Kabete, but was forced to spend ten days in Cairo waiting for his own inoculation against a different virus to take effect: he needed a yellow fever vaccine to get into Kenya. He spent the extra time in Cairo studying the records of its researchers' attempts to get Grosse Île seed virus and Kabete seed virus to grow in eggs. The "history and results of these attempts," he noted, "were very similar to the experience we have had in China, where each strain has either been dead, or lost in propagating on eggs." A similar story awaited him in Kabete. The avianized strain that the Grosse Île researchers had sent there in the spring of 1946 to ensure that it was not lost had not survived the journey. Kabete researchers had been forced to start an avianized strain of their own. They failed eleven times before finally being able to produce two strains that would grow in egg membranes

[54] Ibid.

and yolk sacs.[55] Kesteven brought eight frozen samples of them and some frozen Kabete-O infected bovine tissue with him back to China, hoping for the best.[56] It worked. One of the strains proved amenable to egg passaging in the lab. Nanjing researchers announced success in June.[57] Meanwhile, teams in the field continued their efforts with the lapinized vaccine. China's relief and rehabilitation from rinderpest was in full swing.

The UNRRA rinderpest program was a success. Vaccination teams administered vaccines to hundreds of thousands of cattle and buffalo; researchers established small laboratories around the county. UNRRA laboratory equipment went to "30 different institutions in 20 provinces to assist in the establishment of 5 grade-A and 10 grade-B serum laboratories, and 11 diagnostic and epizootic prevention teams."[58] Critically, researchers also specifically rehabilitated the National Research Bureau of Animal Industry branch in Nanjing. By December 31, 1947, the date UNRRA activities in China officially ended, researchers there were producing two different kinds of rinderpest vaccines – lapinized and avianized – which meant that the vaccination program did not have to depend on foreign vaccines to keep going. The international machinery for "relief and rehabilitation" had been replaced by domestic machinery for development.

The rinderpest program had begun as biological relief, but had become, along the way, biological development. Aid efforts started with the

[55] K.V.L. Kesteven to W.J. Green (October 8, 1947), Correspondence Documents, Rinderpest Meeting – Nairobi – 1948, Animal Production and Health Division, 10AGA407, FAO Archives; Cheng, S.C., T.C. Chow, and H.R. Fischman. "Avianized Rinderpest Vaccine in China," in *Rinderpest Vaccines*, ed. Kesteven, 41–42.

[56] K.V.L. Kesteven to W.J. Green (October 8, 1947), Correspondence Documents, Rinderpest Meeting – Nairobi – 1948, Animal Production and Health Division, 10AGA407, FAO; K.V.L. Kesteven, Transport of Avianized Rinderpest Virus from Kabete, Kenya to Nanking, China (October 6, 1947), Correspondence Documents, Rinderpest Meeting – Nairobi – 1948, Animal Production and Health Division, 10AGA407, FAO; Agricultural Rehabilitation in China, Operational Analysis Report (31 December 1947), Agricultural Rehabilitation Reports, 1944–1949, Folder S-1121–0000-0011, Box S-0528–0019, China Office, AG-018–001, UNRRA, UN; Cheng, S.C., T.C. Chow, and H.R. Fischman. "Avianized Rinderpest Vaccine in China," in *Rinderpest Vaccines*, ed. Kesteven, 32.

[57] Weekly Staff Conference, Agricultural Rehabilitation Division (October 21, 1947), Agricultural Rehabilitation Meetings 1944–1949, Folder S-1121–0000-0010, Box S-0528–0018, China Office, AG-018–001, UNRRA, UN.

[58] Appendix I: Agricultural Rehabilitation in China During the UNRRA Period and Personnel Assistance Needed from FAO, Annual Report: China, 1947–50, Annual Reports from Governments, O1OV2, FAO.

physical transfer of actual animals, but had quickly morphed into the creation of laboratories that produced cutting-edge vaccines. The technology turned "relief" into "rehabilitation" at laboratories that had been either damaged by the war and/or seized from retreating Japanese scientists. It did not stop there, however, for it also fostered the creation of an expansive field vaccination apparatus that had not previously existed. It moved the science out of the laboratories and into the fields, literally, via refrigerated vials of vaccine and non-refrigerated crates of rabbits. It brought the vaccines to the farmers who did not know the vaccines existed; it altered the immune systems of individual cows, protecting them from their deadliest foe. The program had begun reshaping the environment and had created the machinery necessary for China to continue that reshaping on its own once UNRRA left. It did so to great success: the last reported case of rinderpest in China came in 1955.[59]

The rinderpest program had worked, but it was, admittedly, only a small part of a massive operation. The larger story of the UNRRA effort in China, which was primarily economic and political, was far less clear-cut. The plans of the international operatives had not always meshed easily with the goals of the Nationalist government as they had in the rinderpest program. Those tensions spoke to a larger problem of the postwar world: balancing nationalist and internationalist economic visions of the future. The world order that Washington and London were trying to create did not necessarily coincide with the plans of leaders of recipient nations. The differences went to the heart of the debate about relief versus rehabilitation.

J. Franklin Ray, Jr., who had been the UNRRA Chief of Far Eastern Affairs, concluded in the fall of 1947 that "UNRRA did great good in China, though some of its benefits went into unintended channels and others did not reach their targets. The supply program was substantially fulfilled; thousands of workers of all nationalities rendered devoted and effective service to millions of Chinese who needed and deserved help." But it had run into problems in spaces where UNRRA policies did not match up with the primary goal of the Chinese government "to establish unified civil and military control over territorial China."[60] The Nationalists cared more about the second "R" than the first, because they perceived it to be crucial to building a strong, centralized

[59] Spinage, *Cattle Plague*, 489.
[60] Ray, J. Franklin, Jr. "UNRRA in China" (Tenth Conference of the Institute of Pacific Relations; International Secretariat, Institute of Pacific Relations, September 1947).

government. As the historian C.X. George Wei has argued, "For the Chinese government, the more imperative goal was rehabilitation – the solution to China's economic problems – rather than relief, the humanitarian concern. It saw no reason to sacrifice China's task of economic recovery for temporary humanitarian achievements."[61]

This is not to say that the Nationalists did not ask for relief, they did. The point is that they were far more concerned with rehabilitation efforts that would help them legitimize their claims to rule. Rana Mitter explained in *Forgotten Ally* that "the Nationalists had always had an interest in projecting the image of a modernizing, active state."[62] Doing so was an effort to live up to a vision that Sun Yat-sen had shared with the world back in 1921, insisting that China's "vast resources" needed to be "developed internationally" for the "good of the world in general and the Chinese people in particular." China, with its expansive population, he predicted, would, if developed, "be another New World in the economic sense." International cooperation in the effort, ideally in the form of capital investment, but "at least" in Western "machinery," he argued, "will culminate to be the keystone in the arch of the League of Nations." He composed a detailed plan about how it could happen and sent it to governments around the world. The US Secretary of Commerce wrote back that while he agreed that "the economic development of China would be of the greatest advantage, not only to China, but to the whole of mankind," the plans Sun envisioned would "take billions of dollars" and China was already burdened with debts that it could not pay.[63] The United States did not answer Sun's call for assistance, but the League of Nations did create a Program of Technical Cooperation that provided the Nationalist government with a limited amount of technical and financial assistance. UNRRA provided far, far more – over 500 million dollars' worth of supplies – but it still was nowhere near enough.[64]

---

[61] Wei, C.X. George. *Sino-American Economic Relations, 1944–1949*. Westport, CT: Greenwood Press, 1997, 68.

[62] Mitter, Rana. *Forgotten Ally*. Boston: Houghton Mifflin Harcourt, 2013, 355–359. See also Mitter, Rana. "Imperialism, Transnationalism, and the Reconstruction of Post-war China: UNRRA in China, 1944–7." *Past and Present*, Supplement 8 (2013): 51–69; Mitter, Rana. and Helen M. Schneider. "Introduction: Relief and Reconstruction in Wartime China." *European Journal of East Asian Studies* 11 (2012): 179–186.

[63] Sun Yat-sen. *The International Development of China*, 2nd ed. New York: G. P. Putnam's Sons, 1929, xi, 8–9, 198, 257–258.

[64] Zanasi, Margherita. "Exporting Development: The League of Nations and Republican China." *Comparative Studies in Society and History* 49:1 (2007): 143–169; *12th Report*

UNRRA did not "develop" China, but its operations there played a significant role in convincing officials that "relief" and "rehabilitation" were inadequate goals. The shift was evidenced in the changing vocabulary of assistance. A 1944 report to Congress on the agency's activities explained that China had been so depleted by its war with Japan that "she will be helpless to undertake the gigantic task of postwar reconstruction unless she is first given adequate help from outside in the form of relief and rehabilitation supplies." The government had already requested forty tons of medical supplies, five tons of veterinary supplies, three tons of seeds, and one ton of agricultural hand tools to help it get through the "emergency period."[65] By 1947, the description of the China program had fundamentally changed. A report to Congress that summer explained that the effort there "was different in character" from the European programs "because of the depressed state of the Chinese economy before the war." In consequence, "greater effort was therefore expended to provide rehabilitation through new economic development in the hope of alleviating the chronic need for relief." UNRRA activities in China had shifted from a focus on bringing in supplies to establishing "projects," such as those "for the local manufacture of pharmaceutical supplies" and "for the development of small-scale rural industries." Eight million dollars went into building farm-tool shops "to help the Chinese rural population to provide their own supplies and equipment." In addition, "the most spectacular single project in China was the recovery of over 2,000,000 acres of inundated agricultural land through the flood control of the Yellow River." It and the other projects were initiated in the name of meeting "the need for new economic development."[66]

That shift in vision and vocabulary was not limited to UNRRA officials alone. At a September 1947 conference in England hosted by the Institute of Pacific Relations (IPR) on UNRRA activities in China, participants devoted a great deal of attention to the language of postwar assistance. Poeliu Dai concluded that UNRRA had "pursued a limited objective," engaging in a number of r-words: "relief," "rehabilitation," "restoration," "rebuilding," "reclaiming," "rediversion," "repatriation," and

to *Congress on Operations of UNRRA, as of June 30, 1947*, 7. See also Zanasi, Margharita. *Saving the Nation: Economic Modernity in Republican China.* Chicago: University of Chicago Press, 2006.

[65] *2nd Report to Congress on United States Participation in Operations of UNRRA, as of December 31, 1944.* Washington, DC: United States Government Printing Office, 16–17.

[66] *12th Report to Congress on Operations of UNRRA, as of June 30, 1947*, 9–12.

"resettlement." UNRRA's work, however, he continued, "represents only a first step on a long road of rehabilitation, reconstruction, and economic development."[67] Fellow participant H. Belshaw seconded the point, arguing that reconstruction "in the narrow and more precise sense ... is used to connote the rebuilding of material equipment and the reestablishment of the economic and political organizations that have been destroyed or damaged by war." But reestablishment, he argued, was no longer an acceptable goal. "Reconstruction policies which express too great a nostalgia for the status quo ante bellum are likely," he warned, to "do damage to the prospect of an expanding world economy." It is now clear, he continued, that "the expansion of production, purchasing power and trade in other, poorer countries is conducive to their own welfare, and to the peace of the world." When we talk of "reconstruction" today, he insisted, we should understand it to mean "the continuous process of economic progress and development, as it is with this long-run objective, covering at least several decades, that we shall mainly be concerned." It would, Belshaw admitted, be "more accurate ... to use the phrase 'economic, social, and political development or progress,'" but "reconstruction" was "easier to handle" and more "in common use," so he used it "at the expense of somewhat straining its content."[68]

Sun Yat-sen had been ahead of his time when he constructed his plan for China's economic development, but the international community had, by 1947, largely caught up with him, agreeing that development was an international issue. Sun's "development" had a long history that stretched back to eighteenth-century arguments about "improvement" and "progress."[69] The concept spread around the world in the nineteenth century as states created ministries specifically devoted to

---

[67] Dai, Poeliu. "Summary Report of UNRRA Activities in China" (Tenth Conference of the Institute of Pacific Relations; International Secretariat, Institute of Pacific Relations, September 1947). For more on IPR, see Hopper, Paul F. "The Institute of Pacific Relations and Origins of Asian and Pacific Studies." *Pacific Affairs* 61:1 (Spring 1988): 98–121. See also, Akami, Tomoko. *Internationalizing the Pacific: The United States, Japan and the Institute of Pacific Relations in War and Peace, 1919–1945.* London: Routledge, 2002.

[68] Emphasis original. Belshaw, H. "Agricultural Reconstruction in the Far East" (Tenth Conference of the Institute of Pacific Relations; International Secretariat, Institute of Pacific Relations, September 1947).

[69] Arndt, *Economic Development*, 9–48; McVety, *Enlightened Aid*, 5–37; Brewer, Anthony. "Adam Ferguson, Adam Smith, and the Concept of Economic Growth." *History of Political Economy* 31:2 (1999): 237–254; Brewer, Anthony. "The Concept of Growth in Eighteenth-Century Economics." *History of Political Economy* 27:4 (1995): 609–638.

national economic improvement, usually through the employment of science and technology. The idea of spreading development was also gradually co-opted as part of the West's imperial "civilizing mission," with a focus on technical assistance. As Emily Rosenberg has argued, the perceived "benevolence of technology and the transferability of techno-logical solutions formed a major part of the new credo."[70] This sense of development's usefulness informed the international response to Sun's proposal: agreement that the economic development of China would be beneficial to the world as a whole, but not so much so that they would pay for it.

Development took new form and urgency during the 1930s and 1940s, however, in response to the birth and spread of the idea of "the economy," which radically transformed the perceived stakes of global poverty. As Timothy Mitchell has shown, the idea of "the economy" as we think of it today – "a self-contained structure or mechanism" that has the definite article in front of it – only emerged in the 1930s in response to the compilation and publication of new information about global incomes, consumption, production, and more that the League, the ILO, and individual economists began producing in the 1920s. "Standards of living" data made the economy a tangible entity that could be measured. The Great Depression and the outbreak of World War II made it the pivotal factor in the quest for prosperity and security.[71] Once created, the economy could not be ignored and the Allies would construct a great deal of international machinery during

---

[70] Rosenberg, Emily S. *Spreading the American Dream: American Economic and Cultural Expansion, 1890–1945*. New York: Hill and Wang, 1982, 27. See also Lurtz, Casey Marina. "Developing the Mexican Countryside: The Department of Fomento's Social Project of Modernization." *Business History Review* 90:3 (Autumn 2016): 431–455; Cribelli, Teresa. "'These Industrial Forests': Economic Nationalism and the Search for Agro-Industrial Commodities in Nineteenth-Century Brazil." *Journal of Latin American Studies* 45:3 (August 2013): 545–579; Adas, Michael. *Machines as the Measure of Men: Science, Technology, and Ideologies of Western Dominance*. Ithaca: Cornell University Press, 1990); Adas, Michael. *Dominance By Design: Technological Imperatives and America's Civilizing Mission*. Cambridge: Harvard University Press, 2006; Curti, Merle and Kendall Birr, *Prelude to Point Four: American Technical Missions Overseas, 1838–1938*. Madison: University of Wisconsin Press, 1954.

[71] Mitchell, *Rule of Experts*, 81–83. See also, Mitchell, Timothy. "Econo-mentality: How the Future Entered Government." *Critical Inquiry* 40 (Summer 2014): 479–507; Clavin, Securing the World Economy, 159–197; Schmelzer, Matthias. *The Hegemony of Growth: The OECD and the Making of the Economic Growth Paradigm*. Cambridge: Cambridge University Press, 2016, 75–92; Speich, Daniel. "The Use of Global Abstractions: National Income Accounting in the Period of Imperial Decline." *Journal of Global History* 6:1 (March 2011): 7–28; McVety, *Enlightened Aid*, 38–61.

the war specifically for it, most notably at Bretton Woods. The question remained, however, how expansive their vision of it – and their efforts to nurture it – needed to be.[72]

In 1939, Eugene Staley, an economist at Tufts, insisted that humanity was "well into the era of planetary economy," which made poverty anywhere a problem everywhere. Staley called for "a development program designed to carry modern capital equipment and technical knowledge into parts of Asia, South America, and Africa."[73] That same year, Harry Dexter White, who worked in the Treasury Department, sent a memo to Treasury Secretary Henry Morgenthau warning that without US action the arm of Axis power would reach Latin America, preventing the Allies from accessing the region's many valuable resources. To stop it, he argued for a "bold program" of financial aid. "Only capital and technical skill are needed to develop the area," he insisted, "so that it could provide for a much larger population, for a higher standard of living and a greatly expanded foreign trade."[74] Staley and White recommended funding technical assistance projects, not just sending supplies. It was a message increasingly seconded by political leaders around the world.

In the summer of 1943, Ethiopia's Vice Minister of Finance, Yilma Deressa, sent Roosevelt a message explaining that his country's development could help Europe's reconstruction. "Ethiopia can be of aid in supplying foodstuffs to needy European and Eastern nations during the post-war period of reconstruction," he wrote, "If help could now be given her in the form of technical aid and equipment, the temperate climate and fertile soil ... would lend themselves favorably to the production of important crops and cattle for food." Ethiopia was much closer to those countries than the United States, which would allow savings in "transportation and time."[75] The US responded with a technical advisory

[72] Helleiner, Eric. *Forgotten Foundations of Bretton Woods: International Development and the Making of the Postwar Order*. Ithaca: Cornell University Press, 2014; Staley, *The Birth of Development*, 8–34; Borgwardt, *A New Deal for the World*, 88–140; Zeiler, Thomas W. *Free Trade, Free World: The Advent of GATT*. Chapel Hill, NC: University of North Carolina Press, 1999.

[73] Staley, Eugene. *World Economy in Transition*, reissue. Port Washington, NY: Kennikat Press, 1971), 19, 283, 332–333. UNRRA would send Staley to China a few years later to help prepare a survey of that nation's "relief and rehabilitation needs" (Staley, Eugene. "Relief and Rehabilitation in China." *Far Eastern Survey* 13:20 [October 4, 1944]: 183–185).

[74] Helleiner, *Forgotten Foundations of Bretton Woods*, 40–51.

[75] Deressa, Yilma. Aide-Mémoire to President Roosevelt (July 12, 1943). *Foreign Relations of the United States, 1943*, v. IV. Washington, DC: United States Government Printing Office, 1964, 103–106.

mission that "drew up a ten-year economic development plan" for the country but Washington declined to fund any of it. Ethiopia got $5,152,000 in lend-lease aid, some arms and ammunitions, and a recommendation that it direct further inquiries for help to UNRRA.[76]

Ethiopia did so, specifically asking UNRRA for the "technical aid and equipment," as Yilma had phrased it, that it believed most useful "to help develop the natural resources" of the country.[77] The most important of those resources was cattle. At around 20 million head, Ethiopia had one of the largest, if not the largest, cattle populations in Africa.[78] Unfortunately, it also had endemic rinderpest, which was preventing it from being able to turn that particular resource into an export commodity. It had heard about UNRRA's campaign in China, however, and it promptly requested assistance with "the reorganization of the Veterinary Laboratory, including a complete unit for the manufacture of rinderpest vaccine."[79] The Imperial Ethiopian Government (IEG) did not just want vaccines, it wanted the ability to produce them on its own. UNRRA responded with "relief" via vaccines: over fifty cases of "Rinderpest Laboratory" (which was likely lapinized virus vaccine from Nanjing), but the agency was pessimistic about Ethiopia's ability to manufacture its own vaccines, which is what the government really wanted. "The present [national laboratory] service is well-intentioned," UNRRA officials reported, "but is weak in personnel, apparatus, drugs, glassware and nearly everything required for vaccine manufacture on a scale which is necessary for this country."[80] UNRRA agreed that cattle could become a valuable export

[76] Wallace Murray to Yilma Deressa (August 12, 1943) in ibid., 112; State Department, Ethiopia Policy Statement (March 1, 1951), in *FRUS*, 1952, v. V. Washington, DC: United States Government Printing Office, 1982, 1242; State Department, Ethiopia Policy Statement (March 1, 1951), in *FRUS*, 1952, v. V. Washington, DC: United States Government Printing Office, 1982, 1242; Marcus, Harold. *A History of Ethiopia*, updated ed. Berkeley: University of California Press, 2002, 159.

[77] Deressa, Yilma. Aide-Mémoire to President Roosevelt (July 12, 1943). *Foreign Relations of the United States, 1943,* v. IV. Washington, DC: United States Government Printing Office, 1964), 103–106.

[78] Gray, David F. "The Ethiopian Farmer." *Western Mail* (February 2, 1950): 8–9.

[79] A. G. Sandoval to Hugh G. Calkins (September 19, 1946); Agricultural Programme (May 12, 1945 – March 5, 1947) 1944–1949, Folder S-1347–0000-0018, Box S-0527–0475, Ethiopia Mission, UNRRA, Record Group AG-018-019, UN.

[80] Ibid; Description of Equipment, UNRRA Addis Ababa (1947), Ministry of Agriculture – Receipted Requisitions – Commodity Accounting Copy 1944–1949, Folder S-1349–0000-0020, Box S-0527–0478, Ethiopia Mission, UNRRA, Record Group AG-018-019, UN; Signed delivery receipt for UNRRA Addis Ababa (September 6, 1947), Ministry of Agriculture – Receipted Requisitions – Commodity Accounting Copy 1944-1949, Folder S-1349–0000-0020, Box S-0527–0478, Ethiopia Mission, UNRRA, Record Group AG-

commodity, noting in 1946 that "as the stock position improves" Ethiopia should build slaughterhouses and factories for canning meat, tanning hides and more, but it provided only the recommendation, not the funding.[81]

In total, UNRRA provided Ethiopia with a little less than $700,000 in supplies, most of which went to programs that targeted improving *human* health.[82] Unlike in China, in Ethiopia the agency never moved past its focus on relief, but its actions there highlighted the point that technical assistance to fight both human and animal diseases could, without much money, help lay a foundation for economic development. In both its largest and in one of its smallest programs, UNRRA's efforts reinforced the growing sense that international technical assistance was both possible and vital – not just for trying to repair the damages caused by the war, but for trying to build something better. The perceived stakes of that effort only grew higher as the mid-1940s became the late-1940s and global freedom from want continued to be an aspiration instead of an achievement.

***

In his memoirs, Dean Acheson reflected that "UNRRA would have done its work and passed away before we were to know what 'rehabilitation' really required from us." It was left to General Marshall, he wrote, "to outline the task at Harvard."[83] Acheson was referring to the Secretary of State's June 1947 commencement address, during which Marshall laid out the principles of what was to become the Marshall Plan. He opened it with a warning that Americans, "distant from the troubled areas of the earth," did not comprehend how much people were still suffering and could not, therefore, comprehend the gravity of the threat it posed to peace. The Allied powers had not been prepared for the extent of the damage. "In considering the requirements for the rehabilitation of Europe," Marshall explained, "the physical loss of life, the visible destruction of cities, factories, mines and railroads was correctly estimated, but it has become obvious during recent months that this visible destruction was

018–019, UN; Cheng, S. C., and H.R. Fischman. "Lapinized Rinderpest Virus," in Kesteven, *Rinderpest Vaccines*, 53.

[81] Formation of a Veterinary Division in Ethiopia (October 12, 1946) (Agricultural Programme 5–12-45-3-5-47) 1944–1949, Folder S-1347-0000-0018, Box S-0527-0475, Ethiopia Mission, UNRRA, Record Group AG-018-019, UN.

[82] 12th Report to Congress on Operations of UNRRA, as of June 30, 1947, 7.

[83] Acheson, *Present at the Creation*, 69.

probably less serious than the dislocation of the entire fabric of European economy." This *real* rehabilitation will take more time and more effort than we ever imagined, he admitted, and "any assistance that this Government may render in the future should provide a cure rather than a mere palliative." And it needed to provide it. The "consequences to the economy of the United States" of not doing so, Marshall warned, "should be apparent to all."[84]

Marshall insisted that this new assistance would be "directed not against any country or doctrine but against hunger, poverty, desperation and chaos," but that was no more the truth than was Acheson's statement that UNRRA had "done its work and passed away." The United States shut UNRRA down and created the Marshall Plan to fight more than hunger. The month before Marshall's speech, Acheson had announced that "free peoples who are seeking to preserve their independence and democratic institutions and human freedoms against totalitarian pressures, either internal or external, will receive top priority for American reconstruction aid." The United States was now openly agreeing to tackle "reconstruction." It was doing so because it had decided that "not only do human beings and nations exist in narrow economic margins, but also human dignity, human freedom and democratic institutions."[85] Fighting for principles, as opposed to just fighting against want, necessarily required a change in tactics – a shift, as Marshall had phrased it, from "palliatives" to "cures." It also, the Truman administration argued, required a shift away from solely relying on UN machinery to do it.

The previous year, Herbert Hoover, appointed by Truman to head a new Famine Emergency Committee, returned from a round-the-world investigation of global food security with ominous tales of American aid being co-opted by its opponents. Hoover, who had argued back in 1919 that the United States could defeat Bolshevism by defeating hunger, was particularly disturbed by what he found.[86] In Italy, for example, he learned that the communists, "whenever they were able to seize local

---

[84] George C. Marshall, Address at Harvard (June 5, 1947), available at http://www.marshallfoundation.org/library/MarshallPlanSpeechfromRecordedAddress_000.html.

[85] Acheson, Dean. "The Requirements of Reconstruction." *The Department of State Bulletin* 16:411 (May 18, 1947): 994. On the politics of the decision to terminate UNRRA see Reinisch, "Internationalism in Relief." The United States, Melvyn Leffler has shown, "launched the Marshall Plan to arrest an impending shift in the correlation of power between the United States and the Soviet Union" (Leffler, Melvyn P. *A Preponderance of Power.* Stanford: Stanford University Press, 1992, 163).

[86] Patenaude, Bertrand M. *The Big Show in Bololand: The American Relief Expedition to Soviet Russia in the Famine of 1921.* Stanford: Stanford University Press, 2002, 28–34.

FIGURE 3.6 Huang Sung-ling of the Provincial Bureau of Agriculture and Forestry, Robert C. Reisinger of FAO (working as a JCRR veterinary consultant), and others investigate a case of suspected rinderpest during an outbreak in Taiwan.
Photograph courtesy of National Archives, College Park, MD.

governments, they misused UNRRA supplies for their purposes." In Czechoslovakia, he heard that trains bearing American-donated grain bore pictures of Stalin and were met by Communist party officials, who distributed the food "with ceremonies of thanks to Uncle Joe."[87] Additional reports of UNRRA aid going to the "wrong places" and being used by the "wrong people" convinced the Truman administration that it needed to alter its strategy.[88] Under Secretary of State for Economic

See also, Irwin, Julia F. "Taming Total War: Great War-Era American Humanitarianism and its Legacies." *Diplomatic History* 38:4 (2014): 763–775.

[87] Herbert Hoover, My Impressions of Italy (March 22–24, 1946): 2–3, Herbert Hoover Diaries, Round World Trip, PPS – Famine Emergency Committee, HHPL; Hoover, My Impressions of Czechoslovakia (March 27–28, 1946): 4, Herbert Hoover Diaries, Round World Trip, PPS – Famine Emergency Committee, HHPL.

[88] Reinisch, "'Auntie UNRRA' at the Crossroads." 87–89. See also, Harder, Andrew. "The Politics of Impartiality: The United Nations Relief and Rehabilitation Administration in the Soviet Union, 1946–47." *Journal of Contemporary History* 47:2 (April 2012): 347–369.

Affairs William L. Clayton insisted in a memo to Marshall about the proposed new assistance program that "we must avoid getting into another UNRRA. *The United States must run this show.*" The White House agreed.[89]

Congress brought the Marshall Plan to life with the creation of the Economic Cooperation Administration (ECA) in April of 1948 to aid Europe's recovery. At the same time, it created the Joint Commission on Rural Reconstruction (JCRR) for China to help "the Chinese farmer to help himself" through "the introduction and fostering of improved agricultural practices, health and sanitary measures, dike building and maintenance, control of animal and plant diseases," and more. JCRR built on the foundations of what UNRRA left behind.[90]

JCRR administrators immediately began rinderpest vaccination campaigns in those areas of China that were still open to them, vaccinating around 50,000 cattle and buffalo in Canton (its base of operations) with lapinized vaccine in 1948.[91] Efforts went beyond just vaccination, however, to stress the importance of teaching people how to produce and administer the vaccines themselves. JCRR officials made "a technical rinderpest film" that explained "how to keep their instruments sterile and how to do the necessary inoculating work with the rabbits; how to take care of the rabbits, and how to use the rabbits for inoculating the cattle." They would go on to make other filmstrips, but the rinderpest one was the first produced, because UNRRA efforts had already proven how successful the campaign could be at eliminating the virus.[92] The same was not true of the larger JCRR campaign to fight communism. In 1949, the organization moved from Canton to Taipei. It would continue to fight rinderpest in Formosa, but the vaccination campaign in mainland China was now firmly under communist control. They had offered China's

---

[89] [Emphasis original] Clayton to Marshall (May 27, 1947) in *Foreign Relations of the United States, 1947*, vol. III. Washington, DC: United States Government Printing Office, 1972, 233; Zeiler, Thomas W. "Opening Doors in the World Economy," in *Global Interdependence*, ed. Akira Iriye. Cambridge, MA: Harvard University Press, 2014, 219.

[90] Cleveland, Harlan. "Economic Aid to China." *Institute of Pacific Relations* 18:1 (January 12, 1949): 2, 5. For more on the JCRR, see Ekbladh, *The Great American Mission*, 116–121.

[91] *International Technical Cooperation Act of 1949, Hearings before the Committee on Foreign Affairs House of Representatives*, 81st Cong. 1st Session on H.R. 5615, September 27, 28, 30, October 3–7, 1949. Washington, DC: United States Government Printing Office, 1950, 171–172.

[92] Ibid., 173.

farmers more. While the United States had slowly moved from supporting agricultural relief to rehabilitation to reconstruction, the communists had instituted an "agricultural *revolution*," "redistributing 43 percent of land under cultivation to 60 percent of the peasantry." Revolution won.[93]

The JCRR had not achieved its ultimate goal, but, in 1949, American officials viewed it not as a failure, but as "a prototype" for a new kind of aid machinery. Truman had announced in his 1949 inaugural address that the United States needed to "embark on a bold new program for making the benefits of our scientific advances and industrial progress available for the improvement and growth of underdeveloped areas."[94] At congressional hearings on the proposed "cooperative technical-assistance program" that fall, the Acting Secretary of State noted that it offered "a unique opportunity to make available to those peoples who seek them and without loss to ourselves the scientific and technical skills which have flourished here and helped to build our strong democratic institutions." The subsequent Point Four Program made "the development of economically underdeveloped areas of the world" – a designation that the government said covered "two-thirds of the world's population" – US foreign policy.[95]

Point Four would not be a global Marshall Plan; it was far more modest.[96] But its creation signaled the extent to which the idea of technical

[93] Loth, Wilfried. "States and the Changing Equations of Power," in *Global Interdependence*. ed. Iriye, 43. In a 1951 article, Melvin Conant, Jr. wrote that the JCRR put "prime emphasis" on "the technical aspects of the Chinese agrarian problem. (This was, of course, in contrast to the Communist view and tactic, which emphasized the social aspects of the agrarian problem.)" Conant, Melvin, Jr. "JCRR: An Object Lesson." *Far Eastern Survey* 20:9 (May 2, 1951): 89. For information about JCRR activities in Taiwan, see Shen, Tsung-han. *The Sino-American Joint Commission on Rural Reconstruction.* Ithaca: Cornell University Press, 1970.

[94] *International Technical Cooperation Act of 1949, Hearings before the Committee on Foreign Affairs House of Representatives*, 81st Cong. 1st Session on H.R. 5615, September 27, 28, 30, October 3–7, 1949. Washington, DC: United States Government Printing Office, 1950, 54; Harry S. Truman, Inaugural Address (January 20, 1949), available at http://www.trumanlibrary.org/whistlestop/50yr_archive/inagural20jan1949.htm.

[95] *International Technical Cooperation Act of 1949, Hearings before the Committee on Foreign Affairs House of Representatives*, 81st Cong. 1st Session on H.R. 5615, September 27, 28, 30, October 3–7, 1949. Washington, DC: United States Government Printing Office, 1950, 1, 5.

[96] For more on Point Four, see McVety, *Enlightened Aid*, 83–120; McVety, Amanda. "Pursuing Progress: Point Four in Ethiopia." *Diplomatic History* 32:3 (June 2008): 371–403; Macekura, Stephen. "The Point Four Program and U.S. International Development Policy." *Political Science Quarterly* 128:1 (Spring 2013): 127–160; Black, Megan. "Interior's Exterior: The State, Mining Companies, and Resource Ideologies in the Point Four Program." *Diplomatic History* 40:1 (January 2016): 81–110.

assistance for development in the pursuit of "a prosperous and expanding world economy" had grown in popularity during the 1940s.[97] It had done so in part at the insistence of the so-called "underdeveloped countries" who had, a Brookings Institution study argued, expressed "dissatisfaction … with the American thesis that priority for reconstruction and recovery in Europe was justified because the rest of the world would benefit from the revival of world trade and by the provision of European capital for economic development." Governments such as Ethiopia kept pushing for more aid, partially out of dissatisfaction, but also out of a growing confidence in humankind's ability to transform the world through technology. They wanted the aid because they believed it would work.[98]

Small victories in the immediate postwar years encouraged people to want bigger ones. The UNRRA rinderpest program in China was an excellent example. That program, and others like it, had proven that "better productive techniques could markedly alleviate … misery."[99] Point Four gave full expression to a growing faith that technology could help bring development, and it was not the only program that did so. The UN General Assembly had, in December 1948, passed a resolution calling for the allocation of funds for teams "of experts" to provide "technical assistance for economic development."[100] In March, the US delegate to the UN proposed expanding that limited commitment with the creation of "a comprehensive plan for an expanded cooperative programme of technical assistance for economic development through the United Nations and its specialized agencies."[101] The General Assembly,

---

[97] *International Technical Cooperation Act of 1949, Hearings before the Committee on Foreign Affairs House of Representatives*, 81st Cong. 1st Session on H.R. 5615, September 27, 28, 30, October 3–7, 1949. Washington, DC: United States Government Printing Office, 1950, 4.

[98] Brown, Jr. and Opie, *American Foreign Assistance*, 389.

[99] Samuel P. Hayes, Jr. "Truman's 'Bold New Program,'" draft attached to letter from Hayes to Carter (26 August 1949), Point 4 Memoranda – Miscellaneous, Box 109, NEA-Memoranda, Notes, etc. to UNESCO, 1946–1950, Subject Files of the Chief, 1945–1951, Bureau of Public Affairs, Office of Public Affairs, Division of Public Liaison, Lot File 53D387, RG 59, NARA.

[100] UN General Assembly, 3rd session, Resolution 200 (III). Technical Assistance for Economic Development (4 December 1948); available at http://www.un.org/documents/ga/res/3/ares3.html.

[101] Samuel P. Hayes, Jr. "Truman's 'Bold New Program,'" draft attached to letter from Hayes to Carter (August 26, 1949), Point 4 Memoranda – Miscellaneous, Box 109, NEA-Memoranda, Notes, etc. to UNESCO, 1946–1950, Subject Files of the Chief, 1945–1951, Bureau of Public Affairs, Office of Public Affairs, Division of Public Liaison, Lot File 53D387, RG 59, NARA.

"impressed with the significant contribution to economic development that can be made by an expansion of the international exchange of technical knowledge through international cooperation among countries," adopted Resolution 222 that fall, establishing the United Nations Expanded Program of Technical Assistance for Economic Development of Underdeveloped Countries.[102] In consequence, a US official wrote at the time, "economic development ... replaced post-war economic recovery and reconstruction as the subject receiving the greatest attention on their agenda and in their activities."[103] Faith in the power of science and technology to permanently secure freedom from want had never been higher. The rinderpest campaigns had aided the conversion process and they were only just getting started.

[102] Committee on Expenditures in the Executive Departments, *Research Summary on Technical Assistance Proposals of the Food and Agriculture Organization of the United Nations, House Report no. 670*, 82nd Cong. 1st session. Washington, DC: United States Government Printing Office, 1951, 1–2, 13.
[103] Samuel P. Hayes, Jr. "Truman's 'Bold New Program,'" draft attached to letter from Hayes to Carter (August 26, 1949), Point 4 Memoranda – Miscellaneous, Box 109, NEA-Memoranda, Notes, etc. to UNESCO, 1946–1950, Subject Files of the Chief, 1945–1951, Bureau of Public Affairs, Office of Public Affairs, Division of Public Liaison, Lot File 53D387, RG 59, NARA.

4

The Machinery of Development: FAO's Rinderpest Campaigns

In his 1949 Nobel Lecture, Lord John Boyd Orr, who had served as the first director general of the United Nations Food and Agriculture Organization (FAO), insisted that "the road to peace lies only through the cooperation of governments in developing the vast potential wealth of the earth for the benefit of all." The United Nations and its specialized agencies could make it happen, he continued, for "here at last mankind has the machinery through which governments can join in eliminating hunger, poverty, and disease."[1] His fellow directors shared that vision. Julian Huxley, then the head of the United Nations Educational Scientific and Cultural Organization (UNESCO), had argued a few years earlier that UNESCO could "both on its own account and in close relations with other U.N. agencies such as the F.A.O. and the World Health Organization, promote the international application of science to human welfare," in part by educating the public about the new possibilities for "harnessing micro-organisms as the chemical servants of man" and "banishing germ-causing disease."[2] Orr and Huxley firmly believed that such tasks were both possible and desirable – that they were, indeed,

[1] John Boyd Orr, "Nobel Lecture" (December 12, 1949), available at http://www
.nobelprize.org/nobel_prizes/peace/laureates/1949/orr-lecture.html.
[2] Huxley, Julian. *UNESCO: Its Purpose and Its Philosophy.* 1947, London: Euston Grove
Press, 2010, 5–14. For more on Huxley and his vision of human ecology, see Deese, R. S.
"The New Ecology of Power: Julian and Aldous Huxley in the Cold War Era," in
Environmental Histories of the Cold War, ed. McNeill and Unger, 279–300;
Deese, R. S. *We are Amphibians: Julian and Aldous Huxley on the Future of Our
Species.* Oakland: University of California Press, 2015; Sluga, Glenda. "UNESCO and
the (One) World of Julian Huxley." *Journal of World History* 21:3 (September 2010):
393–418. For more on UNESCO's early years, see Graham, S. E. "The (Real)politiks of

primarily what the UN was about. The organization's 1945 charter had included a vow "to employ international machinery for the promotion of the economic and social advancement of all peoples," and they intended to hold its member nations to that promise.[3]

The machinery of development took several forms in the 1950s. The UN specialized agencies were joined by bilateral government agencies and colonial development agencies, just to name a few. Sometimes this machinery worked together toward common goals; sometimes it worked in competition. The politics of development became more complicated in the 1950s as development became institutionalized.[4] Efforts in the name of development focused on the sharing of technology and technological knowledge in pursuit of economic growth, in the confidence that such growth would benefit both individuals and nations.[5] The emphasis upon technology was a hallmark of the period. The postwar world was, as J. R. McNeill and Corrina Unger have argued, "in many respects, a scientific world – one in which political, social, and cultural problems were viewed through the lens of science and in which science was believed to offer solutions to the challenges both of everyday life and of international politics."[6] Development was one of those challenges. To many at the time it seemed that humans now had both the bureaucratic and the technological "machinery" necessary for "eliminating hunger, poverty, and disease," and, if it could be done, then it should be done. The idea of development-via-technical-

Culture: U.S. Cultural Diplomacy in UNESCO. 1946–1954," *Diplomatic History* 30:2 (April 2006): 231–251.

[3] Preamble of the Charter of the United Nations (June 26, 1945); available at http://www.un .org/en/documents/charter/preamble.shtml. On the influence of the International Labour Organization in the wording, see Maul, Daniel. "'Help Them Move the ILO Way': The International Labor Organization and the Modernization Discourse in the Era of Decolonization and the Cold War." *Diplomatic History* 33:3 (June 2009): 387–404.

[4] For a recent discussion of this, see Engerman, David C. "Development Politics and the Cold War." *Diplomatic History* 41:1 (January 2017): 1–19.

[5] The UN Expanded Technical Assistance Program described its stated purpose as "to strengthen the national economies of the member nations of the UN through the development of their industries and agriculture, with a view to promoting their economic and political independence in the spirit of the Charter of UN, and to ensure the attainment of higher levels of economic and social welfare of their entire populations" (Bok, Bart J. "The United Nations Expanded Program for Technical Assistance." *Science* 117:3030 [January 23, 1953]: 67). For comments on the framework of technical assistance, see Mitchell, *Rule of Experts*, 19–53.

[6] McNeill, J. R. and Corrina R. Unger, "Introduction: The Big Picture," in *Environmental History and the Cold War,* ed. McNeill and Unger, 13.

assistance, as shown in the last chapter, predated the war, but victories during the war heightened enthusiasm for it.

UNRRA's disease control efforts offered the most persuasive example of success. The war "provided a new set of tools, making it possible – for the first time – to greatly reduce the global burden of disease through the application of biomedical technologies."[7] UNRRA had fought mankind's microbial enemies with those technologies. A 1947 report to the United States Congress noted that "large quantities of 10-percent DDT powder" had been shipped out to combat typhus, "supplies of vaccine for mass immunization and of chemicals for water purification" had gone to China to "check cholera," and "great quantities of diphtheria toxoid ... and antitoxin" to Europe. In addition, "the new method of spraying with DDT solution was employed in many areas." And they worked. Prime Minister Clement Attlee wrote that year that "no one who has considered the comparative freedom of all liberated countries from the epidemics which ravaged Europe after the first world war can fail to pay tribute to the work which UNRRA has done."[8] That success helped to persuade people that technical assistance for disease control was an excellent starting point for development. During the 1949 hearings on Point Four, the Assistant Secretary of State for Economic Affairs noted that an Institute for Inter-American Affairs program had dramatically reduced cases of malaria in a region of Venezuela that "it is very clearly known what to do to eliminate malaria. But the things that can be done by having some experts on the spot and advising what shall be done by the local officials are very different from just the fact that the information exists in the world." Technical assistance was more than just shipping the technology; it was the "expert" administration of that technology and the training of locals.[9]

The new technologies played a key role in focusing attention on disease control, but they were not the only reasons that it was so appealing. Disease control was perceived to be a fundamental starting point for

---

[7] Packard, *A History of Global Health*, 108.
[8] 12th Report to Congress on Operations of UNRRA, as of June 30, 1947, 16, 50.
[9] *International Technical Cooperation Act of 1949, Hearings Before the Committee on Foreign Affairs House of Representatives*, 81st Cong., 1st Session on H.R. 5615, September 27, 28, 30, October 3–7, 1949. Washington, DC: United States Government Printing Office, 1950), 15. For more on the role of "experts," see Mehos, Donna C. and Suzanne M. Moon, "The Uses of Potability: Circulating Experts in the Technopolitics of Cold War and Decolonization," in *Entangled Geographies*. ed. Gabrielle Hecht. Cambridge: MIT Press, 2011, 43–74.

changing living standards in underdeveloped countries: it could be sold as a politically neutral intervention, and it was already understood to be a task that required international collaboration for success. George Marshall explained in 1948 that "little imagination is required to visualize the great increase in the production of food and raw materials, the stimulus to world trade, and above all the improvement in living conditions, with consequent social and cultural advances, that would result from the conquest of tropical diseases."[10] And who could argue with the ambition? As a US congressman asked in 1949, "If nations cannot cooperate on such matters as combating malaria and plague and cholera and tuberculosis, on what can they be expected to cooperate?"[11]

That had been exactly Huxley's point when he had described one of UNESCO's chief tasks to be educating people about the new possibilities for "harnessing micro-organisms as the chemical servants of man" and "banishing germ-causing disease" in order to bring them into greater international cooperation. "As the benefits of such world-scale collaboration become plain (which will speedily be the case in relation to the food and health of mankind)," he wrote, "it will become increasingly more difficult for any nation to destroy them by resorting to isolationism or to war."[12] By the mid-1940s, the international community already had a long history of cooperation in the name of disease control. What was changing now was the form that cooperation took and the sense of what it could achieve: ambitions were soaring, and they were not limited to microbes that attacked humans.

UNRRA's campaigns against rinderpest did not receive the amount of public attention that its campaigns against human diseases did, but members of the international community tasked with continuing the fight

---

[10] Marshall quoted in Packard, *A History of Global Health*, 95. For a history of this connection made between malaria control and economic development, see Packard, Randall M. "'Roll Back Malaria, Roll in Development': Reassessing the Economic Burden of Malaria." *Population and Development Review* 35:1 (March 2009): 53–87; Packard, Randall M. "Malaria Dreams: Postwar Visions of Health and Development in the Third World." *Medical Anthropology* 17:3 (May 1997): 279–296; Zimmer, Thomas. "In the Name of World Health and Development: The World Health Organization and Malaria Eradication in India, 1949–1970," in *International Organizations and Development, 1945–1990*. ed. Marc Frey, Sönke Kunkel, and Corinna R. Unger. New York: Palgrave MacMillan, 2014, 126–149.

[11] *International Technical Cooperation Act of 1949, Hearings Before the Committee on Foreign Affairs House of Representatives*, 81st Cong. 1st Session on H.R. 5615, September 27, 28, 30, October 3–7, 1949. Washington, DC: United States Government Printing Office, 1950, 77.

[12] Huxley, *UNESCO*, 14.

against want watched closely. By the late 1940s, rinderpest had become an ideal target for disease control. The new vaccines turned the virus into a "chemical servant of man," mutating it into a weapon that would fight off its non-mutated brethren, and opening the door to the possibility of eradication, the most global of all ambitions. Under Orr's leadership, FAO would make the eradication of rinderpest one of its central missions. That decision, just like the World Health Organization's subsequent decision to launch a global program for the eradication of malaria, had political origins and political consequences as well as technological and environmental origins and consequences. Both campaigns were launched in the name of economic development and expanding the global food supply (malaria eradication was also sold as a public health measure), and in the hope of demonstrating the supreme value of the United Nation's international machinery for development. Both faced numerous obstacles.[13] In FAO's case, those obstacles often took the form of competing machinery for development. FAO took the lead in the international effort to eradicate rinderpest, but it was far from the only entity that was excited by the possibilities of the new vaccines. The vaccines – combined with the development aspirations of the day – opened the door to many different kinds of imaginings, not just internationalist ones.

***

FAO began with a small budget and a limited mandate: five million dollars and a charge to "collect, analyze, interpret, and disseminate information relating to nutrition, food and agriculture." Its founders additionally stated that it "shall promote and, where appropriate shall recommend national and international action" on those topics and also "furnish such technical assistance as governments may request."[14] Such assistance, however, would have to be of a very limited nature, given its budget. How limited is evident in suggested remarks sent to Truman from the Department of Agriculture in September of 1945, which explained, "FAO will be an expert consulting agency operating on a modest budget, with

---

[13] Packard, *A History of Global Health*, 133–179; Packard, "Roll Back Malaria, Roll in Development," 56–62; Packard, "Malaria dreams," 279–296; Mitchell, *Rule of Experts*, 19–53. For more on the malaria campaign and on the idea of eradication in general, see Stepan, Nancy Leys. *Eradication: Ridding the World of Diseases Forever?* Ithaca, NY: Cornell University Press, 2011.

[14] United Nations Interim Commission on Food and Agriculture, Facts about FAO (September 11, 1945), United Nations Food and Agriculture Organization, 121–8, President's Secretary's Files, Papers of Harry S. Truman, HSTL.

nothing to give away free except advice." But, that advice would be "the kind that gets results. It will stimulate research and development and the rapid spread of use of knowledge in all its various fields of work."[15] This was the vision of the organization's founders, but it was not the vision of John Boyd Orr. He wanted FAO to do more – to *be* more. The new rinderpest vaccines helped him do it.

Animal health was not part of FAO's initial mission, but it was clearly an important part of food security. In the summer of 1946, Orr formed an ad hoc committee of participants from France, India, Iran, South Africa, the UK, the US, the UNRRA, and the Office Internationale des Epizooties to "study international veterinary questions" and make a recommendation to FAO about what its role should be in that area. The committee met in London and talk immediately turned to disease. Participants "emphasized the essential need for action on the widest international lines" and insisted that FAO could enable "collaborative action in research into veterinary problems, and cooperative trials, and use of new products for the control of disease." It could also facilitate "joint action between countries concerned in any particular campaign." And participants had a particular campaign in mind. The committee "reviewed in some detail the subject of rinderpest," decided that it "still represents a major menace to the world's supply of food," and concluded "that every effort should be made to eradicate the disease completely where it exists in enzootic form." They had reason to think it possible to do so. Participants "noted with the liveliest interest, the cooperative research into this disease carried out by the United States and Canadian Scientists which resulted in the preparation of a vaccine of a new type which is now being tested in the field in Africa and China."[16] The Grosse Île vaccine gave them hope, and a preferred target.

The new vaccine made eradication seem possible, but it was not just a question of technology. It was also a question of organization and administration. Eradication could not be accomplished without permanent machinery for sustained international coordination. Committee members "strongly recommended" that FAO establish a Standing Committee on Animal Health, precisely because the rinderpest case held

---

[15] Suggested Remarks, G.H. (September 24, 1945), United Nations Food and Agriculture Organization, 121–8, President's Secretary's Files, Papers of Harry S. Truman, HSTL.
[16] Notes on FAO Veterinary Meeting held in London (August 13–15, 1946), London 1946 Report, ADHOC Committee on Animal Health, Animal Production and Health Division, 10AGA407, FAO.

so much potential for action and for success. FAO followed their recommendation.

At the first meeting of the new Animal Health Subcommittee in the spring of 1947, Orr "particularly stressed that FAO would now concentrate on the expansion of animal food production," noting that "the reduction of waste due to loss caused by disease was of particular importance." Participant conversations about rinderpest centered on helping nations get the new vaccine. They recommended that "FAO, in consultation with officials of the Canadian Government and with officials in countries such as Egypt, Kenya and Siam, where the vaccine might be established, should take steps to assist in the transfer of the frozen virus to these new locations. FAO might provide for the expenses of a courier to transfer the virus, as possibly also for technical assistance, if needed, in getting the virus established." The subcommittee also discussed the possibility of establishing a Center for the Study of Rinderpest, but recommended that the topic be deferred as "it was understood that the Office International des Epizootics would discuss in detail, at its next meeting in May, the problem of rinderpest research." Participants concluded their conversation with a note that FAO could help facilitate a "meeting of specialists from those countries specifically interested in rinderpest" in the near future.[17] The recommendations centered on what FAO could reasonably do based on its limited budget: it could move frozen virus between labs in different countries and host meetings that brought nations together to discuss the new vaccines – it could, in those ways, create international networks for fighting the disease where none existed.

Everything centered on the vaccines (primarily, but not only, the new avianized), because everyone involved believed that the vaccines were the key. R. Daubney, who had participated in the 1946 London meeting, had sent a note to the subcommittee in February, insisting, "it is upon universal immunization of infected cattle populations that we must depend for the control, and even for the eradication" of rinderpest in those countries where it was enzootic. "Few of them," he argued, "are suited to the employment of veterinary police measures that are adequate to deal with introduction of the disease into more highly developed and closely administered areas, such as Western Europe."[18] Daubney's colonial

---

[17] Minutes of Meeting of Subcommittee on Animal Health (March 31–April 4, 1947), Sub-Committee on Animal Health, Animal Production and Health Division, 10AGA407, FAO.

[18] R. Daubney, "Recent Developments in the Control of Rinderpest" (February 1947), Correspondence-Documents, Rinderpest Meeting – Nairobi, 1948, Animal Production and Health Division, 10AGA407, FAO.

administrator resume was clearly on display, but he was not alone in his perception that widespread vaccination was necessary in areas where the government had not so fully domesticated the environment as in Western Europe. Non-colonial administrators in other parts of the world were arguing the same thing.

Thailand had dealt with endemic rinderpest for years, primarily in its northern and northeastern provinces, but the end of the war had seen its spread throughout the entire country, raising the annual loss of stock to rinderpest from 15,000 to 200,000.[19] The problem had begun with the retreat of the Japanese army from Burma, the government explained, and had continued "by the agency of infected animals amongst the ox-drawn transport." The Thai government initiated a nationwide campaign – "not of reduction, but of elimination" – against the virus in June 1946. It used the tools it had: inactivated buffalo spleen vaccine (Kelser's wet tissue method), serum, and goat-attenuated vaccine (the Kabete caprinized strain).[20] Water buffalo could not tolerate the caprinized vaccine, which is why the Thai still had to make the expensive (it took one buffalo to make around ninety doses of vaccine) and short-immunity-granting inactive tissue vaccine, and why they wanted the avianized vaccine. They turned to FAO to get it.

Thai veterinarian Dr. Charas Suebsaeng, who worked for the Ministry of Agriculture, sent a letter to FAO in early February 1947 asking for assistance, and was promptly informed that "the matter of setting up a committee to study the problem of Rinderpest and Foot and Mouth disease was under consideration."[21] The new Animal Health Subcommittee had not yet met. When it did, it recommended exactly the course of action that Charas Suebsaeng desired.[22] The FAO technical mission to Thailand arrived in the beginning of January 1948 listing its first "specific concern" as "production of avianized serum for the control of rinderpest."[23] FAO advisers, working with their counterpart Keith

---

[19] FAO, Report of the Mission for Siam, 7122G3, FAO.

[20] Ministry of Agriculture, Thailand, Annual Report to the Food and Agriculture Organization, 1949, Thailand 1947–65, Annual Reports from Governments, 010V2, FAO.

[21] R.P. Jones to Orr (January 22, 1948), Outgoing Correspondence, Mission to Thailand, 7122A1, FAO.

[22] Minutes of Meeting of Subcommittee on Animal Health (March 31–April 4, 1947), Sub-Committee on Animal Health, Animal Production and Health Division, 10AGA407, FAO.

[23] FAO Mission to Siam; Memorandum of Agreement, Mission to Thailand, 710M22, FAO.

Kesteven in China (the same man who had the summer before, while still working for the UNRRA, traveled from Nanjing to Kabete in search of avianized vaccine) soon imported seventeen strains of rinderpest from Nanjing. These included both Nakamura III (in its 772nd passage) and Chinese-avianized strains, along with the equipment necessary for continuing their development.[24] Work began immediately.

As the transfer of the virus from Nanjing to Pak Chong demonstrated, FAO was in a unique position to coordinate the movement of frozen virus around the world. The Thai government recognized that it was also in a unique position to bring experts together from around the world to discuss the necessity of international collaboration not only in the creation of vaccines, but in their use. Eradication was impossible without it. As the Thai Ministry of Agriculture explained, "Eradication of rinderpest is a species of warfare, and has to be planned accordingly."[25] A victorious campaign would depend primarily on the weapons, but it also required a coordinated strategy. Everyone knew that FAO did not have expansive funds, yet the organization wielded critical power. It could move frozen virus around the world and it could host international conferences that could lay the groundwork for international cooperation in this "species of warfare." Its leadership did both, anxious to demonstrate that FAO was working – that internationalism was working.

In a July 1948 address in Indiana, Norris Dodd, the new Director General of FAO, told his audience that "FAO has great promise of success" in its effort to address the "most fundamental ... problem of getting enough of the basic necessities of life for the world's people." This is, he admitted, "a big assignment ... and so far FAO is a small organization with a budget of less than five million dollars a year." Dodd insisted, however, that he did not "worry too much about that," for "I believe we will grow in proportion to our success with what we are doing." And they were doing a lot. He went on to explain that first and foremost, "we foster the development of the underdeveloped or so-called backward areas of the world," by providing "advice and technical assistance on a wide range of problems." "I could give you many interesting examples of this kind of work," he told his audience, "but I shall have to confine myself to a few." The first of his examples was the war against rinderpest in China.

[24] Transcript report of the activities of the mission to Siam (March 4, 1948), Speeches, Statements, and Radio/TV Broadcasts, Mission to Thailand, 7122H1, FAO.
[25] Ibid.

Equating FAO activities with those of the UNRRA before it, Dodd wove a tale of science defeating want,

Well, during the war a method was developed in North America for mass production of a vaccine that gives immunity against rinderpest. FAO technicians flew a strain of this virus into China and set to work helping the Ministry of Agricultural to start mass production. A campaign is now beginning which aims at vaccinating all of the cattle and water buffalo south of the Yangtze River – about 15 million head all told. That would be a big operation in any country. It is an outstanding example of the true internationalism of science. Discoveries made in laboratories on one side of the world are being applied for the benefit of people on the other side.[26]

And the benefits do not stop there, Dodd implied, but come back again to the other side of the world – all the way back to Indiana, in fact, because the "gate marked Freedom from Want," he insisted, is "also the gateway to peace."

Dodd ended his story with a plea that the audience encourage "your government" to support FAO. He was worried about FAO's fate. The organization needed more money to provide greater evidence of its potential. John Boyd Orr had been happy to leave FAO when his term expired; he was frustrated over the lack of funding and the lack of faith in the value of internationalism.[27] Dodd shared his predecessor's conviction that FAO could make a difference even beyond its specific programs, if only it were provided the necessary funding. If FAO succeeds, he told his audience, "we can have greater hope for the success of international efforts in other fields."[28] FAO was a proving ground for internationalism itself and the struggle against rinderpest was its most important action. Dodd pitched his story accordingly.

FAO's international status was the source of its limited budget and blurry mandate, but it was also the key to its potential success. It could do more than a single nation could, provided it were given the necessary backing. FAO could not exist without nations: they funded it, they implemented its plans. FAO could transport frozen rinderpest in glass vials, it could send in advisers, it could coordinate meetings and the acceptance of foreign students in veterinary colleges around the world, but it could not

---

[26] Norris Dodd, Food, Farming and Peace (July 19, 1948), Box 1, Folder 7, Norris Dodd Collection, HIA.

[27] Staples, *The Birth of Development*, 95; Vernon, James. *Hunger: A Modern History.* Cambridge: Harvard University Press, 2007, 152–156.

[28] Norris Dodd, Food, Farming and Peace (July 19, 1948), Box 1, Folder 7, Norris Dodd Collection, HIA.

independently wage the war it so desperately wanted to win. In this, FAO was not alone. A 1950 article about the World Health Organization (WHO) noted that "notwithstanding their recognition that collective action is essential to combat and prevent disease, the nations of the world community have not been willing to yield enough power of decision to make any international health organization more than a collaborative agency."[29] Nationalist agendas regularly warred with internationalist ones for both dominance and access to resources, and they were not the only agendas competing for influence in the postwar world. Both had to also deal with imperial ones.

***

In 1947, the OIE, noting "great interest" in the US and Canadian work that "provides new ways of immunization," resolved that research on rinderpest be undertaken at the "international level," and committed itself to coordinate with FAO *and* with Great Britain's Colonial Office (CO) in that effort.[30] The two entities shared an interest in rinderpest eradication, but they pursued it through different administrative networks for very different overarching purposes.

FAO was already working with specific nations throughout Asia and Africa, encouraging them to work more closely with their neighbors, sharing information and technological know-how. The Colonial Office had been pursuing its revised regional control efforts in Africa since 1940 and had begun networking more closely with other imperial powers – France, Belgium, and Portugal – in pursuit of eventual continental eradication. French and British veterinary offices met in Dakar in May of 1946 to discuss rinderpest control, immediately after the reports of the work undertaken at Grosse Île were published. It was the first in a series "planned for Anglo-French Colonial Collaboration in West Africa" and was, the British Consul-General reported, "a success in developing technical contacts and in reaching conclusions as to the combined treatment of various diseases." (Along with rinderpest, the participants also discussed pleuro-pneumonia.) Participants agreed that "the ultimate goal at which we aim is the total eradication of rinderpest ... from the whole of the African continent," but acknowledged that they currently lacked the

[29] Allen, Charles E. "World Health and World Politics." *International Organizations* 4:1 (February 1950): 31.
[30] OIE, Résolution no° 1, 15éme Session Général, 1947, available at http://www.oie.int/doc/en_document.php?numrec=556503.

necessary personnel and equipment. They focused, instead, on "intensify-[ing] control of the disease" through actions "synchronized on both sides of International frontiers." The approved press release stressed that the ultimate aim of this coordination was "to raise the standard of living of the people."[31] Rinderpest control had been co-opted into colonial development.

Britain's Africa program depended on *imperial* international coopera-tion. There was plenty of room for the OIE, which did not "interfere," in that plan. Whether or not there was room for FAO, which seemed from its public statements likely to "interfere," was open to debate.[32] The question became a subject of debate when Britain decided, "as a result of discus-sions between the Colonial Office and certain European Governments with African Colonial interests," to host a "Pan-African Conference to discuss the problem of eradication of rinderpest in Africa" in 1948. This conference had a larger vision than the 1946 one, which had focused solely on West Africa. That one had demonstrated that technical coopera-tion could prove a beneficial field for colonial cooperation in a region; it now seemed logical to expand the effort to include an entire continent. The decision for the new conference also came hot on the heels of the Ministry of Agriculture and Fisheries' "refusal ... to allow the importa-tion of refrigerated beef from rinderpest infected territories." This was an expensive ban. The CO may also have been concerned about the recent "introduction of rinderpest into Malta through the importation of cattle from the Sudan." The virus was on its radar. So, too, were the new vaccines.[33]

Letters (in English or French) to potential participants explained that "the main object of the meeting will be to consider the latest developments in the production of biological preparations and the control and possible eradication of rinderpest in all countries afflicted."[34] They scheduled it for October in Nairobi, so that participants could visit Kabete and witness

[31] British Consulate-General to H.M. Principal Secretary of State for Foreign Affairs (May 31, 1946), FO 371/60000, NAUK; Agreed Conclusions of Anglo-French Veterinary Conference held at Dakar (May 11–15, 1946), FO 371/60000, NAUK; British Representatives Draft of Item 1 of Conclusions of Anglo-French Veterinary Conference, FO 371/60000, NAUK; Sir. H. Blood to S. of S. (May 17, 1946), FO 371/60000, NAUK.
[32] Notes for Mr. Eastwood of the C.C.T.A., Working Party on Rinderpest (August 29, 1950); CO 852/1228/7, NAUK.
[33] Mr. McGilvray, Rinderpest Conference in Nairobi (July 26, 1948), MAF 35/740, NAUK.
[34] Draft of letter from Herbert Broadley (August 16, 1948), Animal Production Branch, 1946–58, Animal Production and Health Division, 10AGA407, FAO.

vaccine production. There were exciting things happening there. In August of 1948, the same month the letters went out, K. L. V. Kesteven sent Kabete lapinized "dried virus by air" from Nanjing. This was "Nakamura Strain III which had been serially passaged in rabbits for 795 generations." By the time the conference began, researchers at Kabete were experimenting with caprinized, avianized, and lapinized vaccine.[35] For most participants, it would be their first exposure to the last two. The "latest developments in the production of biological preparations" were, indeed, waiting for them in Kenya.

The CO imagined the conference to be a meeting of fellow colonialists. To its dismay, however, John Boyd Orr found out about it early on and wrote to the Foreign Secretary asking that FAO and all of its African members be allowed to participate. The CO was not pleased. A cable from a member of the British Food Mission in the United States reported back to London that "Orr is very resentful at Colonial Office attitude." "Understand," the memo continued, he is "likely to write to the Colonial Office asking whether they would be willing to enlarge the Kabete conference so as to enable FAO to bring to its experts or representatives from other areas such as the Middle East." Well aware that the CO would not look favorably on Orr's letter, the memo's author asked his recipient in the Ministry of Food to intervene with the CO "to secure a favourable and friendly reply." The United Kingdom, he insisted, needs to "show readiness to cooperate with FAO in conferences on a wider scale."[36]

The memo's recipient was no more welcoming than his CO counterparts, writing in his message to them, "I need hardly say that I thoroughly agree with the idea that when a limited number of countries 'get together' to solve some mutual difficulty F.A.O. should not immediately step in and inflate the whole question on to the international scale." But it made no sense to alienate the organization with a blanket refusal. Doing so was bad politics. "While I am all opposed to turning the Kabete conference into an international circus on the usual F.A.O. lines," he insisted, "I wonder whether there is not some way in which we can meet Sir John Orr's main point of making the conference immediately available to other countries who are practically concerned in the

---

[35] Brotherston, J. G. "Lapinised Rinderpest Virus and Vaccine: Some Observations in East Africa." *Journal of Comparative Pathology* 61 (1951): 263–288.

[36] Note by the Joint Secretaries, F.A.O. National Official Committee for the U.K., African Conferences (April 28, 1948), with annexes, FO 957/16, NAUK.

Rinderpest problem." In the end, the CO made two concessions to FAO: it could send "observers" to the African Rinderpest Conference in Nairobi and FAO could host its own rinderpest "meeting" in Nairobi immediately afterward.[37]

The imperial conference opened first. If FAO was going to make a "circus," the CO wanted it to be of side-show size. It cared about securing cooperation with France, Belgium, and Portugal, and, in that, it was successful. The attending colonial powers agreed that "in spite of any agricultural, sociological, or administrative repercussions, control of rinderpest with a view to its complete eradication is desirable and necessary in the interests of Africa as a whole."[38] Gone were the days of viewing the virus as a helpful check on the size of African-owned herds. It was now politically expedient to support eradication for both economic and image purposes. It was defending colonialism by different means, but it was still all about that defense.

Participants at the African Rinderpest Conference resolved that "rinderpest can be eradicated from Africa with the biological immunizing agents already at our disposal." The technology was there, as the trip to Kabete made tangibly clear. Now, it was all about organization for success. Participants recommended the creation of an African Rinderpest Organization to coordinate information between members and liaise with FAO and the OIE and an African Scientific Committee to monitor vaccine production. Participants additionally agreed to "the creation of game barriers," recognizing that the movement of the disease across Africa could not be controlled without them. This emphasis on the spread across borders, however, brought them to the "problem" of Africa's independent countries, and that brought them straight to FAO.[39]

In a third resolution, participants acknowledged that they recognized "that in the initiation, co-ordination and prosecution of rinderpest eradication certain territories, namely, Ethiopia, Somalia, Eritrea, and the Anglo-Egyptian Sudan, may require assistance in finance, provision of personnel, and provision of prophylactics, and arranging for the marketing of surplus stock." These did not have an imperial "benefactor" to

[37] Ibid.
[38] Roeder, Peter and Karl Rich. "The Global Effort to Eradicate Rinderpest." IFPRI Discussion Paper 00923 (November 2009), 18; available at http://www.ifpri.org/sites/default/files/publications/ifpridp00923.pdf.
[39] Secretary of State's Visit to Paris and Brussels, Anglo-French Cooperation, African Rinderpest Conference, CO 852/1228/7, NAUK; Resolutions of African Rinderpest Conference (October 1948), CO 852/1228/7, NAUK.

provide it for them. Lest there be any confusion about the last country mentioned, British correspondence before the meeting had firmly noted that "Sudan is not a British Colony." Sudan was complicated, existing in a gray area between independence and imperialism, exactly the kind of realm where FAO could work. Participants of the conference also noted at the meeting that they were worried about "overstocking" in areas outside of their direct control. They "considered that FAO would be the most suitable organization to consider the solution of these problems."[40] They did not want the responsibility.

Resolution 3 was a grudging admission, born more of desperation than of desire. Continental eradication required that every nation participate, but involving FAO challenged the entire point of that campaign from the imperial perspective. The colonial powers hoped to use rinderpest eradication in Africa as proof of the benefits of imperialism. Outside aid to neighboring independent countries would demonstrate that international aid did not have to flow through imperial networks to work; development could come through other channels. Even after agreeing to involve FAO, the conference's imperial participants soon wanted to rescind it, as some revealing notes from a 1950 meeting make clear: "Although this resolution was passed by the conference it subsequently appeared that France, Belgium, the Union of South Africa and Portugal were not happy about its implications. They did not like the idea of FAO interfering in an African campaign against rinderpest," preferring, instead, that "all possibilities of the African Powers handling this problem between themselves should be explored" first. Participants "suggested therefore, that the attention of F.A.O. should not be drawn specially to this resolution (they were represented by an observer at the conference and have not themselves raised the point)" until "information should first be obtained as to the amount of assistance the territories concerned would require, with a view to deciding whether the other African powers should assist the United Kingdom in the provision of personnel and finance to undertake rinderpest campaigns in these territories."[41] The British Colonial Office was not happy about it either. The imperial powers wished to maintain their

---

[40] Resolutions of African Rinderpest Conference (October 1948), CO 852/1228/7, NAUK; Animal Production Branch, 1946–58, 10AGA407, FAO, Economic Relations Department to Chancery (September 18, 1948), FAO Conference for the Control of Rinderpest; FO 371/69127, NAUK.

[41] Notes for Mr. Eastwood of the C.C.T.A., Working Party on Rinderpest (August 29, 1950), CO 852/1228/7, NAUK.

stranglehold and believed FAO could threaten it.[42] They were right to worry. FAO had been contemplating that very issue since its own Nairobi meeting two years earlier – hot on the heels of the imperialists' conference.

***

As had been the case at the African Rinderpest Conference, FAO's 1948 Nairobi Meeting on Rinderpest also focused on "latest developments in the production of biological preparations and the control and possible eradication of rinderpest."[43] In an October press release for the meeting, FAO explained that discussion at Nairobi will "center around" the vaccines. The press release noted that R. Daubney will present on the caprinized vaccine and S. C. Cheng on the lapinized vaccine and avianized vaccine. "There will also be a discussion of tissue vaccine," it read in a deliberately vague statement that is explained later in this chapter. After four days "devoted to consideration of the good and bad points of each vaccine," it continued, and the meeting would spend its final day discussing the practicalities of eradication: the "uses of vaccine under various circumstances," "costs," and "the possibilities of international cooperation." "Stress will be laid on the fact that rinderpest is no respecter of international boundaries, and that complete control is possible only if governments work together in a coordinated program."[44] The Nairobi meeting was FAO's first significant effort to help governments achieve that kind of coordination by sharing vaccine technology, information about vaccination programs, and just simply getting veterinarians from the different countries talking to each other – developing relationships that reached across borders.

Some delegates stayed on from the Nairobi conference to attend the Nairobi meeting, which lasted from October 28 to November 1. Frustration in London and Paris was evident. FAO asked Sir Daniel Cabot, the United Kingdom's Chief Veterinary Officer from 1938 until 1948 and the person who had assigned Edwards to engage in secret rinderpest research at Pirbright during the war, to chair the meeting. Cabot, who also represented the OIE at the meeting, wrote afterward

[42] For more on this see Kent, John. *The Internationalization of Colonialism*. Oxford: Clarendon Press, 1992.

[43] Draft of letter from Herbert Broadley (August 16, 1948), Animal Production Branch, 1946–58, Animal Production and Health Division, 10AGA407, FAO.

[44] FAO, "Governments Meet in Africa to Plan New Attacks on Rinderpest" (October 27, 1948), Animal Production Branch, 1946–1958, 10AGA407, FAO.

that the "French delegation found itself in an embarrassing situation owing to 'political considerations.'" A French researcher had been asked and had agreed to present a paper on inactive tissue vaccines, but he backed out at the "last minute," neither attending, nor sending his paper. After inquiring as to the reason for the absence, Cabot "was informed that the French African delegation would disperse after the first conference, leaving only their leader (M. Feuntun) to attend the F.A.O. meeting as an observer." When Cabot confronted Feuntun about the decision, he "explained that their view was that there was not sufficient time to digest the findings of the first conference before embarking on the second." Cabot, unimpressed, "told him that view was not valid as the programmes were quite distinct. M. Feuntun then admitted that there were political aspects and that he was bound by his written instructions." In the end, after "telegraphic appeals" to Paris, the French delegation remained, "though only as observers, taking no part in the discussions."[45] Despite "some feeling in Paris" about the meeting, it went on as planned. Orr was not about to let France dictate FAO's agenda. A familiar delegate from the United States stepped in to offer what the French government had denied. "In the absence of Dr. Vittoz (of French Indo-China) who had promised a paper on tissue vaccines," Richard E. Shope, formerly at Grosse Île, now at the Rockefeller Institute, "opened the discussion on this subject."[46]

The biggest names in rinderpest vaccine research attended the Nairobi meeting: Daubney, Cheng, and Shope were responsible for some of the most important advances in the field. They shared their expertise with veterinarians from the Belgian Congo, Portuguese West and East Africa, Egypt, Ethiopia, India, the Union of South Africa, Southern Rhodesia, Nigeria, Gold Coast, Northern Rhodesia, Nyasaland, Kenya, Uganda, Tanganyika, Zanzibar, the British Military Administration Somalia, the East African High Commission, France, Great Britain, the United States (Shope), and FAO headquarters (Kesteven). Some were national representatives; some were colonial representatives. Some countries, such as Ethiopia, were represented by an FAO technician who was stationed in their country.[47] Discussion revolved around the vaccines: creating them, producing them, and utilizing them in the field.

---

[45] Sir Daniel A. E. Cabot, F.A.O. Nairobi Conference on Rinderpest (October 28–November 1, 1948), MAF 35/740, NAUK.
[46] Ibid.
[47] List of Delegates, Observers, etc., F.A.O. Nairobi Conference, MAF 35/740, NAUK.

Those present rejected the serum-simultaneous method out of hand. Shope, who had not had time to prepare a "formal paper" on the inactivated vaccines, spent half of his presentation talking about the Grosse Île research that had produced the avianized vaccine. There was discussion about the inactive vaccines, but participants agreed that it was "not recommended for general use." The uncertainty of the length of time of the immunity it conferred made it too risky.[48] The formal papers on the attenuated vaccines made for far more exciting conversation. Daubney's paper on the caprinized vaccine contained much that was already known to the meeting's participants, as it was the oldest of the attenuated vaccines and the one that had been "used with great success on a large scale for certain types of stock." All agreed that it was still highly recommended as the vaccine of choice for cattle, such as most zebu, with "a degree of natural resistance" to the virus. There were, at present, three strains of Daubney's original KAG vaccine available: Kabete had one in the 750th passage, Vom in Nigeria had one in the 599th passage, and Cairo in Egypt had one in the 690th passage. Participants agreed that it would be wise for each laboratory to get each strain "to provide strains of varying potency for use on cattle with different levels of susceptibility." They also agreed that it would be "unwise to passage the virus in goats more frequently than necessary once a satisfactory passage has been obtained for a particular type of cattle." The virus needed to be altered only so far and no more.[49]

Everyone acknowledged that the caprinized vaccine was important, but it was not the star of the Nairobi meeting. Participants were far more excited about the presentations on the avianized and lapinized virus which made the work that S. C. Cheng and his colleagues had done at Nanjing over the past two years publicly available for the first time. Cheng and company had just that summer announced their successful creation of an avianized vaccine using two of the avianized strains that Kesteven had brought to Nanjing from Kabete in the fall of 1947. "We have one strain ready for use," Cheng reported, and another still "quite virulent for calves" currently being passaged into greater attenuation. This was an exciting achievement, because so many people had been unable to replicate the success of Shope and company's Grosse Île research. That being the case, "much of the discussion which followed the paper centered

---

[48] Shope, Richard E. "Inactivated Virus Vaccine," in *Rinderpest Vaccines*, ed. Kesteven, 23–28.
[49] Daubney, R. "Goat-Adapted Virus," in *Rinderpest Vaccines*, ed. Kesteven, 6–22.

around the difficulty in adapting the virus to grow on eggs." Indeed, Kabete researchers, who had successfully passaged the virus in eggs, had since lost its strains, their first either "somewhere between the 35th and 51st passages, or alternatively, it had lost its ability to immunize." They were now having trouble repeating their own initial success, which made the details of Cheng's research – and the successful strains that he had created – that much more important. Everyone wanted to know exactly what he had done.[50]

Cheng's paper on avianized vaccine generated a great deal of discussion, but all the participants had already known from the Grosse Île research that it was possible to create a vaccine via passaging through eggs. Cheng's paper on the lapinized vaccine, however, was altogether new – the first exposure many had to the discovery Nakamura had made years earlier – and it aroused "considerable interest."[51] Cheng readily gave credit to Nakamura for the initial strain and then went on to describe both the ongoing research with the strain that was taking place in Nanjing and how the vaccine was being administered in the field. It was fast becoming the vaccine of choice in China. Participants agreed that the lapinized virus "presents outstanding possibilities for furthering the study of rinderpest control." Cheng reported that researchers were already trying to passage it through other animals to both get around the problem of a limited rabbit supply in certain areas and to even further reduce its virulence. Thus far, he wrote, passages through chicken embryos, the most appealing option, had failed. Research continued in Nanjing and abroad as researchers there had already sent Nakamura III to Egypt, Thailand, India, Kenya, Pakistan, and Ethiopia.[52]

Research also continued in one place they had not sent the virus. Junji Nakamura had not been invited to the meeting because he was Japanese and, as such, still the enemy. He was now at the Nippon Institute of Biological Science in Tokyo, since Japan had lost its imperial laboratories in the war. He had been busy trying to attenuate his lapinized virus "by serial passages through chicken embryos, to attenuate it for cattle of the highest susceptibility." He had first tried to do it between 1942 and 1945 – the exact time that Shope and his team had been engaged in their own

[50] Cheng, S. C., T. C. Chow, and H. R. Fischman. "Avianized Rinderpest Vaccine in China," in *Rinderpest Vaccines*, ed. Kesteven, 29–45.

[51] Scott, G. R. "Adverse Reactions in Cattle after Vaccination with Lapinized Rinderpest Virus." *Journal of Hygiene* 61 (1963): 193.

[52] Cheng, S. C. and H. R. Fischman. "Lapinized Rinderpest Virus," in *Rinderpest Vaccines*, ed. Kesteven, 47–65.

efforts with the straight bovine virus – but "the study was then discontinued," no doubt because of the progress of the war. He was, at the time of the Nairobi meeting, trying again. He was also paying attention to all the research being published on rinderpest in the West. Nakamura was currently down, but decidedly not out. He was too vital to the cause for the international community to isolate him for long simply for being Japanese. The larger goal – eradication – was too important.[53]

When Cheng had announced on June 16 that his team at Nanjing had successfully produced their first batch of avianized vaccine, he had told reporters, "While this is only the first step, it means that large-scale production of avianized rinderpest vaccine can now commence in the near future. Once large-scale production gets under way, it is not too optimistic to say that our goal, the complete eradication of rinderpest from China, and indeed from the whole world, will be in sight."[54] Participants at the Nairobi meeting shared his enthusiasm, concluding that "the prophylactics now available" had made eradication "a practical possibility," but they warned that the technology alone was not enough to achieve success. It also mattered a great deal where, when, and how it was used. Eradication would depend on "the closest international co-operation," they insisted, "particularly in regard to boundaries."[55] In order to spread the word, participants recommended that FAO publish the papers presented at the meeting and organize a meeting for Asia and the Far East. It promptly did both.[56]

The emphasis on international cooperation was critical, not only because it was true that rinderpest paid no attention to political boundaries, but because FAO's entire existence depended on showing that human prosperity could be best achieved through global – not local or imperial – efforts. FAO was constructing its own network. Its utility became clear when, toward the end of 1948, "the political disturbances in China" led FAO to transfer the avianized virus, equipment for manufacturing it, and the chief FAO veterinarian, Dr. Harvey Fischman,

[53] Nakamura, Junji and Takeshi Miyamoto. "Avianization of Lapinized Rinderpest Virus." *American Journal of Veterinary Research* 14 (1953): 307–317.

[54] Leo H. Lamb to P.W.S.Y. Scarlett (June 21, 1948), FAO Conference for the Control of Rinderpest, FO 371/69127, NAUK.

[55] Summary of Conclusions, F.A.O. Nairobi Meeting on Rinderpest (October 28–November 1, 1948), Animal Production Branch, 1946–1958, 10AGA407, FAO.

[56] See Kesteven, K.L. *Rinderpest Vaccines: Their Production and Use in the Field*, 1st ed. Washington, DC:FAO, 1949. Original copies of the reports can be found in MAF 35/740, NAUK.

from Nanjing to Thailand "so that work on the avianized vaccine could continue." Fischman (who had worked closely with Cheng), the seized equipment, and the "rescued" avianized virus set up shop at Pak Chong laboratory, to the northeast of Bangkok. The Thai government approved the expansion of the facilities, "so that the invaluable work done in China should not be lost to the world."[57] Or, at least, lost to the "free world."

Thailand also, of course, had more selfish reasons for readily approving the vaccine transfer. The avianized "embryo material" sent there from Nanjing in 1948 had "proved to be dead, and another strain started made no headway." By the end of 1948, Thailand had produced 285,677 doses of inactivated buffalo spleen vaccine (compared to 149,647 the year before); 608,244 doses of wet caprinized vaccine (compared to 188,660 the year before); 27,216 doses of dried caprinized vaccine (none the year before); and 20,841 doses of lapinized vaccine (none the year before). In 1949, however, the number of inactivated vaccines produced dropped to 77,440 and that of wet caprinized to 266,200, while the lapinized doses rose to 68,880. A general shortage of rabbits was proving troublesome, but it was hoped that the new "rabbit breeding unit" would alleviate the problem. It was also hoped that the new avianized strain would survive the journey. "If all goes well," its Ministry of Agriculture reported to FAO in 1949, "rinderpest should be completely eradicated in Thailand in the course of a year, after which it will be urgently required that neighboring countries co-operate to prevent re-infection. It is hoped," it concluded, "that the FAO will initiate proposals in regard to this subject."[58] Eradication needed cooperation across borders. Thailand needed more from FAO than just viral strains. So did everyone else struggling with the virus.

This was why rinderpest was such a good opportunity for FAO. The organization's Regional Representative of the Director General for Asia and the Far East argued as much in a letter to headquarters in January of 1949. Noting that there had been talk lately about encouraging coordination between neighboring countries through plant breeding efforts, he countered that "it is highly desirable that we should achieve success in attacking one specific problem before attempting numerous projects

---

[57] Ministry of Agriculture, Thailand, Annual Report to the Food and Agriculture Organization, 1949, Thailand 1947–65, Annual Reports from Governments; 010V2, FAO.
[58] Ibid.

which have a diversity of features which are not of the same magnitude of interest to each of the member countries." Rinderpest was the obvious choice: "I feel that greater progress can be made in rinderpest control than in any other agricultural problems which face us on a regional basis." He concluded with a request that a regional meeting be organized "as soon as possible."[59]

FAO leaders made rinderpest a key target both because the vaccines had made eradication "a practical possibility" and because the disease could unite countries from all over the world around one specific problem. True success in the war on rinderpest would mean the global eradication of the virus. What other campaign (besides the eradication of a virus that directly attacked humans, which was outside FAO's scope of work) could provide stronger evidence of the utility of international organization for the betterment of human beings everywhere?

While the ultimate goal was bold, in practical terms, FAO's assault on rinderpest had modest funding and a modest plan of action. At its rinderpest meeting for Asia and the Far East in Bangkok that June, FAO organizers "stressed that it was planned to consider the situation as it is at the present time, the best means of combating the disease, and finally to arrive at a definite agreement between countries on the problems of research, and control and eradication plans." It had limited, clear goals, the primary one being to encourage countries to work together against "the common enemy."[60] The FAO adviser to the Thai delegation offered a "stirring call" for "international action in eradicating the disease throughout the rinderpest belt," insisting that participants needed "to follow the disease wherever it might be found, regardless of national barriers." He noted that "such a campaign would cost money," but that it would end the losses currently accruing from dead cattle and reduced food production. Lest there be any confusion about where that money would come from, K. V. L. Kesteven promptly "pointed out that, while it might be possible through the International Bank or through the United States plan for assistance to under-developed areas (Point Four) to provide additional technical assistance, the nations themselves would have to provide local

---

[59] W.H. Cummings to Ralph w. Phillips (January 21, 1949), FAO Rinderpest Conference for Asia and the Far East – Bangkok, 1949, Animal Production and Health Division, 10AGA407, FAO.

[60] K.V.L. Kesteven, Report on FAO Rinderpest Meeting for Asia and the Far East (August 2, 1949), FAO Rinderpest Conference for Asia and the Far East, Bangkok, 1949, Animal Production and Health Division, 10AGA407, FAO.

financing."[61] FAO would not be footing the bill for eradication; it would, however, help nations with their programs by ensuring they got the technological know-how and administrative advice they needed. These measures, though limited in scope, were novel in scale, dispersed throughout Asia and Africa. FAO was helping humanity as a whole by helping individual nations, which is exactly what Orr had wanted it to do from the beginning.

The countries involved prided themselves on these being *national* campaigns. They were very happy to have them connected internationally via FAO in the form of vaccine transfers and international meetings and, when they could get it, additional laboratory technology and field work tips, but they were fighting rinderpest for their own development. They desired coordination, but not control. Nowhere was this more true than India, which had only just won its independence from Britain in 1947. Rinderpest was a serious problem in the country – it had been for a long time. Edwards had insisted in 1928 that the "losses from the great cattle plagues" were "the cause of great – some might even say, with much reason, that they are the greatest single cause of – economic impoverishment to the country." The situation had not improved in the interim.[62] India had sent a representative to the Nairobi meeting to find out about the new vaccines. They had the caprinized vaccine, of course, which had been originally created there, but they were interested in something milder for more susceptible breeds.

In the aftermath of the Nairobi conference, they sent a request to FAO for "strains of rinderpest which had been adapted to eggs." Those strains were now, after FAO's retreat from Nanking, in Thailand. Kesteven, who had been in Bangkok for the regional meeting, left there with the strains on June 29. He arrived at Mukteshwar a few days later with a thermos of virus packed in dry ice. The researchers at Mukteshwar hoped to begin production immediately and were disappointed to learn that it was impossible to do without a "reliable deep-freeze unit," because the fluids harvested from the yolk sac needed to be kept below –20 degrees centigrade. Kesteven taught the Indian researchers how to keep the strain alive by passaging it through egg membranes until such time as they could secure the equipment necessary to begin producing the avianized vaccine. He

---

[61] FAO Press Release No. 4 (June 23, 1949), FAO Rinderpest Conference for Asia and the Far East, Bangkok, 1949, Animal Production and Health Division, 10AGA407, FAO.

[62] Edwards, "The Problem of Rinderpest in India," 1. For a more extensive history of rinderpest in India, see Spinage, *Cattle Plague*, 447–484.

then talked with them about their production of the lapinized vaccine and their general plan for eradication, which would involve the vaccination of "approximately 320 million head of stock." It was a daunting project. He recommended trying to accomplish the vaccinations within a single five-year period, which would cost more money and would require a delay while they built up their vaccine reserves, but would be more likely to achieve success. It was better to attack the virus everywhere at one time. He also noted that the plan would "very likely … have to be coordinated with neighboring countries."[63]

Kesteven's trip exemplified the nature of FAO assistance in 1949. FAO provided strains of the virus, technical instruction on how to produce vaccines, and help with the development of an eradication plan. It was doing something similar in Ethiopia. It had sent R. C. Reisinger to Ethiopia the year before to pick up where the UNRRA had left off. Since the UNRRA had provided "considerable equipment and supplies," Reisinger had been able to continue work on Nigerian caprinized vaccine and began work on a lapinized vaccine via strains sent in from Nanking. Ethiopia's Veterinary Laboratory at Goulele even had a deep-freeze unit, courtesy of the UNRRA, but it had broken and required spare parts to be shipped in from the United States. Its imminent repair did not immediately open the door to work on the avianized vaccine, however, because the power was not reliable at the laboratory. "We have," one of the FAO advisers reported in the spring of 1949, "been obliged to keep a constant supervision, day and night," bringing in "a small petrol driven generator" every time the power fails. The key was not just having the right equipment, but being able to ensure that it could reach its full potential.[64] Despite the setbacks, researchers at Goulele had produced 75,000 doses of Nigerian goat-virus vaccine by February of 1949 and the Ethiopian government began planning a wide-scale campaign. FAO technicians asked for a delay to allow for further testing of the caprinized vaccine around the country. When FAO reported satisfactory results, the Imperial Ethiopian Government (IEG) announced the campaign's commencement in December 1949.[65]

[63] Kesteven, Report on Trip to India, June 29 to July 14, 1949, Animal Production and Health Division, Tours and Trips, 1949–60, 0440GI, FAO.

[64] Ethiopia and FAO Technical Services (undated), Reports by Experts, 1948–1953, Mission to Ethiopia, 07120G3, FAO Archives; David F. Gray, Veterinary Report from Ethiopia (October 1948–March 1949), Reports by Experts, 1948–1953, Mission to Ethiopia, 07120G3, FAO Archives.

[65] *FAO Report No. 497, Report to the Government of Ethiopia on the Control of Diseases of Livestock.* Rome: FAO, May 1956, 4–7.

FAO technicians provided the virus, assistance with equipment repair, vaccine production expertise, and advice to the Ethiopian government in 1948 and 1949, but the campaign was to be Ethiopia's, not FAO's. There were strict limits to what FAO could offer; its assistance took specific form: a thermos of virus, the expertise of its operatives, technical meetings, and world-wide statistical reports, for example. That changed, however, in 1950, in part because of a program that Kesteven had mentioned at the Bangkok meeting: Point Four.

***

In 1954, former UN Secretary General Trygve Lie, reflecting on the changes that had happened at the organization under his watch, noted that following Truman's 1949 inaugural address, "technical assistance suddenly leaped into prominence as a major factor in international life and captured imaginations everywhere."[66] It was not, as was shown in the last chapter, so abrupt. It was true, as Francis Wilcox, the Chief of Staff of the Senate Committee on Foreign Relations, pointed out in 1950, that Truman's speech "gave the U.N. its real impetus toward an expanded technical assistance program," but "technical assistance for economic development was by no means a new concept in the United Nations."[67] It had, indeed, been adopted by the League of Nations as part of its mission. What changed in 1950 was the specific commitment, not the concept. It came from the United States, and, as Lie later noted, it gave the UN specialized agencies "new life."[68]

In a 1950 article about the new UN Program for Technical Assistance, Wilcox acknowledged that "some U.N. supporters have raised the question as to why the United States should persist in conducting its own program." Why Point Four, they asked. Why "not put *all* our technical assistance eggs in one basket?" The United States government, Wilcox responded, had offered several justifications: existing programs and experience in the field and the efficiency of a "compact task force" from

---

[66] Lie, Trygve. *In the Cause of Peace; Seven Years with the United Nations.* New York: Macmillan, 1954), 146.

[67] Wilcox, Francis O. "The United Nations Program for Technical Assistance." *Annals of the American Academy of Political and Social Science* 268 (March 1950): 45–46.

[68] Lie, *In the Cause of Peace*, 146. Joseph M. Jones echoed this idea in *The United Nations at Work: Developing Land, Forests, Oceans … and People* (Oxford: Pergamon Press, 1965), writing, "in 1950 came the beginning of change. Powerful world currents and events disclosed critical urgent needs for development help in three underdeveloped continents, and led by the United States … the United Nations and its family of organizations stepped up its response to the need" (109).

one country rather than many among them. The most "persuasive argument," however, Wilcox admitted, was "the financial one." Very few nations could make significant contributions to the new UN fund and it "would be most unwise for us to encourage the United Nations to launch activities which its members cannot pay for on an equitable and democratic basis." To avoid such a calamity, the United States would "shoulder a part of the burden bilaterally, at least until the other members of the organization are in a sounder financial position." This should not pose any problem, Wilcox concluded, because "there is no inherent conflict between bilateral and multilateral activities of this character."[69] Everyone wanted the same thing, in part.

During the 1949 debate in the General Assembly about creating a technical assistance fund, nineteen countries had declared their intention to contribute: Argentina, Australia, Belgium, Chile, Denmark, France, India, Liberia, Mexico, the Netherlands, New Zealand, Norway, Pakistan, Peru, Sweden, the United Kingdom, the United States, Uruguay, and Venezuela. These were not assumed to be the only nations that would contribute, but the ones who would contribute the vast majority of the funding.[70] Notably absent from the list were any of the Eastern European nations. The Soviet Union had never joined FAO or UNESCO and it had just recently left WHO. There was a reason the United States was happy to encourage a fund to help those specialized agencies in their efforts. A focus on the Cold War divide, however, masks deeper divisions within the "free world" itself. As the African Rinderpest Conference had revealed, France and the UK were eager to limit FAO's influence in Africa, viewing the organization as a threat to their dominance over the continent. They were, in consequence, a bit wary about this new push. It showed in their contributions.

In 1950, the UN held a Technical Assistance Conference in which fifty countries pledged twenty million dollars for the effort. The United States provided sixty percent. It was not willing to wait to until all nations could "pay ... an equitable and democratic basis." The total was small compared to what it had spent and was spending on bilateral international assistance programs, and it kept the United States in a position of dominance.[71] The additional funding proved very useful for FAO, but it also further

---

[69] Wilcox, "The United Nations Program for Technical Assistance," 47–48.
[70] Ibid. 51.
[71] Owen, David. "The United Nations Expanded Program of Technical Assistance – A Multilateral Approach." *Annals of the American Academy of Political and Social Science* 323 (May 1959): 28.

complicated its mission. No "Iron Curtain country" offered to send money to the UN expanded program.[72] There was a great deal of politics wrapped up in the new burst of enthusiasm for technical assistance.

FAO was delegated twenty-nine percent of the first twenty million dollars to expand its technical assistance programs. The year before, it had had only 1.2 million dollars for that work. In 1952, it would get 6.2 million dollars.[73] At the end of that year, Dodd reported that "he believed that during the past year FAO had made more progress in its work than in the previous five years of its existence, adding that this was mainly because nations were 'moving ahead more rapidly in the program of technical assistance for economic development.'"[74] The money was "a godsend," Dodd acknowledged, but it also complicated the organization's internationalism.[75] China, Hungary, Poland, and Czechoslovakia withdrew from FAO in protest.[76] The organization was finally getting closer to the kind of funding it had wanted from the beginning – the kind of funding that would enable it to do more than simply offer advice and transport vials of viruses – but it was not getting it for the reason it wanted.

Huxley and Orr had argued for a peace based on expanding human welfare, countering US and British efforts to secure peace via their own national strength. The United States government was now more committed than ever before to the concept of expanding human welfare, but this was primarily because it believed it would contribute to its national security. This was international action in the service of nationalism and it was not what Huxley and Orr had set out to achieve. Indeed, Orr's resignation from the leadership of FAO had been partially based on his frustration with the US decision to not work with the UN to distribute Marshall Plan aid.[77] The United States was creating more machinery for

[72] Sharp, Walter R. "The Institutional Framework for Technical Assistance." *International Organization* 7:3 (August 1953): 349.

[73] Hambidge, Gove. *The Story of FAO*. Toronto: D. Van Nostrand Company, Inc., 1955, 83–84; Jolly, Richard, Louis Emmerij, Dharam Ghai, and Frédéric Lapeyre. *UN Contributions to Development Thinking and Practice*. Bloomington, IN: Indiana University Press, 2004, 68-73; Yates, P. Lamartine. *So Bold an Aim: Ten Years of International Co-operation Toward Freedom from Want*. Rome: FAO, 1955, 119–121; Wilcox, Francis O. "The United Nations Program for Technical Assistance." *Annals of the American Academy of Political and Social Science* 268 (March 1950): 45–53.

[74] "Food and Agriculture Organization." *International Organization* 7:1 (February 1953): 131.

[75] Staples, *The Birth of Development*, 99.

[76] "Food and Agriculture Organization." *International Organization* 6:3 (August 1952): 431.

[77] Vernon, *Hunger*, 154–155.

international cooperation and development, but it was doing so with the plan of using that machinery to pursue its international agenda.

In the aftermath of the burst of new funding, Dodd struggled to maintain FAO's identity. "Nations," he wrote in 1951, "seem at last to be preparing for a genuinely large-scale world war against want," which will "prevent another war of nations against nations." FAO wanted to take the lead in a struggle that pitted humans, not against each other, but against their nonhuman enemies. We are in a fight, Dodd continued, "that will defeat no one and nothing except starvation and malnutrition and social deterioration." It is "a truly international and non-partisan effort, without regard to race or creed."[78] Dodd did not want to be distracted by the Cold War. He wanted to focus on "the concrete and effective things that will produce more food and bring it to the world's hungry."[79] That meant he also did not want to have to deal with infighting in the so-called "free world," where Britain and France worked both with FAO and against it at the same time. They donated money, but they also tried to limit FAO's presence in and around their colonial territories, fearful that its mission threatened their own. FAO was already working closely with their former colonies; they did not want it working with the ones they still had.

The French were particularly disturbed by the new technical assistance push. A British official reported in 1949 that there was "a new fever in Paris carried by the 4th Point bug," that was producing "successive moods of elation at United States money being lavished on the colonies and acute depression with visions of the Specialized Agencies passing round the hat to finance a broader and wilder campaign of sticking noses into colonial affairs." They loathed the idea of UN "interference" above all. A British official responded by calling for a stronger commitment to raising living standards in Africa, warning that without it "we will be overwhelmed by the combination we know so well in Asia, local nationalism and American technical and financial intrusions." Paris agreed.[80]

In January of 1950, at the Commonwealth Conference on Foreign Affairs in Colombo, Ceylon, Britain grudgingly agreed to work with the combination of "local nationalism and American technical and financial intrusions" in Asia by launching the Colombo Plan for Cooperative Economic and Social Development in Asia and the Pacific. In Africa,

[78] Quoted in Hambidge, *The Story of FAO*, 87.
[79] Staples, *The Birth of Development*, 99–100.
[80] Quoted in Kent, *The Internationalization of Colonialism*, 198–199.

Britain fought against it, joining that same year with France, Belgium, South Africa, Portugal, and Southern Rhodesia to launch the Combined Commission for Technical Co-operation in Africa south of the Sahara (CCTA). The French hoped the new organization would serve as a "barrier" to UN and US "indiscretions" in Africa. British governors were informed by London that "one of the primary purposes" of the Commission was to "secure effective arrangements for the co-ordination of action in the technical field between countries having responsibilities in the area, as a substitute for the setting up of bodies for this purpose by the United Nations."[81] Inter-imperial cooperation was escalating as an effort to ward off American-dominated UN action.

Rinderpest played an important role in that effort. Colonial officials were still regretting Resolution 3 from the Nairobi meeting and the door that it had opened to FAO in Africa. A March 1950 letter in the Colonial Office reported, "when we discussed this resolution with the other Powers in Paris a year ago it was apparent that they all viewed it with misgiving. They were all most anxious not to allow F.A.O. a footing in Africa except on the technical assistance [unreadable]." Most of the African territories concerned were, at the time, under British influence, but they were not colonies. London weighed the financial versus the political costs of shouldering the burden of rinderpest eradication programs within them. So, too, did its fellow colonial powers. After much back-and-forth, the letter continued, "the principle was accepted that the campaigns were primarily the United Kingdom's responsibility, secondarily that of the other African Powers and only more remotely that of F.A.O."[82] The following year, the CCTA created the Inter-African Bureau of Epizootic Diseases to coordinate its animal disease efforts. This type of independent action was exactly what FAO and the United States were hoping to avoid by putting more money into FAO's technical assistance budget.

The Truman administration liked the idea of FAO taking the lead in rinderpest eradication. It perceived, as FAO did, that the effort had the potential for success that would be beneficial to the world *and* to the organization, demonstrating the importance of the United Nations and its specialized agencies, which were now more closely associated with the

[81] Kent, *The Internationalization of Colonialism*, 264–267. See also, Gruhn, Isebill V. "The Commission for Technical Co-Operation in Africa." *The Journal of Modern African Studies* 9:3 (October 1971): 459–469; The Colombo Plan, History; available at http://www.colombo-plan.org/index.php/about-cps/history.

[82] Letter to Sir G. Clauson from unknown (March 17, 1950), CO 852/1228/7, NAUK.

United States than ever before. In a November 1949 speech to FAO, Truman had emphasized the point, insisting, "In the beginning, the greatest advance will probably result from the most elementary improvements. The control of animal diseases ... would greatly increase production and better the lot of millions of small farmers in many parts of the world."[83] His listeners, of course, had heard that before. It was the central reason that the organization had originally decided to expand its mission to include animal disease control back in 1946. It mattered, however, that the president of the United States also now thought that way. The Truman administration was eager to help FAO use animal disease control to its greatest advantage, which just happened to also be the US government's greatest advantage. That eagerness led to an expanded budget in 1950, but that was not all.

In May of 1950, representatives from FAO and the OIE met in Paris at FAO's initiative to discuss the possibility of creating a "single international system for the collecting and disseminating of information" about transmissible animal diseases. The OIE was represented by members from France, the United Kingdom, and Venezuela, joined by Sir Daniel Cabot, the president of the OIE. FAO had invited representatives from three non-OIE member nations to attend: the United States, Canada, and Brazil. The American member, H. W. Schoening, Chief of the Pathology Division at the United States Department of Agriculture (USDA), ended up attending with only his adviser. They were joined by K. V. L. Kesteven, who officially represented FAO. Brazil never responded to the invitation and the Canadian representative died a few days before the meeting and was not replaced. It was, however, "understood" that the Canadian position was "in accord with that of the U.S." And the US had a very specific view. It had been planning for this meeting.[84]

On the first day of the meeting, the US representative submitted an official proposal for a "Draft Constitution for an International Office of Epizootics of the Food and Agriculture Organization," explaining that the United States felt "that all animal disease reporting and control work

---

[83] Truman, Speech to be delivered at a session of FAO (November 21, 1949), Food and Agriculture Organization of the United Nations [2 of 2], OF 85–N, Box 532, Official File, Papers of Harry S Truman, HSTL.

[84] Report of the United States Delegate to the Meeting of the Joint Veterinary Committee of FAO and OIE (May 1–5, 1950), US Del/Joint Veterinary Committee/FAO/OIE, Box 4: UN Conf. Adoption Conv. On Narcotic Drugs to Report – USDEL/Indo-Pacific Fisheries Council, Office of International Conferences (OIC), Lot File 71D221, Supplemental Office Files, 1948–1966, RG 59, NARA.

should be centered in one international agency and that FAO is that agency." The French representative was furious, insisting that "he saw no point in continuing the discussion," because the proposal made the OIE subordinate to FAO. "He pointed out that OIE is much older than FAO; while the League of Nations dies, OIE has lived and if FAO died OIE will live on." The Venezuelan representative seconded the point, explaining that "FAO's primary job is to increase food production in backward areas, that job will be done sometime and FAO will go out of business, but animal diseases will go on forever, so ... there will always be a need for an OIE." The two representatives also criticized the proposal's description of membership in the new office, which was open to all member countries of the OIE and FAO "except for such governments as may be temporarily barred from membership in the United Nations agencies by resolution of the General Assembly." At the present moment, they pointed out, that would include Spain. Both insisted that the OIE "would never accept an arrangement, which required them... to refuse to work with countries on political grounds." The French and Venezuelan representatives insisted that the OIE needed to maintain its autonomy. The organization would "be committing suicide to accept this proposal." The UK representative disagreed, arguing that FAO was already so strong that, if the OIE refused the proposal, FAO would "go ahead anyway and set up a world-wide body on animal disease problems and ... the OIE will be left out." The final vote was three for the proposal and two against. Cabot abstained.[85]

The vote itself did not really matter. The committee acknowledged that it did not "have the power of decision"; it simply submitted its report to FAO and the OIE (minus signatures from the French and the Venezuelan representatives, who refused). It was "for the OIE and FAO to decide on the question." In his report to Dean Acheson on the meeting, Schoening warned that the United States needed to "make sufficient advance preparations so that member governments of FAO will have time to instruct adequately their representatives to FAO and to arrange, where necessary, for identical instructions to be given to the representatives of these same countries in the OIE." Despite these efforts, the OIE did not, unsurprisingly, decide to give up its independent existence. Unlike FAO, it included many nations over whom the United States wielded little influence. The episode, however, visibly demonstrated the level of US interest in

[85] Ibid. The French and the Venezuelan representatives presented a counter proposal that was defeated 3 to 2 with the same abstention.

solidifying FAO's dominance in the field of international animal disease control. The OIE, which had shown itself to be "uninterested in a program of technical assistance for the control of animal diseases, along the lines proposed by OEEC [the Organization for European Economic Co-operation, set up under the Marshall Plan] for participating countries and the proposed expanded technical assistance program of FAO," could, in the US government's mind, only "hamper" FAO.[86] The OIE had committed the sin of not taking the American assistance bait not just once, but twice. Therein lay the real roots of the US frustration with the organization: its global machinery operated outside American control.

FAO did not take over the OIE, but it did become the dominant player in the effort to coordinate the international effort to eradicate rinderpest. FAO used its new technical assistance budget to begin providing countries with laboratory equipment (particularly expensive freeze-driers that would allow the replacement of "wet" vaccines with more stable "dry" ones), along with the now-standard thermoses of viruses, technical advisers, and reports. Its efforts became both more concrete and more effective. The change did not happen overnight, but occurred slowly in different locations between 1950 and 1953, as FAO expanded its operations on the ground. Throughout it tried, as Dodd maintained, to stay "a truly international and non-partisan effort."[87] It was not an easy task.

***

On January 24, 1950, a letter arrived at the British Foreign Office's South East Asia Department from the British embassy in Kabul, passing along the message that the American ambassador had told them that "there are a number of Russian veterinary experts actually in Afghanistan at the present time." Embassy officials had not been able to confirm his information, but they reported it: "Some time ago there was a serious outbreak of rinderpest in the Herat region. The Afghans tried to get serum from Tehran but were unable to do so because the Persians had a simultaneous outbreak in Khorassan. The Afghan government then turned to Russia, engaging Russian veterinarians on the grounds that they could thereby be sure of getting supplies of serum."[88] FAO

---

[86] Ibid.   [87] Staples, *The Birth of Development*, 100.

[88] Russian Veterinary Experts in Afghanistan in Connection with Outbreak of Rinderpest (January 14, 1950), FO 371/83096, NAUK. For the larger story of Cold War development in Afghanistan, see Nunan, Timothy. *Humanitarian Invasion: Global Development in Cold War Afghanistan*. Cambridge: Cambridge University Press, 2016.

"veterinary expert" E. Pierson later explained the situation in more detail, writing that throughout 1950 in Herat, Afghan veterinarians Mirak Shah Khan and Mohammed Asiam Khan (the only two in the country), "assisted for a few months by a Russian veterinarian, were endeavoring to control the spread of rinderpest by the use of formalized tissue (cattle) vaccine, produced in Afghanistan and augmented by a small supply from Russia."[89] Pierson expressed no concern about the Russian presence in his report, just worry about the potential success of the Afghani campaign against the virus without access to better vaccines. His were technical concerns, not Cold War ones.

When it signed an agreement with FAO in 1950, Afghanistan was told it would receive an FAO veterinarian "to advise the Government on problems of animal health and parasite control and, at the outset, to give particular attention to the control of rinderpest" and "to advise on the organization and execution of a field program to bring the current outbreak under control." But it got more than that. The limitations of the inactivated vaccine were readily apparent, and FAO brought in strains of both caprinized and lapinized virus. The caprinized proved the most efficient and officials promptly established a temporary laboratory for its preparation in Herat. By the end of March 1951, teams had vaccinated over 30,000 cattle. None of this fell outside of the pattern of earlier FAO activities, but its subsequent actions did.[90]

Determining that the laboratory at Bini Hissar "was poorly equipped" and "unsuitable ... even for vaccine production," FAO equipped it. By December of 1953, the resident veterinarian could write that "all the necessary apparatus and equipment for the establishment of a satisfactory laboratory have now been provided by the Food and Agricultural Organization of the United Nations and the necessary structural alterations have been made." FAO had even supplied "an Edwards freeze-drying unit ... for the preparation of vaccines, by modern techniques."

---

[89] *FAO Report No. 204, Report to the Government of Afghanistan on the Control of Animal Diseases*. Rome: FAO, December 1953, 7. Kesteven noted that "a Russian D.P. veterinarian employed by the Ethiopian Government has been taught to manufacture the pleuro-pneumonia vaccine, using the Australian v-5 strain," in a December, 1950, letter to the Colonial Office (CO 852/1228/7, NAUK). More information also available in OIE, 2ème conférence de l'OIE pour l'Asie – Bangkok, Thaïlande (Février 13–20, 1954), available at http://www.oie.int/doc/en_document.php?numrec=3177003 and in several documents in FO 371/92112, NAUK.

[90] *FAO Report No. 204; Report to the Government of Afghanistan on the Control of Animal Diseases*. 1953, 5–8.

Freeze-drying the vaccines made them more stable, enabling longer periods of storage.[91] FAO equipped Afghanistan to be at the front lines of vaccine production technology.

The freeze-drying units were symbols of FAO's expanded technical assistance program. FAO reported in 1952 that it had provided Thailand's Department of Livestock with Nakamura III and "considerable equipment to enable the production of this vaccine." Thailand had declared rinderpest eradicated on December 31, 1949, following the vaccination of 617,000 cattle and buffalo, but it was "reinvaded" by, it believed, "the smuggling in of infected animals from Cambodia." A new vaccination campaign began with the lapinized vaccine, which was "very popular" with farmers, because it produced such a mild reaction in their buffalo that they could get right back to work. "A great trouble arose in obtaining sufficient rabbits," but this was solved by bringing a new animal into the attenuation process. Rinderpest attenuation via passage through pigs had not worked in the past, but a British researcher brought in by the Thai government found that he could inject Nakamura III into domestic Thai pigs and then harvest vaccine from its spleen without any change in the vaccine's impact on buffalo. Pigs were more plentiful than rabbits and produced more vaccine (about 800 doses per pig). They were useful transports. In 1952, utilizing the new equipment FAO had given them, researchers at the Pak Chong laboratory produced 510,120 doses of lapinized-pig virus.[92]

Rinderpest's "reinvasion" of Thailand highlighted the point that eradication could not be a national project alone, but required coordination at the regional, and, ultimately, global level. Thailand's particular struggles with the disease were representative of the larger struggle to coordinate the war against it. FAO had intended to hold another Asian Rinderpest Meeting in the fall of 1952 "mainly for the purpose of receiving reports on the progress that had been made towards establishing control of the disease," but the decided *lack* of that control had forced it

---

[91] Ibid.

[92] Ministry of Agriculture, Thailand, Annual Report of the Government of Thailand for the Year 1951–1952 to FAO, Thailand 1947–65, Annual Reports from Governments, 010V2J, FAO; Ministry of Agriculture, Thailand, Annual Report of the Government of Thailand for the year 1955–1957 to FAO, Thailand 1947–65, Annual Reports from Governments, 010V2J, FAO; *FAO Report No. 455, Report to the Government of Thailand on the Control of Animal Diseases.* Rome: FAO, February 1956; Richard Hudson, Second Report on an Attempt to Obtain Improved Prophylactic Agents for the Control of Rinderpest (March 20, 1950), Tours and Trips, Animal Production and Health Division, 1949–60, 0440G1, FAO.

to change its mind. "It seemed clear that very little advance had in fact been made" in the region. FAO technicians argued that the source of the problem lay in the kind of vaccines that the countries were producing, specifically "wet" instead of "dry." FAO officials decided "that in many countries no great progress would be possible until the standardized production of suitable living virus vaccines in the freeze-dried state had been established on an adequate scale." Instead of a meeting, then, FAO decided to hold a training workshop at the Indian Veterinary Research Institute in Izatnagar the following year.[93]

FAO prescribed a technical solution to what it deemed a technical problem: the continued lack of success against rinderpest. "The success of any mass-immunization campaign that may be planned as part of the drive to eradicate the disease from countries in Asia," an FAO veterinarian explained, "will be mainly dependent upon the availability of efficacious vaccines, endowed with sufficient viability to enable them to be transported to the area of operations in the field."[94] It was no longer enough just to be able to produce living attenuated virus vaccines, they had to be freeze-dried. The vaccines were still the answer, they just needed to be distributed in an improved form. This decision, made possible by the expanded budget that could now include freezers, did not require any serious change in thinking about the campaigns. The focus remained on the technology – on the vaccines – as the primary solution to the problem of rinderpest. They had, after all, been the reason that FAO had taken on the issue in the first place. A lot was riding on them.

In early 1953, H. R. Fischman (the FAO veterinarian who had retreated from Nanjing to Bangkok in 1949) reported from Pakistan on his efforts "to advise and assist in the production and use of Rinderpest vaccines." Pakistan needed the help. It was regularly losing about 2.5 million cattle, out of a population of around 24 million, to disease, primarily rinderpest. The goat-attenuated version (Kabete strain) was in widespread use throughout the country, Fischman noted, produced in "most provincial laboratories and issued in a 'wet' state to the field workers." It was having a less-than-desirable impact, however, because "considerable amounts of anti-rinderpest serum are used in conjunction with the vaccine . . . to offset the 'expected reaction.'" This led to a "partial or passive immunity of short duration," which was not helpful to the eradication campaign. His

---

[93] *FAO Report No. 149, Report on the International Training Centre on Living Virus Vaccines (Veterinary)*. Rome: FAO, August 1953, 1.
[94] Ibid., 8.

own subsequent field trials of both the caprinized (KAG) and lapinized vaccines revealed both "to be perfectly safe for almost all types of animals" in Pakistan. Since KAG was cheaper to produce, Fischman dubbed it the vaccine of choice. To aid its production, FAO sent an Edwards Centrifugal Freeze-Dryer to the chief laboratory at Peshawar. Pakistan supported that effort with "two deep freezers, one Latapie mincer, several Waring blenders," and additional equipment "essential to the production of desiccated vaccine." The laboratory was currently stockpiling doses – it had over 110,000 desiccated caprinized vaccine and over 10,000 desiccated lapinized vaccine – in preparation for the forthcoming eradication program.[95]

Pakistani researchers already knew how to freeze-dry their vaccines, but they sent researchers to the Training Center at Izatnagar that February anyway. They were joined by researchers from Afghanistan, Burma, Ceylon, Japan, Malaya, Thailand, and, of course, India. There were thirty-two participants in all, "the maximum number that could be accommodated." Most of the FAO personnel involved came from the United States and the United Kingdom. Also in attendance was a Point Four adviser who worked at the Institute.[96] The course lasted three weeks. It focused on rinderpest, but there were also classes on vaccine production for Newcastle disease, sheep pox, and fowl plague.[97] It was an ideal, well-stocked location. FAO had recently provided the Institute with several freeze-drying units; the United States government, via Point Four, had also provided "technical advice" and "modern equipment."[98] Izatnagar was not the only place such overlap was occurring.

FAO's expanded technical assistance program regularly came into contact with the United States' technical assistance program. FAO had moved its headquarters from Washington, DC to Rome, closer to the so-

[95] *FAO Report No. 103, Report to the Government of Pakistan on Control of Animal Diseases.* Rome: FAO, February 1953, 2–7; E.B. Evans, Report on a Trip to Pakistan, Egypt and Ethiopia in the Interest of Rinderpest Eradication and Animal Disease Control (January 26, 1953), Tours and Trips, Animal Production and Health Division, 1949–60, 0440G1, FAO.

[96] *FAO Report No. 149, Report on the International Training Centre on Living Virus Vaccines (Veterinary).* Rome: FAO, August 1953, 2–4.

[97] Syllabus, International Training Center, February 16 – March 7, 1953, FAO International Training Center on Manufacture of Rinderpest and Other Virus Vaccines – India 1953, Animal Production and Health Division, 10AGA407, FAO.

[98] Speech by Dr. Panjabrao S. Deshmukh, Minister for Agriculture, India, at Izatnagar (U.P.) to inaugurate the International Training Center (February 16, 1953), FAO; International Training Center on Manufacture of Rinderpest and Other Virus Vaccines – India 1953, Animal Production and Health Division, 10AGA407, FAO.

called "underdeveloped" nations that were its chief targets of concern, but it could not escape the looming presence of its chief donor. FAO technicians found themselves working side by side with American Point Four technicians in fields and laboratories around the world. It only made sense to do so. The technical assistance programs that grew out of Truman's speech were framed as cooperative endeavors from the beginning.[99] In addition, because Point Four ended up with a vastly more expansive budget than the UN's Technical Assistance Fund (about 300 million dollars versus 20 million dollars in 1953), FAO was hardly in a position to reject Point Four collaboration, even if such collaboration made its claims of political impartiality sound more hollow.[100]

FAO and Point Four worked together on rinderpest eradication campaigns in a number of countries throughout the 1950s.[101] Their technical assistance took almost identical form: equipment, "expertise," training, reports, surveys, and support to send students abroad for advance schooling. The grassroots goals and approaches were the same. Point Four operatives just had access to more money, although how much more varied greatly from country to country. American aid had an overt political purpose, but there was enough to go around so that even nations deemed non-strategic could get funding provided they were "free." By early 1953, thirty-five nations were receiving Point Four assistance, some of it for rinderpest eradication.[102]

Point Four technicians did not start their own rinderpest eradication programs; they offered to help FAO with the national ones that it was helping support throughout Asia and Africa. An FAO report out of Burma in April of 1953 noted that "FAO have provided an Edwards Centrifugal Freeze-Dryer and some other items, but a vast quantity of capital goods, equipment, and expendable stores have been procured through Point Four funds." In the appendix, the report explained that FAO and Point Four officials met "every two weeks . . . at the US Embassy" to discuss "projects and problems."[103] They also had to discuss them with the Burmese

---

[99] Wilcox, "The United Nations Program for Technical Assistance," 45–48.
[100] Bok, Bart J. "The United Nations Expanded Program for Technical Assistance." *Science* 117:3030 (January 23, 1953): 67.
[101] Point Four was originally administered through the Technical Cooperation Administration, then the Foreign Operations Administration, and then the International Cooperation Administration, so using "Point Four" avoids needless confusion.
[102] "Point Four Promotes Better Life in 35 Nations, Survey Finds." *New York Times* (January 12, 1953): 1.
[103] *FAO Report No. 105, Report to the Government of Burma on Control of Animal Disease.* Rome: FAO, April 1953, 6, 18.

government, because it was still expected to contribute the most funding. Technical assistance depended, above all, on cooperation and collaboration with the home government. It was *help*. Recipient governments were expected to do most of the eradication planning and most of the eradication work. This led to tension, at times, when government officials did not organize projects the way that FAO technicians wanted them organized. They were, after all, the "experts," and they wanted their advice heeded, even as they insisted that they were not in charge. Technical "assistance" was a complicated "gift."

In December 1951, FAO's senior veterinarian in Ethiopia, H. T. B. Hall, answered a request from Rome about "why the disease control program, although functioning to some extent in a reasonable manner, has not progressed at a rate which could be normally expected with the presence of four FAO veterinarians." Ethiopia was still losing over one million cows a year to disease, primarily rinderpest. A frustrated Hall wrote back that officials at the Ministry of Agriculture "have neither cooperated nor assisted" our efforts. "So little" of our advice "has been taken," he fumed, that we might as well not have been here. He had some specific complaints, the most important being that the Ethiopian government was trying to pay for the campaign by charging farmers fifty cents a vaccination. The result, another FAO official explained, was that only "a fantastically small percentage of cattle" were receiving vaccinations, to the point where the campaign "cannot even be considered at present a public service."[104] They eventually turned to Point Four for help.

FAO officially asked for Point Four help with the rinderpest campaign in the spring of 1953. The head of Point Four operations in Ethiopia reported that the UN organization had, "for some time ... been providing the technical talent for a vaccination program against rinderpest." It supplied "technicians but no funds," and the program was "financially handicapped."[105] The funding was, of course,

---

[104] H.T.B. Hall, Statement on the Present Position of the Disease Control Program in Ethiopia (December 28, 1951), Animal Production Branch, 1949–1958, Agricultural Division; 10AGA407, FAO; E.B. Evans, Report on a Trip to Pakistan, Egypt and Ethiopia in the Interest of Rinderpest Eradication and Animal Disease Control (January 26, 1953), Tours and Trips, Animal Production and Health Division, 1949–60, 0440G1, FAO; Nels Konnerup, Survey of the Rinderpest Program in Ethiopia (December 27, 1951), Animal Production Branch, 1949–1958, Agricultural Division, 10AGA407, FAO.

[105] Marcus J. Gordon, Summary of Point 4 Operations in Ethiopia for Mr. Black (March 16, 1953), Administration 1953, Box 3, Folder 1, Mission to Ethiopia Subject Files,

supposed to come from the Ethiopian government, but it, too, turned to Point Four for help. The Ministry of Agriculture, Point Four, and FAO subsequently created a Technical Advisory Committee for Animal Disease Control to coordinate the eradication campaign. Point Four offered to defray the cost of the vaccinations for farmers, along with providing additional funding for vehicles, laboratory facilities, refrigerators, and other equipment.[106] Free vaccinations began in May of 1954.

The United States had multiple reasons for paying for those vaccinations. Ethiopia's location, "so near the restless Arab states," made it essential that "the Western World" strengthen its "ties of friendship" with it. Such ties were not only useful for keeping the communists out (there were "indications" that they wanted in), but also for helping expand global food security.[107] The eradication program was "concentrated in localities where cattle are most readily available for export."[108] In this, FAO and Point Four were decidedly on the same page. Reisinger had written back in 1948 that it was "useless, as far as Ethiopia is concerned, to eradicate rinderpest or any other animal disease, or to save any animals at all, unless full use be made of such animals." Such "full use" would require "a meat-export program which includes the construction and use of modern abattoirs, canning and freezing plants, and improved transportation."[109] But eradication had to come first, ensuring that Ethiopia was no longer "a central distribution point" for the

1951–1954, General records of the Agency for International Development and Predecessor Agencies, 1948–1961, 469/250/80/31/01, NARA.

[106] US Operations Mission to Ethiopia, The Point 4 Program in Ethiopia (September 1954), Box 9, Folder 5, Mission to Ethiopia Subject Files, 1951–1954, General records of the Agency for International Development and Predecessor Agencies, 1948–1961, 469/250/80/31/01, NARA.

[107] Marcus J. Gordon to E. Reeseman Fryer (December 24, 1952), Ethiopia: Field Submissions and Country Desk Writeup FY 1945, Country Files 1950–53, Near East and Africa Development Service, Technical Cooperation Administration, 469/250/78/29/06–07, NARA; Information for Evaluation TCA Program at Country Level, Program – 1952, Mission to Ethiopia Subject Files, 1951–1954, General records of the Agency for International Development and Predecessor Agencies, 1948–1961, 469/250/80/31/01, NARA.

[108] US Operations Mission to Ethiopia, The Point 4 Program in Ethiopia (September 1954), Box 9, Folder 5; Mission to Ethiopia Subject Files, 1951–1954, General records of the Agency for International Development and Predecessor Agencies, 1948–1961, 469/250/80/31/01, NARA.

[109] R. Reisinger, "Veterinary Services Ethiopia," in FAO Ethiopia Report (August 18, 1948), Reports by Experts, RG 0 7120G3, FAO.

virus.[110] At present, Ethiopia's rinderpest problem was its neighbors' problem as well. In this, it was not alone.

In May of 1955, Bangkok hosted another rinderpest conference, this time for its immediate neighbors, "with a view of putting into practice a Rinderpest Program in South East Asia." Bangkok hoped to end rinderpest's periodic incursions across its borders by establishing an "international plan" for regional eradication. Records of the meeting in United States Operations Mission files list participants as Laos, Cambodia, Thailand, and Vietnam, with "South" crossed out – it might be an acceptable modifier for FAO, but it was not for the Americans. Also present were representatives from FAO, the US Foreign Operations Administration (Point Four), the OIE, and "observers" from the Colombo Plan, Hong Kong, and the Philippines. Thailand had received extensive outside help from the United States and FAO, the other countries less so, but that was changing, and it was why the nations could come together to discuss an "international plan" – they finally had the funding to seriously pursue their national ones.

Cambodia had reported earlier in the year that the American economic mission in its country had promised money to help fund its production of vaccines for the control of animal diseases, principally rinderpest. Work continued there on both inactivated tissue vaccine and on Nakamura III (932nd passage!) which it had received from "the laboratory of Tachikava" in 1953. This presumably refers to Tachikawa, which was at that time the location of an important US air base in Japan.[111] Meanwhile, the United States Operations Mission had drafted a "Plan for Eradication of Rinderpest in Vietnam," promising an expansive amount of help from the Point Four office in Saigon to make it happen. It was scheduled to begin

[110] Formation of a Veterinary Division in Ethiopia (October 12, 1946), Agricultural Programme (May 12, 1945 – March 5, 1947) 1944–1949, Folder S-1347-0000-0018, Box S-0527-0475, Ethiopia Mission, UNRRA, Record Group AG-018-019, UN.

[111] Rapport Biennal 1953–1954, destiné à la F.A.O. au titre de l'article II de l'acte contitutif de la F.A.O., Mission to Country Files – Cambodia, 010V2, FAO; Nghiem to P. H. Allen, Report: Peste Bovine, 1955 – Rinderpest Information, Box 2, Unclassified Subject Files, 1955–1960, Agriculture and Natural Resources Division; ICA US Operations Mission to Vietnam, RG 469, Records of the US Foreign Assistance Agencies, 1948–1961, NARA; Stevenin, Huard, and Goueffon, Vaccination Trials of Asiatic Buffalo Against Rinderpest By Using Lapinized Virus NakamuraIII, 1955 – Rinderpest Information, Box 2, Unclassified Subject Files, 1955–1960, Agriculture and Natural Resources Division, ICA US Operations Mission to Vietnam, RG 469, Records of the US Foreign Assistance Agencies, 1948–1961, NARA.

on October 1, 1955.[112] In addition to that from the Americans, technical assistance also came from FAO and from the Colombo Plan. As an FAO report later explained, in Vietnam "excellent assistance in the matter of funds, transport vehicles and other equipment was given by USOM. A set of freeze-drying equipment was furnished by the Colombo Plan (United Kingdom) and another by FAO."[113]

By the mid-1950s, this kind of cooperation for rinderpest eradication had become the norm, not the exception. The international machinery for development existed now in many forms as development itself became an ever more integral part of international relations. People remained confident about the ability of humans to use technology to expand their dominion over other living and non-living things in order to advance human welfare. Rinderpest eradication efforts were particularly encouraging. Following a tour of laboratories and field vaccination programs in Ethiopia, Egypt, and Pakistan in the winter of 1952–1953, an FAO consultant reported back to headquarters that FAO and Point Four "are probably the most effective means by which the free world can assist many of the under developed countries. I am thoroughly convinced that the practical work observed in connection with these agencies is far-reaching and will go a long way in helping these countries remain part of the free world."[114] Just a few years later, a history of FAO's first ten years insisted, "It would be fair now to say that another major plague of rinderpest is unlikely. The countries have the means and the knowledge to produce vast quantities of vaccine – enough to control any outbreak, and ultimately to achieve complete eradication."[115] The end goal seemed in sight.

In 1957, Binay Ranjan Sen, Director General of FAO since 1956, sent a letter to ministers of agriculture around the world regarding the question, "what concrete results FAO can show for all the work that it has been doing during the eleven years of its existence." He turned to animal disease control for his answer.

[112] Plan of Campaign for the Eradication of Rinderpest in Vietnam (undated), 1955 – Rinderpest Information, Box 2, Unclassified Subject Files, 1955–1960, Agriculture and Natural Resources Division, ICA US Operations Mission to Vietnam, RG 469, Records of the US Foreign Assistance Agencies, 1948–1961, NARA.

[113] *FAO Report No. 1202, Report to the Governments of Burma, Cambodia, Laos, Thailand and Vietnam on Animal Disease Control*. Rome: FAO, 1960, 6.

[114] E.B. Evans, Report on a Trip to Pakistan, Egypt and Ethiopia in the Interest of Rinderpest Eradication and Animal Disease Control (January 26, 1953), Tours and Trips, Animal Production and Health Division, 1949–60, 0440G1, FAO.

[115] Hambidge, *The Story of FAO*, 151.

Ten years ago it was estimated that some two million cattle were dying in the Far East every year from rinderpest alone. Today's figure is probably less than 50,000. This progress can be attributed principally to the activities of FAO. Afghanistan is now completely free from rinderpest, so are Iran and Thailand. In India, Burma, Vietnam, Laos, Cambodia and Pakistan the disease is under control and mass vaccination programs are planned which are aimed at nothing less than complete eradication. Within another ten years rinderpest may well be expected to have disappeared from the whole of the Far East. In Africa, for long Ethiopia acted as a reservoir of infection for other countries and rendered control work in the rest of Africa largely ineffective. As a direct result of FAO's intensive work in Ethiopia, Kenya, Uganda and Tanganyika have had no outbreaks of rinderpest to report during the last two years. In Ethiopia itself eighty FAO-trained teams, under the supervision of FAO experts, are in the field using vaccines produced according to methods elaborated by FAO specialists. Eradication of the disease from Ethiopia is in sight, and there is a confident hope that it will subsequently be driven out of the whole of Africa.[116]

The rinderpest eradication program had made readily identifiable, concrete progress, just as its originators had hoped it would. Rinderpest had not yet been eradicated from the planet, but it had been eliminated in many countries. It was under human dominion in a way that it had never been before. It was not a complete victory, but it was a measurable success story. The vaccines had made it possible. FAO, with help (noticeably unmentioned by Sen) from UNRRA, the United States, the Commission for Technical Cooperation in Africa South of the Sahara (CCTA), the OIE, and national governments throughout Asia and Africa, had made it happen.

FAO interpreted this success as a vindication of its original mission – proof of the utility of international cooperation in the name of a common humanity. The opening of *So Bold an Aim: Ten Years of International Cooperation toward Freedom from Want*, a 1955 retrospective on its activities, insisted, "man is a striving animal," hungry for "material advancement and moral improvement." FAO, it continued, "had to be created as a vehicle for this striving, because the radically changed condition of the world makes old vehicles as obsolete as the stage-coach. New times, new institutions." The international agencies connected to the United Nations – that machinery for global development – were not simply useful, but *necessary* to humankind: "Men have found that they need these additional administrative instruments in their attempt to regulate and control their environment." FAO still

---

[116] B.R. Sen, Copy of letter, Sent to Ministers of Agriculture for your information and advice, No. 58 (April 1957), Director General's Monthly Letter File, Office of the Director General, FAO.

embraced the ideology that had defined its opening years: that human dominion over nonhumans could best be secured through international cooperation under internationalist direction.[117]

Its first director general continued to embrace it as well. John Boyd Orr wrote in 1953 that "a world which, to-day, has to deal with atomic energy and with the new and equally powerful biological forces, is so different from the world of yesterday that the economic and political ideas with which the older politicians were familiar are obsolete." They cannot address the complexity of the situation, he insisted. "Nineteenth-century economics and politics cannot carry twentieth-century science." Humanity needed "a World Government," but, since that was not possible right away, it could settle for "international co-operation."[118] A few years earlier, the British historian Arnold J. Toynbee had written that he guessed that the current age would be remembered in the future as having been the first "in which people dared to think it practicable to make the benefits of civilization available for the whole human race." Science had made it possible. This "sudden vast enhancement of man's ability to make nonhuman nature produce what man requires from her has, for the first time in history, made the ideal of welfare for all a practicable objective instead of a mere utopian dream," he wrote.[119]

Confidence in the possibilities of science and technology dramatically changed international society in the postwar years. The ability "to make nonhuman nature produce what man requires from her" became a central part of the growth and expansion of the machinery for development. That was not, however, its only consequence. There were other ways to employ the biological forces that had been unleashed during the war. Here, too, the rinderpest vaccines played an important role, for the same "extraordinary biological plasticity" that made the virus an ideal candidate for attenuated vaccine experimentation also made it a tempting candidate for biological weapons experimentation.[120]

---

[117] Yates, P. Lamartine. *So Bold an Aim: Ten Years of International Co-operation Toward Freedom from Want.* Rome: FAO, 1955, 5, 13.

[118] Orr, John Boyd. *The White Man's Dilemma.* 1953, New York: British Book Centre, Inc., 1954), 7, 96.

[119] Toynbee, Arnold J. "Not the Age of Atoms But of Welfare for All." *New York Times* (October 21, 1951), 168.

[120] Zinsser, *Rats, Lice and History*, 64.

5

Back to Grosse Île: Biological Warfare in the Postwar World

In November of 1947, the Canadian government declared Grosse Île a "prohibited place." It did so for several reasons, one being that the "Dominion of Animal Pathology considers the actual soil of the island to be toxic in certain areas" and another being that "due to the unique nature of the installations [it] is being maintained for such use in the future."[1] The island, the graveyard of thousands of cholera victims whose dreams of immigrating to North America ended there, had long been a place of separation, of isolation, and of disease. Parts of its soil were toxic; parts of its history were, too. Yet, in 1947, the island was now famous in certain circles for the work that had taken place there during the war. The avianized rinderpest vaccine inspired hope, not despair. The mutated virus that left the island for China and Kenya in sealed vials was a biological weapon – a living organism that programed cattle bodies to kill other strains of that organism – but it was an internationally sanctioned and shared biological weapon. Its success at that job, however, had gotten scientists thinking about what else they could do with it.

Earlier that year, the *Journal of Immunology* published a long report titled "Bacterial Warfare" by two scientists from Columbia University, Theodor Rosebury and Elvin A. Kabat. They had actually written the report in 1942, as members of a subcommittee of the War Effort Committee of the American Association of Scientific Workers, and had submitted it to the National Research Council. Even though it was based

---

[1] A. D. P. Heeney, P.C. 4728 (November 19, 1947), Diseases of Animals, Rinderpest Control Vaccination Project, Grosse Isle, QC, Record Group 17, Volume 3029, Page 83, File 37–23, LAC.

solely on published works, the report was deemed too hot for publication and kept confidential throughout the war. Rosebury and Kabat wanted it to be made public when possible "for the sake of its value per se, but also as a contribution to an informed discussion of the portentous moral and political issues involved." The science, the authors insisted, cannot be isolated from the politics, but the report itself was concerned with the science. Looking primarily at pathogens that affected humans and animals, they selected "potential infective agents" according to "their suitability for military use estimated, by reference to a series of predetermined criteria": "infectivity, casualty effectiveness, availability and resistance of the agent, means of transmission, epidemicity, availability of means of immunization and therapy, ease of detection and possible retroactivity." Rinderpest made the cut.[2]

In the case of rinderpest, Rosebury and Kabat wrote, "Preparations of virus in animal blood might possibly be used for bacterial warfare." In terms of the "military engineering" of pathogens, the authors explained, "a general principle of some potential importance concerns the selection, during the course of repeated subcultures and animal passages, of highly virulent or drug-resistant variants, or both." Careful passaging might develop "strains with unusual properties." Passaging should, they insisted, "be one of the problems studied by a military experimental unit."[3] The same mutability through animal passaging that made rinderpest *vaccine* production possible, potentially made rinderpest *weapon* production possible. The virus was vulnerable to human manipulation – to "military engineering." Humans had figured out its weaknesses, and they were already exploiting them, but the potential was there for exploitation along other lines, too.

That potential was evident in 1942 when Rosebury and Kabat originally wrote their report. It was the same year Richard Shope and his team traveled to Grosse Île. There, they manipulated the virus into a new kind of vaccine, which, beginning in 1946, encouraged a widespread focus on rinderpest eradication. The Grosse Île research, however, had other consequences as well. During World War II, rumors about Allied work on rinderpest encouraged the German government to pursue biological warfare research. That research, combined with what the Allies later learned about Japan's biological warfare research, helped to make biological

---

[2] Rosebury, Theodor and Elvin A. Kabat, with the assistance of Martin H. Boldt. "Bacterial Warfare." *The Journal of Immunology* 56 (1947): 7, 9.

[3] Ibid., 24–25, 83.

warfare seem a significant threat in the postwar period, something against which countries needed to be able to defend themselves. The existence of biological warfare research made additional biological warfare research probable, just as it did nuclear weapons research, particularly after the disappearance of the cooperative spirit of the alliance that had won the war.

The Canadian government declared Grosse Île a "prohibited place" in 1947 in large part because it feared that it was going to need to use it for rinderpest research again.[4] And, indeed, it did, starting in the summer of 1948, when researchers returned to the island to test the virus that they had left behind two years earlier. Canada, the United States, and the United Kingdom agreed to work together again to stockpile vaccines in case of a biological attack by the Soviet Union. In the process, they also began studying ways to turn rinderpest into a more effective biological weapon. Its mutability made that a possible option; their fears of what their enemy could/would do with it made it a tempting one.

$$***$$

Rinderpest had always been a frightening disease, terrible in its fury. The Allies began preparing against it when they worried it might be turned against them. Rumors of that primarily defensive research, however, were interpreted abroad as primarily offensive. Interviews with captured German scientists and military officials, along with documents secured toward the end of the war by the bacteriological warfare team of the ALSOS Mission (which sent American and British soldiers hunting for Germany's scientific secrets), revealed that the Germans believed the Allies were researching rinderpest, along with several other diseases, as possible weapons of war.[5] This, in turn, roused Germany's interest in biological warfare in general.

Although he never explained why, Hitler banned all offensive work on biological weapons early in the war, perhaps after reading a memo on

---

[4] A.D.P. Heeney, P.C. 4728 (November 19, 1947), Diseases of Animals, Rinderpest Control Vaccination Project, Grosse Isle, QC, Record Group 17, Volume 3029, Page 83, File 37–23, LAC.

[5] There were actually three parts to the ALSOS Mission: Italy, France, and Germany. British and American operatives were charged with obtaining "advance information regarding scientific developments in progress in enemy research and development establishments which are directed towards new weapons of war or new tactics." They were primarily looking for atomic intelligence, because ALSOS was part of the Manhattan Project, but biological weapons were also of interest. See, Groves, Leslie R. *Now It Can Be Told.* New York: Harper & Row, 1962.

the subject by the Director of Hygiene for the Waffen-SS. One of the German officials captured after the war told his ALSOS Mission interviewers that "as far as the Wehrmacht was concerned there had been no interest in BW until 1942 and then only from the defense side." That year, the Germans ostensibly became "aware of considerable B.W. in the U.S.A." This intelligence, the ALSOS interviewers soon realized, had not been reliable, but the Germans had not known that.[6] It contained a lot of rumors and very little evidence. One such 1942 rumor was that the Americans were experimenting with the transmission of rinderpest by dropping it out of aircraft in Texas.[7] Others reported potential attacks on animals, plants, and humans with a variety of pathogens. In response, officers at the Heereswaffenamt decided that it "was absolutely necessary to study methods of defense against potential attacks by B.W. agents." General Field Marshal Keitel approached Hitler on the subject in December of 1942. "Hitler expressed his distaste of B.W. as well as of C.W. and forbid any contemplation of offensive B.W. activity but allowed a Heeres Waaffenamt committee to be set up to study methods and measures of protection against a potential B.W. attack." Keitel subsequently sent out a directive from Field Headquarters in March of 1943 reading, "Our Führer has given the order to complete defensive measures against the use of bacteria by the enemy with all possible zeal, and to hold in readiness the means of defense. All other preparations are forbidden." Keitel also created the Blitzarbleiter Committee.[8]

---

[6] Report on the Interrogation of Oberst Hirsch (May 19–21, 1945), BW file; Security – Classified "ALSOS" Mission Reports and Correspondence Relating to the Progress of German Scientists in Connection with Nuclear Physics, 1944–45, Box 137, Office of the Director of Intelligence (G-2), Subordinate Officers, Foreign Liaison Branch; RG 165, Records of War Department General and Special Staff, NARA. See also Geissler, Erhard. "Biological Warfare Activities in Germany, 1923–45," in *Biological and Toxin Weapons: Research, Development and Use from the Middle Ages to 1945*. ed. Erhard Geissler and John Ellis van Courtland Moon. Oxford: Oxford University Press, 1999, 91–126.

[7] Directives and Correspondence of Prof. Kliewe, Reports and Messages, 1918–1951, ALSOS Mission, Assistant Chief of Staff (G-2), Intelligence Administrative Div., Document Library Branch, Box 3, RG 319, Records of the Army Staff, NARA.

[8] ALSOS Intelligence Report (September 12, 1945), ALSOS Intelligence Report B-C-H-H/ 305, ALSOS Reports, Naval Technical Mission to Europe, Official Files and Reports, Box 8, RG 38, Records of the Office of the Chief of Naval Operations, NARA; Report on the Interrogation of Oberst Hirsch (May 19–21, 1945), BW file, Security – Classified "ALSOS" Mission Reports and Correspondence Relating to the Progress of German Scientists in Connection with Nuclear Physics, 1944–45, Box 137, Office of the Director of Intelligence (G-2), Subordinate Officers, Foreign Liaison Branch, RG 165, Records of War Department General and Special Staff, NARA.

Work, however, had already begun. ALSOS Mission records revealed that, in January of 1943, Dr. Heinrich Kliewe, who had been sent to Paris in 1940 to examine BW research at French laboratories, and who had reviewed all of the incoming intelligence reports about Allied BW activities, "drew up a long memorandum entitled, 'The Preparation of Our Enemies for BW.'" In it, he discussed ongoing Allied research on "anthrax, foot and mouth disease, rinderpest, and glanders, and point[ed] out methods of spreading these agents." He argued, "it is probable that England's bacterial war will concentrate on the use of disease agents against the livestock of the Axis powers," adding "that one of the purposes of BW is to starve the civilian population." He was promptly sent to Berlin to continue his research. "In Kliewe's file next to the directive describe above," the investigators reported, "is a short undated parchment, in which he discusses the possible methods of spreading fowl cholera, Pasteurella infections of cattle in tropical climates, Black Leg . . ., swine cholera, foot and mouth disease and rinderpest."[9] Thinking about how the Allies could spread disease of course gave Kliewe the opportunity to think about how Germany could do it. It was a safe way around the Führer's prohibition on offensive research: it was all theoretical.

Blitzarbleiter Committee chief veterinarian, Professor Nagel, likely thinking about an agent's report of June 1942, which "stated that work on BW in the U.S. was concentrated on rinderpest, rather than on foot and mouth disease, anthrax or glanders," spoke with Kliewe in June 1943 about "the possibility of reprisals against Germany should she start BW by spreading foot and mouth disease" (likely in England), adding "that America might disseminate rinderpest which would have catastrophic results." Meanwhile, "the director of Armed Forces Science Bran W/Wiss stated in 1943 that 'America would have to be attacked simultaneously with different human and animal diseases.'" At the July committee meeting, Nagel, of the Military Veterinary Inspectorate, "stated that work on rinderpest was forbidden in Germany and that neither anti-serum nor vaccines were available." He reiterated the point at the September meeting, noting that "work on rinderpest is not planned and that only 'Veterinary-police' measures will be undertaken should the enemy introduce rinderpest."[10]

---

[9] ALSOS Intelligence Report (September 12, 1945), ALSOS Intelligence Report B-C-H-H/ 305, ALSOS Reports, Naval Technical Mission to Europe, Official Files and Reports, Box 8, RG 38, Records of the Office of the Chief of Naval Operations, NARA.

[10] Translation, Resume of England-American BW Intelligence Authored by German Intelligence Organizations, ALSOS Mission – Undated, Military Intelligence Division,

Nagel's concern about that possibility was no doubt elevated by a September intelligence report that insisted, "especially advanced ... is the cooperative work ... on dissemination methods for rinderpest." The Allies have "rejected" the "dissemination of human diseases," the report concluded, but, "judging from the above the entire German veterinary profession should become well conversant with rinderpest." An October report argued that "the introduction of a definite type of hoof-and-mouth disease organism is likely. But the Russians could justifiably object to the use of rinderpest organisms (as well as of hoof-and-mouth disease organisms). England may concur in the protest," because of short distances between them and Germany. These fears made it unlikely that the Allies would resort to "mass infection," but "the introduction of the said agents, or of poisons, by saboteurs must be reckoned with."[11] Biological warfare was not deemed a dire threat, but an issue of concern.

That concern helps to explain why, at the May 24 meeting, "the representative of the Reich Marshall, Prof. Blome," informed the committee that "a program was started concerning rinderpest which included production of a vaccine."[12] But it took more than fear of sabotage to make that shift, because the fear had existed for months. How did it happen that Germany began studying rinderpest as a biological weapon when, first, Hitler had expressly forbid such research and, second, there was no rinderpest in Germany, the country having given up its stock of the virus years earlier in agreement with its European counterparts to avoid a possible laboratory escape? The answer is the personal involvement of one of the few men powerful enough and confident enough to disobey Hitler: Heinrich Himmler.

Himmler, the ALSOS investigators concluded, "apparently" was "the only influential leader in Germany who expressed any desire to initiate the offensive use of BW." The Americans learned the story from Kurt Blome, whom the United States Counter Intelligence Corps arrested in Munich in May of 1945.

During a conference between Blome and Himmler in February-March 1944, Himmler asked about rinderpest and when Blome pointed out that no strains of

---

"S-C Intelligence Reference PUBS. ("P" File) 1940–1945, Box 162, RG 165, Records of the War Department General and Special Staffs, NARA.

[11] Ibid.

[12] ALSOS Intelligence Report (September 12, 1945), ALSOS Intelligence Report B-C-H-H/ 305, ALSOS Reports, Naval Technical Mission to Europe, Official Files and Reports, Box 8, RG 38, Records of the Office of the Chief of Naval Operations, NARA.

the virus existed in Germany, Himmler said he would take steps to obtain some. Sometime later Blome received a message that a strain of rinderpest had been obtained in Turkey and he was instructed to send an expert to collect it. Prior to this, Blome had mentioned rinderpest to Waldmann and the latter had agreed that it was probable that suspension of the virus could be dried and preserved in the same way as foot and mouth disease. Accordingly, Bloom then asked Prof. [Erich] Traub, the head of the Bacteriology Section of the Institute in Reims, to report to an SS office in Berlin and from there Traub was sent to Turkey and returned with the virus. Blome did not know any of the details as to how or where Traub actually obtained the virus. Traub returned to Reims with the virus and there, in an old building isolated from the main institute, an attempt was made to infect some cattle by rubbing the virus on their nose and mouth. One animal was sick, but recovered and the others were unaffected. Further work was therefore impossible as the strain had proved avirulent. On reporting this to Himmler at a later meeting, Blome was told that attempts would be made to secure another strain but he heard no more concerning the matter.[13]

Blome and Himmler also collaborated on far more sinister work. Blome readily admitted to his ALSOS investigators that Himmler had "ordered him to study the offensive side" of BW back in 1943, and "offered him facilities in a concentration camp to study plague. He promised Blome all the support of the SS in carrying out his task." Notes from the September, 1943, Blitzerbeiter meeting recorded, "Since it is not known under what conditions inhaled aerosols or dispersed droplets of certain pathogenic germs cause disease in man, Prof. Blome suggested experiments on human beings."[14] Clearly, Blome and Himmler thought alike.

In December 1946, Blome was hauled before the Nuremberg tribunal in *U.S.A. v. Karl Brandt et al*, also known as the Doctors' Trial. Blome was one of twenty-three doctors and administrators accused of war crimes and crimes against humanity. The prosecutors charged him with participation in experiments on humans to study malaria, mustard gas and polygal; with executing Polish nationals alleged to have tuberculosis; and with euthanizing "undesirables" throughout Germany. (Another charge, of experimenting with sulphanilamide, was withdrawn.) The trial lasted until August.[15] In the end, the tribunal acquitted Blome, claiming, "It may well be that defendant Blome was preparing to experiment on human beings in connection with biological warfare, but the record fails to disclose that fact, or that he actually conducted

---

[13] Ibid.   [14] Ibid.
[15] Harvard Law School Library, "Introduction to NMT Case 1 U.S.A. v. Karl Brandt et al.," Nuremberg Trials Project: A Digital Document Collection, available at http://nuremberg .law.harvard.edu/php/docs_swi.php?DI=1&text=medical.

experiments. The charge of prosecution cannot be sustained."[16] Two months later, four representatives from the Army Chemical Corps (the former Chemical Warfare Service) at Camp Detrick, Maryland, the headquarters of American biological weapons research, interviewed Blome about his and others' research activities during the war. In 1951, he signed a contract to work for the Army Chemical Corps under "Project 63," "whose purpose was to deny the Soviet Union the German scientists' skills." His immigration required approval by the US consul in Frankfurt; none of the background material presented to the consul by the Americans offering him the job contained any mention of his work during the war, his arrest, or the trial. The consul deemed him inadmissible for immigration anyway, following an "incriminating interrogation report." Blome wound up with the position of camp doctor at the European Command Intelligence Center in Oberusel.[17] It paid to be an unconvicted war criminal in the postwar era, especially if the "alleged" crimes involved biological weapons. This was also true in Japan.

The United States had launched the ALSOS Mission to discover what kinds of progress the Germans had made on new weapons and tactics of war. It launched a similar mission in Japan. Under the direction of Edward L. Moreland, Karl T. Compton, who was president of MIT and a member of Vannevar Bush's Office of Scientific Research and Development, led a team of scientists to Tokyo the first week of September, 1945, to "make a quick, preliminary investigation to determine how the Japanese had organized for scientific war research." They were eager to get there before Japanese scientists had more time to destroy the records of their activities.[18] The central concern, Compton later explained, was that "in this war of such highly developed technical character, how far had Japan gone, either in production or in development or even in conception of new methods of warfare which might prove dangerous to world peace in the

---

[16] Quoted in Hunt, Linda. "U.S. Coverup of Nazi Scientists." *Bulletin of the Atomic Scientists* 2:8 (April 1985): 22.

[17] Ibid., 21–23; Deichmann, Ute. *Biologists under Hitler.* trans. Thomas Dunlap. Cambridge, MA: Harvard University Press, 1996), 288–289.

[18] "Report on Scientific Intelligence Survey in Japan, September and October 1945," Volume I, Scientific Intelligence Survey in Japan, Vol. I, Military Intelligence Division: "S-C Intelligence Reference Pubs. ("P" File) 1940–1945, Box 2045, RG 165: Records of the War Department General and Special Staffs, NARA. For more on the investigation see Sasamoto, Yukuo. "The Scientific Intelligence Survey: The Compton Survey," in *A Social History of Science and Technology in Contemporary Japan*, Vol. I, ed. Shigeru Nakayama. Melbourne: Trans Pacific Press, 2001, 59–72.

future?"[19] The search for the answer led the Americans to seek out officials from many places. The team held a total of 135 interviews with approximately 300 "scientific and technical people, representing some 50 separate institutions or projects," including Noborito.[20]

On September 19, two team members (along with a translator) interviewed two officials from Noborito who had worked on the balloon bombs. The men reported that "all records and plans in connection with this weapon were burned by order of the Japanese General Staff." That order had come down on August 15 and had, to the dismay of the Americans, been widely followed. This made the interviews the primary source of information. The Noborito officers told their interviewers quite a bit about the balloons, describing how many had been made at what cost and when and why they had stopped production. They explained that they had been loaded with "incendiaries and high explosives" and insisted that "it had never been the Japanese intention to alter these loads with CW or other weapons."[21] The Americans took them at their word, as they did most of the men they interviewed. The team's November report explained that, "on the whole," they believed that they had formed "a reasonably complete and reasonably accurate over-all picture of Japanese scientific and technical developments" and that "nothing of military significance has escaped attention." Whether that final statement was true or not, of course, depended on one's definition of "military significance."[22]

At the end of the interview with the Noborito officials, the Americans learned that Major General Sueki Kusaba, the head of the balloon bomb program, had recently returned to Tokyo. The Compton team sent him a request for an interview that was scheduled for October 3. It is unclear if the interview ever took place, because he is not mentioned again in the

---

[19] Coen, *Fu-Go*, 190.

[20] "Report on Scientific Intelligence Survey in Japan, September and October 1945," Volume I, Scientific Intelligence Survey in Japan, Vol. I, Military Intelligence Division: "S-C Intelligence Reference Pubs. ("P" File) 1940–1945, Box 2045, RG 165: Records of the War Department General and Special Staffs, NARA.

[21] "Report on Scientific Intelligence Survey in Japan, September and October 1945," Volume II, Scientific Intelligence Survey in Japan, Vol. II, Military Intelligence Division: "S-C Intelligence Reference Pubs. ("P" File) 1940–1945, Box 2046, RG 165: Records of the War Department General and Special Staffs, NARA.

[22] "Report on Scientific Intelligence Survey in Japan, September and October 1945," Volume I, Scientific Intelligence Survey in Japan, Vol. I, Military Intelligence Division: "S-C Intelligence Reference Pubs. ("P" File) 1940–1945, Box 2045, RG 165: Records of the War Department General and Special Staffs, NARA.

final report.[23] Kusaba had been at the general staff meeting back in September of 1944 where Kuba presented his rinderpest balloon bomb plan, so he was well aware that there had been a plan to weaponize the balloon bombs, for a moment at least, until Tojo quashed the idea. There may well have been other, similar plans, but the Americans appear to have never known about them, just as they appear to have never known about Kuba's rinderpest balloon bomb proposal. Japan's biological warfare research on rinderpest, unlike Germany's, remained secret. It had had time to destroy its records.[24]

Other aspects of Japan's biological warfare activities were not so easily erased, for they involved human beings. A few of their victims survived to tell their stories. At the inauguration of his death factory at Ping Fan, Ishii Shiro told his fellow scientists and soldiers that even though "the research work on which we are now about to embark … may cause us some anguish as doctors," they needed to carry on. "Pursue it," he ordered, "based on the dual thrill of (1) a scientist to exert efforts to probing for the truth in natural science and research into, and discovery of, the unknown world, and (2) as a military person, to successfully build a powerful military weapon against the enemy."[25] Ishii and his men infected their victims (primarily Han Chinese, but also Allied prisoners, Mongolians, Koreans, Europeans suspected of spying for the Allies, and persons with disabilities) with "countless diseases."[26] Along with recording how their victims suffered, how long it took them to die, and how they died, the Japanese scientists researched ways of fighting pathogens and ways of using pathogens to fight. They produced vaccines, keeping chickens and rats on hand for the production. They experimented with delivery systems, exploding bombs filled with bacteria over fields full of human and nonhuman captives. They also let loose rats covered in plague-infested fleas.[27] Some of this became clear

[23] "Report on Scientific Intelligence Survey in Japan, September and October 1945," Volume II, Scientific Intelligence Survey in Japan, Vol. II, Military Intelligence Division: "S-C Intelligence Reference Pubs. ("P" File) 1940–1945, Box 2046, RG 165: Records of the War Department General and Special Staffs, NARA; Coen, *Fu-Go*, 194.

[24] There is no reference to rinderpest in Volume 5 of the "Report on Scientific Intelligence Survey in Japan," which focuses exclusively on biological warfare.

[25] Harris, Sheldon H. *Factories of Death*, revised edition. New York: Routledge, 2002, 56–57.

[26] Ibid., 63, 77–78.

[27] Ibid., 78–88; Guilleman, Jeane. *Biological Weapons*. New York: Columbia University Press, 2005, 83–86; Harris, Sheldon. "The Japanese Biological Warfare Programme: An Overview," in *Biological and Toxin Weapons: Research, Development and Use*

during the Compton team's interviews in the fall of 1945; some came out later.[28]

It is impossible to provide a specific number of the victims of Ishii's Unit 731. Ishii and his men were cagey with their records and they burned most of them before the Red Army arrived. And Unit 731 was only the most notorious of the Japanese biological experimentation stations; others were built in Mukden, Nanjing, and Changcun. Almost immediately following Japan's surrender, allegations of its crimes against humanity came flooding into the Allied headquarters in Tokyo. The Japanese Communist Party played a critical role, releasing information to the United Press that specifically named Ishii and that was subsequently published in the *New York Times*.[29] United States military intelligence (G-2) was also targeting Ishii, but it had different motivations. Living in comfortable house arrest in Tokyo, Ishii promised to trade information for immunity. The United States took his bait.

In March of 1947, the Joint Chiefs of Staff placed all biological warfare investigations under G-2 so that they were untouchable by the US Adjutant General Office in Tokyo that had been officially tasked with the pursuit of war criminals. A few months later, Norbert Fell, from Camp Detrick, and a few colleagues, arrived to interview Ishii. Fell reported that "evidence gathered in this investigation has greatly supplemented and amplified previous aspects of this field. Such information could not be obtained in our own laboratories because of scruples attached to human experimentation." He recommended immunity. G-2 officials agreed, arguing that the information provided by the Japanese was too valuable to be made public in a trial. They did not want the USSR to get it. The United States did grant its former ally access to Ishii and some fellow scientists in 1947, but only "after coaching the Japanese to keep their responses minimal." The Tokyo war crimes trials came and went without mention of biological experimentation on humans.[30] Liberated American POWs were "threatened with court martial if they revealed to

from the Middle Ages to 1945. ed. Erhard Geissler and John Ellis van Courtland Moon. Oxford: Oxford University Press, 1999, 127–152.

[28] "Report on Scientific Intelligence Survey in Japan, September and October 1945," Volume V, Scientific Intelligence Survey in Japan, Vol. V, Military Intelligence Division: "S-C Intelligence Reference Pubs. ("P" File) 1940–1945, Box 2045, RG 165: Records of the War Department General and Special Staffs, NARA.

[29] Harris, *Factories of Death*, 86–87, 242–243.

[30] Ibid., 86–87, 95–104, 263–307; Guilleman, *Biological Weapons*, 76–79; Powell, John W. "A Hidden Chapter in History." *The Bulletin of the Atomic Scientists* 37:8 (October 1981): 44–52.

their families, friends, or the media anything suggesting that the Japanese engaged in BW experiments on American POWs."[31]

The same was not the case in the Soviet Union. A dozen captured Japanese scientists went on trial in Khabarovsk in 1949. Testimony described human experiments that American and British officials publically dismissed as Soviet propaganda. The defendants all went to prison. Western officials interpreted their 1956 release and repatriation as evidence that the Soviets had also traded freedom for information.[32] By that point, both the United States and the Soviet Union were far along their own biological warfare investigative paths.

Blome and Ishii directly profited from the growing tension between the United States and the USSR. That tension made both sides keen to interview captured scientists and loath to allow the other to do the same. It had deep roots and seems, in hindsight, almost unavoidable. There were certainly signs all along – the secrets kept from each other, the ALSOS Missions of American and British soldiers capturing German scientists before the Soviets could get to them. It is misleading, however, to read 1949 onto 1945. In the immediate aftermath of the war, the Western powers viewed much of their biological warfare research as knowledge that could now be put to greater use by being shared freely with the world as a whole.

In October of 1945, George Merck urged the Secretary of War to allow the publication the research the Allies had done during the war that had "contributed significant knowledge to what was already known concerning the control of diseases affecting humans, animals and plants." Merck wholeheartedly supported "the release of such technical papers and reports by those who have been engaged in this field as may be published without endangering the national security."[33] This decision allowed Shope and his team to publish their reports the following spring. The vaccine technology was exactly the kind of research that Merck believed was valuable to the world at large, and not a threat to "national security."

---

[31] Harris, *Factories of Death*, 320–321.

[32] Ibid., 317–321; Guilleman, *Biological Weapons*, 80. For more on this, see *Materials on the Trial of Former Servicemen of the Japanese Army Charged with Manufacturing and Employing Bacteriological Weapons*. Moscow: Foreign Languages Publishing House, 1950.

[33] Report to the Secretary of War by Mr. George W. Merck, Special Consultant for Biological Warfare (October 1945), CCS 385.2 (12-17-43) Sec. 4, Chemical, Biological & Radiological Warfare (JCS 1822), Central Decimal File, 1942–45, Box 375, RG 218, Records of the US Joint Chiefs of Staff, NARA.

Merck was careful to make it clear that the willingness to share the "safe" knowledge gained during the war should not be taken as a cheerful conclusion to the effort as a whole. He believed that the United States could not now halt what it had started. "While it is true that biological warfare is still in the realm of theory rather than fact," he acknowledged, the results of our research demonstrate that "it cannot be discounted by those of this nation who are concerned with the national security." The United States needed the research. "Work within this field," he insisted, "born of the necessity of war, cannot be ignored in time of peace," precisely because other nations were unlikely to ignore it. Unlike atomic weaponry, he warned, biological warfare does not require "vast expenditures of money or the construction of huge production facilities." Scientists simply needed basic labs. It was, therefore, unwise to shut down our own.[34]

Merck justified continued biological warfare research as necessary to "national security." The concept, as Andrew Preston has demonstrated, was one of World War II's most important legacies. Roosevelt was directly responsible. Although the term preceded him, FDR gave it new life by "blending the defense of territory with the protection of core values so thoroughly that the two became inseparable." When Merck used the term "national security," he used it in a World War II framework. It was, after all, only October of 1945 and the war was barely won. His enemies were Japanese and German. The term, however, outlived the circumstances of its creation. FDR had invoked the term to rouse the country; it worked too well. "The war legitimized the doctrine of national security," Preston argued. "In a few short years, it went from being a radical idea to conventional wisdom. And with that, Roosevelt had set up the basic structure of the Cold War."[35] Merck's "national security" warnings were not directed specifically at the USSR, just as Roosevelt's certainly had not been, but the framework made it very easy to slip the Soviets into a "threat" position over the course of the next two years.

It did not happen all at once. Indeed, it seemed that Merck's call for continued work was initially disregarded. In January of 1946, the War Department released part of his report to the public. It did so in the name of transparency, but primarily because it did not believe that it needed to be kept secret. It also encouraged the publishing of the research

---

[34] Ibid.
[35] Preston, Andrew. "Monsters Everywhere: A Genealogy of National Security." *Diplomatic History* 38:3 (June 2014): 492–499.

undertaken under its aegis. The United States government announced that it would end all support for Grosse Île the following month. Later that year, in August, the government turned the Chemical Warfare Service into the Chemical Corps, making it an official branch of the Army headquartered at Camp Detrick, but dropped its budget from a wartime high of $2.4 million to $933,000. The camp had had between 3,000 and 4,000 employees working on research toward the end of the war. By March, it had between 300 and 400.[36] Meanwhile, in Canada, the Department of National Defense turned the Grosse Île station over to the Department of Agriculture in September of 1946. North America was moving away from biological warfare in 1946, readily sharing both research and, as in the case of the vaccines sent to China that year, results.

That movement abruptly changed direction in 1947. The United States War Department "issued a stop order on all references from military sources to the subject of biological warfare" on January 23.[37] That year, the Chemical Corps budget jumped to 2.75 million dollars, and its scientists flew to Europe and Asia to interview (and to offer immunity to some) German and Japanese scientists.[38] The United States, Canada and the UK also restarted their previous strategic wartime alliance, promising to share all of their research with each other in the Tripartite Biological and Chemical Weapons Agreement.[39] Canada's Department of Agriculture returned Grosse Île to the Department of National Defense, which promptly turned its management over to the new Defense Research Board.[40] In Britain, several defense committees recommended that equal priority be given to the development of biological weapons as to nuclear ones.[41] The tide had turned.

[36] Guilleman, *Biological Weapons*, 96; Moon, John Ellis van Courtland. "The US Biological Weapons Program," in *Deadly Cultures*, ed. Mark Wheelis, Lajos Rózsa, and Malcolm Dando. Cambridge, MA: Harvard University Press, 2006, 19.

[37] Report by the Joint Intelligence Committee to the Joint Chiefs of Staff on Proposed Press Release and Proposed Official Statement Concerning the Potentialities of Biological Warfare (April 22, 1948), CCS 385.2 (December 17, 1943) Sec. 6, Chemical, Biological & Radiological Warfare (JCS 1822), Central Decimal File, 1942–45, Box 375, RG 218, Records of the US Joint Chiefs of Staff, NARA.

[38] Guilleman, *Biological Weapons*, 96.

[39] Avery, Donald. *Pathogens for War*. Toronto: University of Toronto Press, 2013), 56.

[40] Defense Research Board, War Disease Control Station (March 6, 1950), Biological Warfare – Rinderpest Virus, RG24, Series F-1, Vol. 4224, LAC.

[41] Balmer, Brian. "The UK Biological Weapons Program," in *Deadly Cultures*. ed. Mark Wheelis, Lajos Rózsa, and Malcolm Dando. Cambridge, MA: Harvard University Press, 2006, 50–51; Balmer, Brian. *Britain and Biological Warfare*. London: Palgrave, 2001, 55–78.

The altered level of attention paid to biological weapons reflected a changing international environment. "By the autumn of 1946," Melvyn Leffler has shown, "there was general agreement in the United States that the Kremlin was an ideological enemy with no legitimate fears or grievances. The cold war had begun."[42] Truman and Stalin had both initially favored cooperation over confrontation, but "the condition of the international system engendered fears and opportunities" that made both less willing to cooperate with the other.[43] Both became more competitive. "Wartime internationalism," Mark Mazower wrote, "was on the wane in policymaking circles by mid-1946."[44] Growing tensions were evident in a variety of ways, some of which have been discussed in earlier chapters. They also manifested in the way the major players approached biological warfare – the attempt "to harness certain forces of nature for purposes of war."[45]

By far the most important force of the era was, of course, nuclear. Governments and the scientists they employed struggled mightily to harness that terrible power. Biological forces were a distant second, but they were there. They had their own peculiar horror: they could wipe out life, yet leave all nonliving structure fully intact. They seemed to have potential not yet imagined, a sense that sparked both fear and anticipation. It all depended on who tapped – you or your enemy. And it now seemed certain that the United States had an enemy eager to do it.

In 1948, Secretary of Defense James Forrestal received a report warning that "the current research and development program for biological weapons is grossly inadequate in view of the potentialities of the present international situation."[46] The following spring, he put together an ad hoc committee to do "a full examination of all the technical and strategic possibilities of biological warfare." Its report, which the committee

---

[42] Leffler, Melvyn. *A Preponderance of Power*. Stanford: Stanford University Press, 1992, 140. For more on the origins of the Cold War, see Leffler, Melvyn P. and David S. Painter. *Origins of the Cold War*. 2nd ed. New York: Routledge, 2005.

[43] Leffler, Melvyn. *For the Soul of Mankind*. New York: Hill and Wang, 2007, 79.

[44] It would pick back up in 1949 in the aftermath of Truman's inaugural address, when the United States committed itself to fighting the Cold War through technical assistance, as discussed in the previous chapter (Mazower, Mark. *Governing the World*. New York: Penguin, 2012) 215, 284–285.

[45] Secretary Forrestal Issues Statement on Biological Warfare Potentialities (13 March 1939); CCS 385.2 (December 17, 1943) Sec. 8, Chemical, Biological & Radiological Warfare (JCS 1822), Central Decimal File 1948–50, Box 206, RG 218 Records of the US Joint Chiefs of Staff, NARA.

[46] Avery, *Pathogens for War*, 57.

presented to his successor Louis A. Johnson in July, concluded that "the offensive and defensive potentialities" of biological weapons (and also chemical and radiological weapons), were "at present potent, but not decisive," for "biological warfare is still in its infancy." It would not stay there, however, for the "medical and biological sciences are on the threshold of great new advances comparable to those in the atomic field." That meant the United States needed to devote far more attention to the issue, because it "may be a particularly important factor in the present balance of power." We know the United States "enjoys nuclear superiority" over the USSR, they wrote, but it "does not necessarily possess a corresponding superiority in the field of biological warfare – in fact, the situation might be the reverse." This was problematic on several levels, but primarily because biological weapons were far more versatile than their atomic counterparts. They could be used "subversively for a long-term economic attrition during a period of 'cold war' without excessive risk of detection." Biological weapons seemed well-suited for the struggle at hand: undeclared war.[47]

The committee deemed this particularly true of biological weapons that attacked plants and animals and were, in that way, less aggressive, less likely to provoke the United States into open war. The Soviets, they argued, were in peacetime far more likely to attack "food resources" than "mass populations." They would aim for economic disruption. The possibility "of enemy action against our meat supply constitute a substantial threat," they continued. "Attacks against animal and plant resources, using such self-spreading diseases as rinderpest and hoof-and-mouth disease and the rusts which affect cereal crops, could be launched by a small number of enemy agents armed with biological weapons of a type which the United States has already developed." Their point was that all it took to make rinderpest a basic biological weapon was a vial of the living, non-attenuated virus. It took very little of a pathogen to commit an act of biological sabotage. What one would be able to do with that same pathogen in the future, in terms of warfare, as opposed to sabotage, remained open to the imagination. What we and our enemies can do today, they insisted, is nothing compared to what we will all be able to do in the future: for better and for worse. In the coming decade, the

---

[47] Report of the Secretary of Defense's Ad Hoc Committee on Biological Warfare (July 11, 1949), CCS 385.2 (December 17, 1943) Sec. 8, Chemical, Biological & Radiological Warfare (JCS 1822), Central Decimal File 1948–50, Box 206, RG 218 Records of the US Joint Chiefs of Staff, NARA.

committee predicted, "man's new knowledge ... may well place in his
hands powers both of healing and of destruction not even contemplated
today." The dichotomy was key; the research that made biological warfare possible also made vaccines and antibiotics possible.[48]

Biological warfare was complicated. It posed "new and difficult problems in international relations and in public understanding of national
defense policies." So, too, of course, did atomic warfare, but that was
getting enormous amounts of attention. Biological warfare was not. This
mattered not only because of the threat of sabotage by small, quiet
weapons that would be infinitely more difficult to trace than their atomic
counterparts, but also because it would be so easy for a nation to claim
that the United States had engaged in biological sabotage. "False allegations that such agents have been used by another nation, or the mere threat
that they will be used against another nation, may have profound political
and psychological effects in periods of 'cold war' as well as in wartime,"
the committee warned.[49] Biological warfare posed a threat to "national
security" on several levels – levels George Merck had not considered in his
1945 report. The Cold War had changed the threat calculus. Weapons
could hurt in new ways when "security" was both "territorial *and*
ideological."[50] On the up side, correspondingly, vaccines could help in
new ways.

On November 12, 1948, Richard Shope submitted a confidential report
to the Committee on Biological Warfare, Research and Development
Board, National Military Establishment, on the recent FAO Meeting on
Rinderpest in Nairobi. Shope included a copy of the official "Conclusions"
of the meeting (discussed in the previous chapter), but also wrote his own,
explaining "a number of things, not included in the formal 'conclusions' of
the Conference, which may have interest for Committee X, were observed
or heard during the Conference proceedings." The US military had different
priorities than FAO. Shope was well aware of them.[51]

Shope opened with a report on the extensive work being done on
rinderpest vaccines around the world. He was, not surprisingly, particularly concerned with the avianized vaccine that he and his team had
created. Work continues, he wrote, in Nanjing, Kabete, and Bangkok,

---

[48] Ibid.   [49] Ibid.

[50] Preston, Andrew. "Monsters Everywhere: A Genealogy of National Security."
Diplomatic History 38:3 (June 2014): 477–500.

[51] Richard E. Shope to Chairman, Committee on Biological Warfare, RDB (November
12, 1948), Biological Warfare – Rinderpest Virus, RG24, Series F-1, Vol. 4224, LAC.

all on strains that came from Kabete. "We have no assurance at all that any other strain of rinderpest virus will ever be found which can be grown and attenuated in embryonating eggs," he warned, which meant that Kabete needed to be protected, especially considering the mounting tensions with the Soviet Union. "It is apparent, I believe, from consideration of the present world political situation that the only currently safe repository of the Kabete virus is the Veterinary Research Institute in Kenya," Shope wrote. The other locations were too vulnerable to be trusted, and, having only one "safe repository" seemed risky. "It will be recalled," Shope continued, "that some months ago the Panel on Programs recommended activation of a rinderpest program, part of whose purpose would be to preserve the egg-adapted Kabete virus for use should we ever need it." And we might. "Should we ever have rinderpest emergency to deal with, I believe that all of the facts point to the avianized virus vaccine as the one of choice for us." The lapinized vaccine might work on our cattle, he acknowledged, but, even if it did, "in all likelihood fertile eggs in sufficient numbers would be easier to obtain in our country."[52] Either way, both vaccines needed further testing before the United States could be confident in their ability to work. But how?

The first step would be to monitor the vaccine work being done in Africa and Asia through FAO's assistance. "In the absence of any direct work with rinderpest in our own country and in view of our interest in the disease in the event of another war," Shope wrote, "it would seem particularly urgent that we maintain as direct contact as possible with workers in the field in other countries." But that was not enough. Shope insisted that the United States also needed to "establish at the earliest possible moment a laboratory under our own control where our own scientists could study rinderpest." This is the only way, he wrote, that we will be able to "best serve our own peculiar purposes."[53] Those purposes were, in Shope's report, defensive in nature. He was clearly concerned about a biological attack on American cattle by the Soviet Union, and he was not alone.

A few months earlier, Canadian scientists under the direction of Charles Mitchell of the Dominion of Animal Pathology, had traveled to Grosse Île to test their existing stock of rinderpest vaccine, which had been left in an ice chest when the station shut down. The results had not been encouraging. The first two specimens of attenuated vaccine proved unable to protect inoculated cattle who were subsequently injected with

[52] Ibid.   [53] Ibid.

"challenging virus" that had also been kept frozen for over two years. In response to these concerning findings, Canada's Defense Research Board (DRB) began contemplating reopening Grosse Île for rinderpest research. They told the Chemical Corps at Fort Detrick as much in February 1949, "indicating that American assistance would be appreciated." The Chemical Corps responded that it was "extremely interested." As Shope had written, it liked the idea of a lab that could serve its "peculiar purposes."[54] The DRB's "contemplations" continued throughout 1949, part of a larger conversation about ABC (atomic, biological, and chemical) warfare in general.[55]

In the meantime, Canadian scientists returned to Grosse Île that summer to resume testing of the frozen vaccine stock. This time, the results were a bit more positive. Mitchell wrote to the DRB from Grosse Île on July 15 that this time his inoculated cattle were demonstrating "a rise of temperature which paralleled that which had been experienced formerly in testing vaccines of known potency." He planned to test them soon with non-attenuated virus. A handwritten note, dated July 25, on Mitchell's letter explains that it should be distributed to a select few people, but "for reasons already discussed we should mark these Top Secret and destroy all copies (not original) after."[56] Canada was not eager for the world to know that it was researching rinderpest again. "It is undesirable," Mitchell would write in February, "to discuss the propagation of this virus in Canada if the information were going outside of this country and this, of course, for obvious reasons."[57] Just testing vaccines in North America could be construed as a politically provocative act, because the virus had never broken out there.

Meanwhile, the Americans engaged in some not-so-subtle prodding, encouraged, no doubt, by their own findings on Soviet weapons research. The Secretary of Defense's Ad Hoc Committee released their report in July. In October, the Office of Strategic Information reported that "information received during this quarter definitely confirms the

---

[54] Defense Research Board, War Disease Control Station (March 6, 1950), Biological Warfare – Rinderpest Virus; RG24, Series F-1, Vol. 4224, LAC; Charles A. Mitchell to Glen Gay (July 15, 1949), Biological Warfare – Rinderpest Virus, RG24, Series F-1, Vol. 4224, LAC; J.C. Bond to Defense Research Board (December 1, 1949), Biological Warfare – Rinderpest Virus, RG24, Series F-1, Vol. 4224, LAC.

[55] Avery, *Pathogens for War*, 66–71.

[56] Chas. A. Mitchell to Glen Gay (July 15, 1949), Biological Warfare – Rinderpest Virus; RG24, Series F-1, Vol. 4224, LAC.

[57] Chas. A. Mitchell to Glen Gay (February 3, 1950), Biological Warfare – Rinderpest Virus, RG24, Series F-1, Vol. 4224, LAC.

supposition that the USSR is engaged in research on BW and shows that Soviet military training in BW is of long standing."[58] This intelligence was, no doubt, on the mind of the head of the Chemical Corps when he wrote to the DRB in December that "we are still greatly interested in undertaking further investigations on diseases of animals, particularly the Rinderpest work."[59] The United States government believed that Moscow was, in the immediate future, far more likely to attack its cattle than its people. London agreed. At a November, 1949, meeting of the Crop Committee of the Advisory Council on Scientific Research and Technical Development, participants from several of the British ministries concluded that "a B.W. attack on livestock seems to be the only risk that need be considered at this stage." A participant "quoted rinderpest as an example of the kind of risk that existed."[60] British scientists at the Pirbright Research Station began new work on rinderpest that year, feeling, like their counterparts in North America, that they needed to be prepared.[61] They also approached Canada for assistance. Grosse Île was too valuable for Washington or London to ignore.

That desirability bothered Mitchell a bit. In a letter to the chairman of the DRB in May of 1950, he wrote that he "opposed" allowing American visitors to join his team at Grosse Île that summer, as he had "opposed" it the previous summer. First, he said, "we have no way of determining in advance whether a virus has been propagated." He was nervous about the limited success he had had the previous summers. Second, he reiterated that it seemed "desirable that no attention should be attracted to this work," particularly not "until we have the vaccine strain established again."[62] The DRB disagreed.

A few days later, members of the DRB met with members of the Chemical Corps to discuss rinderpest research. The director of Camp Detrick announced that he had $150,000 available starting July 1. As "there was no place in the United States where work on rinderpest could be done," all of the money could go toward a "joint project" at

[58] Estimate of the Situation in BW Intelligence in the USSR (October 7, 1949), OSI Situation Reports, Box 218, File 2, President's Secretary's Files, Papers of Harry S. Truman, HSTL.

[59] A.C. McAuliffe to E. L. Davies (December 16, 1949), Biological Warfare – Rinderpest Virus, RG24, Series F-1, Vol. 4224, LAC.

[60] Notes of a Meeting, Crop Committee, Advisory Council on Scientific Research and Technical Development (November 29, 1949), WO 195/10738, NAUK.

[61] Biological Research Advisory Board, Note by the Joint Secretaries (February 27, 1950), WO 195/10780, NAUK.

[62] Chas. Mitchell to Miles Benson (May 30, 1950), Biological Warfare – Rinderpest Virus; RG24, Series F-1, Vol. 4224, LAC.

Grosse Île. The "proposed scope of the programme might be," he continued, "to determine the status of the present avianized strain," "to produce a stockpile of three to five million doses of vaccine," "to undertake research on improved methods of stabilizing and packaging," and "to consider offensive problems such as ease of production, stability, etc." Such were America's "peculiar purposes" for rinderpest research in the spring of 1950.[63] Mitchell might have resented the intrusion, but he did not resent the money, which would enable him to make repairs at Grosse Île that would allow him to expand his research beyond the summer months, to which he had been limited the past two years. Before work could begin, however, he needed new seed material. His tests had demonstrated that his seed virus could still sicken cattle, but it could not survive attenuation in egg embryos. He needed to import some rinderpest virus. He turned to FAO.

Mitchell sent a letter to K. L. V. Kesteven, then at FAO's Washington, DC, headquarters, on June 10, "inquiring as to the possibility of FAO locating and making available ... seed material of the avianized and lapinized rinderpest virus." Kesteven recommended that Mitchell write to R. Daubney in Cairo. It would be the "surest and safest way of obtaining the material." Daubney will be traveling to England in July, Kesteven noted, and I will be there as well. "If you wish," we could have him bring the material with him and I will then bring it on to Canada. "This procedure," Kesteven continued, "would avoid the necessity of committing funds and would at the same time give me an opportunity to discuss with you your work in this field in relation to work going on elsewhere." That last point raised some concern for Mitchell. He passed along Kesteven's letter to DRB with a note that "I should like guidance on our attitude towards him when he arrives."[64] It raised concern with the DRB as well. Its chairman, E. Ll. Davies, promptly sent a top-secret telegram off to the Canadian Joint Staff in London that began, "Urgently require for Doctor Mitchell's BW program this July various preparations of rinderpest virus available in Egypt. Mitchell could obtain through United Nations FAO channels but these are undesirable because they lack security and publicity unwise." Davies then listed the specific desired materials

---

[63] DRB Memo, Rinderpest Research, US Chemical Corps Proposal (June 2, 1950), Biological Warfare – Rinderpest Virus, RG24, Series F-1, Vol. 4224, LAC.

[64] K.L.V. Kesteven to Chas. A. Mitchell (June 15, 1950), Biological Warfare – Rinderpest Virus, RG24, Series F-1, Vol. 4224, LAC; Chas. A. Mitchell to E. Ll. Davies (June 19, 1950), Biological Warfare – Rinderpest Virus, RG24, Series F-1, Vol. 4224, LAC.

(based on what Kesteven had recommended in his letter to Mitchell) and explained that they could "be obtained only from Doctor R. Daubney," who was soon heading to London. "Suggest that British BW authorities be asked to request materials from Daubney and arrange delivery to England. DRB staff now visiting in England could bring them to Canada with clearances arranged to avoid customs inspection." Davies ended the telegram with a stern warning: "Canadian destination should not be revealed to Daubney as information might reach FAO."[65]

Mitchell was willing to sacrifice his desire for secrecy in return for the seed viruses he desperately needed. However, once Kesteven had told him how to get those viruses through Daubney and explained the specific form they should take to best ensure survival on the journey, Kesteven became a complication. The DRB decided that they needed to remove Kesteven from the picture, something that was easy to do since Daubney was heading directly to London. They did not want FAO to know that they had begun researching rinderpest again and urged secrecy from Daubney as to the material's final destination. Considering Mitchell had already written as much to Kesteven, this seems an overreaction on the DRB's part. Its contacts in London, at the Canadian Joint Staff, thought so as well. In his response to Davies's telegram, the Assistant to Defense Research Member explained that "the feeling was that if Daubney was not told the ultimate destination it would arouse unnecessary speculation and comment and that the best plan, on the advice of those who know him personally, was to tell him in confidence and trust to his discretion not to talk about it." They were able to make contact with Daubney via the British Embassy in Cairo. He received the telegram with the request the night before he left Cairo and subsequently packed thirteen tubes of different strains of avianized and lapinized virus to bring with him to London. A member of the DRB carried it on from there to Canada.[66]

Mitchell finally had his new seed viruses. The lapinized strains proved particularly useful. They had always been easier to propagate than the avianized and this time proved no different. On July 17, Mitchell wrote, "I feel hopeful that the lapinized material may be the answer to what we have been seeking for many months." If this proves to be the case, he

---

[65] E. Ll. Davies to CANRESEARCH London (June 23, 1950), Biological Warfare – Rinderpest Virus, RG24, Series F-1, Vol. 4224, LAC.

[66] A.J. Skey to E. Ll. Davies (date ripped off), Biological Warfare – Rinderpest Virus, RG24, Series F-1, Vol. 4224, LAC; R. Daubney (June 29, 1950), Biological Warfare – Rinderpest Virus, RG24, Series F-1, Vol. 4224, LAC.

continued, we need to be making "structural preparations permitting the building up of a large stock pile of vaccines. I am sure many of us would sleep better at night if we knew this were available and likely the same is true of our American friends." On December 4, he wrote to Ormond Solandt, Chairman of the Defense Research Board, that his hopes for the lapinized strains had come to fruition. "We can now say definitely that we have at hand ways and means of defending ourselves against what might be an exceedingly grave attack," he cheerfully reported. You may inform our anxious allies in London and Washington, he continued, that, as soon as we have the "appropriate large-scale drying apparatus," we shall be in production.[67]

Mitchell and his team had solved the DRB's central concern: finding a vaccine that could be mass-produced so that Canada and its Tripartite allies could have immediate access to a stockpile in case of a rinderpest sabotage attack by the Soviets. This was excellent news for all concerned, but it did not end Grosse Île's perceived usefulness; in fact, it expanded it. Now, inspired by the outbreak of war in Korea and growing concern about what the Soviets were planning, the Tripartite allies could turn their attention in new directions. That concern led Secretary of Defense George C. Marshall to issue a directive prioritizing biological weapons research that fall. The Joint Chiefs of Staff agreed.[68]

In January 1951, the Central Intelligence Agency released the top-secret *National Intelligence Estimate 18: The Probability of Soviet Employment of BW and CW in the Event of Attacks upon the US. NIE-18,* which opened insisting, "It is highly probable that the Soviets are carrying on an extensive program to develop BW agents and equipment, and they appear to have given some attention to the possible use of BW agents for sabotage activities." Such attacks "may be employed … any time, even well in advance of D-day." They would try to weaken us first, *NIE-18*'s authors predicted, and they would potentially hit human, animal, and plant targets. "BW attack on animals would be directed primarily at the food

[67] Chas. A. Mitchell to L. W. Billingsley (July 17, 1950), Biological Warfare – Rinderpest Virus, RG24, Series F-1, Vol. 4224, LAC; Chas. A. Mitchell to O. M. Solandt (December 4, 1950), Biological Warfare – Rinderpest Virus, RG24, Series F-1, Vol. 4224, LAC.

[68] G. Marshall to Secretary of the Army, Secretary of Navy, and additional recipients, October 27, 1950, NARA RG 218, Central Decimal File, 1948–1950, Box 207, Folder "Chemical Biological, and Radiological Warfare," quoted in Hamblin, Jacob Darwin. *Arming Mother Nature: The Birth of Catastrophic Environmentalism.* New York: Oxford, 2013, 41–42.

supply. Cattle could be attacked in shipping centers, stockyards, and other concentration areas." Perhaps the report's authors remembered the German attacks on horses in shipyards during World War I. The Soviets have options, they wrote, "US livestock is notoriously vulnerable to foot and mouth disease and rinderpest."[69] The anticipated vaccine stockpile would make it significantly less vulnerable to rinderpest, but it was not yet ready. In the meantime, the Chemical Corps decided that it needed to pursue the study of rinderpest as a weapon, both to better determine how the Soviets might use it *and* to be ready to use it themselves.

The head of the Chemical Corps sent a letter to the DRB in March of 1951, asking if some of its researchers could use Grosse Île "to conduct field trials" of some "anti-animal agents." The DRB agreed, with the stipulation that Mitchell would have authority over the trials.[70] Chemical Corps interest did not stop there. At the end of April, its director sent a letter to the DRB containing "an outline of the investigations which we consider it desirable to institute on the virus of rinderpest." The Americans were, after all, helping foot the bill for Grosse Île; they wanted to put it to use. "You will note that this outline includes studies of both defensive and offensive aspects," the director added, "and that it is rather comprehensive." The outline was over three pages long, most of it dedicated to a list of the offensive research that the Americans wanted conducted. The line between offensive and defensive research was necessarily blurry, for both required expanded knowledge about yields and sustainability. Only the former, however, called for "development towards final form of end item and methods for dissemination:" (1) "Comparison of dry and wet preparations in aerosol experiments" and (2) "Comparison of dissemination units such as feed pellets, feathers, etc. for introducing the agent to farm animals."[71] The Chemical Corps wanted to figure out how to turn rinderpest into a weapon. An updated "list of desired studies" in June (which was regrettably ripped in half at some point, purposefully or no, so it is only partially readable), following a DRB and Chemical Corps meeting, focused on finding "optimal route of infection," along with dosage requirements and "optimal" source, harvesting,

[69] CIA, *NIE-18: The Probability of Soviet Employment of BW and CW in the Event of Attacks Upon the US* (January 10, 1951).

[70] A. C. McAuliffe to O. M. Solandt (March 9, 1951), Biological Warfare – Rinderpest Virus, RG24, Series F-1, Vol. 4224, LAC; E. Ll. Davies to A.C. McAuliffe (March 24, 1951), Biological Warfare – Rinderpest Virus, RG24, Series F-1, Vol. 4224, LAC.

[71] Oram C. Woolpert to E. Ll. Davies (received May 3, 1951), Biological Warfare – Rinderpest Virus, RG24, Series F-1, Vol. 4224, LAC.

and storage information. The DRB responded that those studies were already in progress and "have top priority."[72]

The Chemical Corps list also included a statement of "lower priority" interest "to discover whether certain of these strains can break [missing word] immunity conferred by lapinized or other vaccines, both f[missing] offensive and defensive purposes." The Americans were interested in rinderpest's mutability. The DRB responded that that research was "dependent on your activity, having regard to the agreement that the strains will be collected through American agencies." This referred to an American plan, discussed again at conference between the DRB and several British representatives in early December, "to dispatch an aeroplane round the world, fitted with appropriate instruments, to collect as many different strains of Rinderpest virus as possible." The Americans, a participant pointed out, have "been becoming gradually more interested in [rinderpest]." A Colonial Office representative added that he had recently "met experts from the United States who were flying to Africa with the intention of setting up a laboratory there utilizing and enlarging on British work." The United States was expanding its biological warfare horizons.[73]

By the end of 1951, the Tripartite allies shared a new stockpile of rinderpest vaccines and a growing interest in the virus's offensive capabilities. This was far from the extent of their collaboration on BW research, but it was, in several ways, representative of that research on the whole: collaborative and, in the larger perspective of postwar weapons research, minimal. Overall, US, UK, and Canadian biological weapons research remained limited in the final years of the 1940s and the early 1950s. A 1958 Army report described the period from 1947 to 1952 as "an era of boards, committees, Ad hoc groups, panels, contractors, etc. investigating, evaluating, and advising on various phases of the BW program." The findings tended to be the same: that the United States needed to put more resources into biological weapons research to meet the probable

---

[72] Oram C. Woolpert to E. Ll. Davies (received June 5, 1951), Biological Warfare – Rinderpest Virus, RG24, Series F-1, Vol. 4224, LAC; E. Ll. Davies to Col. Creasy (June 27, 1951), Biological Warfare – Rinderpest Virus, RG24, Series F-1, Vol. 4224, LAC.

[73] Oram C. Woolpert to E. Ll. Davies (received June 5, 1951), Biological Warfare – Rinderpest Virus; RG24, Series F-1, Vol. 4224, LAC; E. Ll. Davies to Col. Creasy (June 27, 1951), Biological Warfare – Rinderpest Virus, RG24, Series F-1, Vol. 4224, LAC; Canadian and Colonial Office Rinderpest Research Programme, Notes on Meeting (December 4, 1951), Biological Warfare – Rinderpest Virus, RG24, Series F-1, Vol. 4224, LAC.

Soviet threat in this area.[74] But biological weapons remained overshadowed by their nuclear counterparts. Funding for the Chemical Corps "rose steadily but not dramatically" in the late 1940s.[75] Britain was more committed to biological weapons research, primarily because it did not yet have nuclear weapons.[76] Canada, as the historian Donald Avery has argued, "despite its unwillingness to become directly involved with weapons of mass destruction," assisted "the offensive BW programs of its British and American allies."[77] The Tripartite countries were in biological weapons research together, but, by the end of 1952, as the rinderpest example attests, no one was in it more deeply than the Americans.

The British noted the shift with alarm, reporting in a March 1952 Chiefs of Staff appraisal of biological weapons that the US military had become more interested in BW and, in consequence, there was now an "intense" push to concentrate on studying their offensive capabilities.[78] The United States government directed increasingly more money toward BW research. The Chemical Corps listed 271 projects in April of 1952 for funding in FY 1953.[79] That expanded pursuit of BW made the United States both stronger and weaker in an international environment where national security had territorial and ideological dimensions. Danger came in many forms. In 1952, an outbreak of foot and mouth disease in Canada highlighted the danger of bacterial sabotage on North America's livestock. That same year, accusations by the Communist world that the United States had used biological weapons against North Korea highlighted the danger of public relations sabotage on a global scale. "Security" was a complicated goal: strength in one area revealed vulnerabilities in another. This was true of biological weapons research in general and of research into rinderpest in particular.

Sharing a vaccine with the world in the name of humanity left a nation less able to use the pathogen it protected against as a biological weapon against its enemies. But what if that pathogen could be mutated to get around those vaccines, to break down immunity? If that were possible,

---

[74] Wright, Susan. "Evolution of Biological Warfare Policy: 1945–1990," in *Preventing a Biological Arms Race*. ed. Susan Wright. Cambridge, MA: MIT Press, 1990, 29.

[75] Moon, "The US Biological Weapons Program," in *Deadly Cultures*. ed. Wheelis et al., 9, 21.

[76] Balmer, "The UK Biological Weapons Program," in *Deadly Cultures*, ed. Wheelis et al., 47–61.

[77] Avery, *Pathogens for War*, 57.   [78] Balmer, *Britain and Biological Warfare*, 134.

[79] Moon, "The US Biological Weapons Program," in *Deadly Cultures*. ed. Wheelis et al., 21.

then a country could bolster its reputation by funding an international war on a disease at the same time as it bolstered its security by funding secret research on how to use that disease as a weapon. That was exactly what the United States decided to do. Meanwhile, the Soviet Union decided to fight it on two fronts: security *and* reputation. Injuries could take a variety of forms in the Cold War.

***

The history of the Soviet pursuit of biological weapons is still, as scholars of the subject readily admit, "shrouded in mystery." Its offensive program "appears to have officially commenced in 1928," following a report by Yakov Fishman, the director of the Military Chemical Agency, which had been founded three years prior. Fishman, after some experiments with anthrax and botulin, concluded, "the bacterial option could be success-fully used in war."[80] When World War II began, only Japan and the Soviet Union possessed "significant offensive BW programs," but the Soviet's had been damaged by the purges of the late 1930s. Its BW researchers spent the war focused on developing vaccines and penicillin to protect troops. The program revitalized, however, in the late 1940s. Biological weapons scholars Milton Leitenberg and Raymond Zilinskas provided three reasons for this move in their 2012 book, *The Soviet Biological Weapons Program*. First, they argued, the Soviets learned the extent of Japan's program, "which likely gave Soviet leaders an indication of how powerful biological weapons could be," along with the data they gained from captured records and scientists. Second, Soviet officials believed that Western countries were developing biological weapons. "In particular, they drew such conclusions from a 1947 article by Theodor Rosebury and Elvin A. Kabat" – the same article, "Biological Warfare," discussed in the opening of this chapter. Rosebury followed that article up with a 1949 book, *Peace or Pestilence: Biological Warfare and How to Avoid It*. Leitenberg and Zilinskas reported that "some of those interviewed for this book believe that the Rosebury article and book were the main determinant of the Soviet government's decision to bolster its BW pro-gram." Third, if they had any doubts about Western intentions, those

---

[80] Bojtzov, Valentin. and Erhard Geissler. "Military Biology in the USSR, 1920–45," in *Biological and Toxin Weapons*. ed. Geissler and Moon, 153–157; Leitenberg, Milton and Raymond A. Zilinskas with Jens H. Kuhn. *The Soviet Biological Weapons Program*. Cambridge, MA: Harvard University Press, 2012, 21. See also, Hart, John. "The Soviet Biological Weapons Program," in *Deadly Cultures*, ed. Wheelis et al., 132–156.

were quashed when they learned about the Tripartite agreement. They knew that their former allies were working together again, and leaving them out of the loop, just as they had during the war.[81]

The Soviets had reason to be suspicious, but they were also happy to be suspicious. Rumors of biological weapons research in the United States, Canada, and the UK supported their vision of their former allies as corrupt capitalists whose existence threatened the security of the USSR. It also gave them an additional weapon to wield against the United States, in particular, which still had not signed the 1925 Geneva Protocol, which prohibited "the use of bacteriological methods of warfare."[82] Biological warfare was not a comfortable subject. Many people had moral qualms about it. At the same time, the intelligence that suggested that the United States and its allies were pursuing offensive biological weapons research encouraged the Soviet Union to do so as well; it upped the ante, stoking the fires at the heart of the Cold War, driving both nations deeper into the intractable conflict. While the Soviet Union secretly expanded its BW arsenal, it vocally attacked the Western powers for their similar efforts. It waged an international public relations take-down campaign.

The BW component of that campaign began in May of 1947, at a meeting between Stalin, Viacheslav Molotov, Andrei Zhdanov, and three secretaries of the Union of Soviet Writers. Stalin informed the writers that he wanted them to write about the "theme of Soviet patriotism," particularly regarding the need for expanded patriotism among the "scientific intelligentsia." Stalin ordered one of the secretaries to read aloud a document that Zhdanov had written a few days earlier which accused two Soviet scientists and the former minister of public health and the former academician-secretary of the medical academy of allying themselves "with the Americans" by sharing biological secrets. The document referred to the trial, about to get underway, of two Soviet scientists, Nina Kliueva and Grigorii Roskin, who had discovered a way to use the South American parasite, *Trypanosoma Cruzi* (which causes Chagas disease) to "dissolve" tumors in mice. KR, as they called it, had the possibility to cure cancer. News of the discovery went public in March of 1946 and was quickly picked up by the Associated Press. A cure for cancer was global news. The new US ambassador, General Walter Bedell Smith, had taken note, in large part after being besieged by letters from

[81] Leitenberg and Zilinskas, *The Soviet Biological Weapons Program*, 38, 43–45.
[82] 1925 Geneva Protocol, available at http://www.un.org/disarmament/WMD/Bio/1925 GenevaProtocol.shtml.

American cancer patients and their families, and had worked to get American access to the research. Smith's efforts were, in the end, successful. The minister of public health told the academician-secretary of the medical academy to share the research and the "KR vaccine" with the Americans, and that is why they were both now "former."[83]

Hoping to use this international interest in their work to their advantage, Kliueva and Roskin wrote to Zhdanov, asking for additional resources and warning that the Americans would pursue this work and "we, Soviet specialists, being now in the leading position, undoubtedly will be left far behind." It would have been wiser to keep silent. Zhdanov seized on the issue as an opportunity to enhance his power. Although he knew that Molotov had given ultimate approval to the transfer of the research to the Americans, he composed a narrative that shifted the blame onto the scientists themselves and the officials immediately involved in the transfer. Stalin decided the narrative was "a theme for a literary work." He instructed one of the secretaries at that May meeting, Konstantin Simonov, a famous poet and novelist, to write one. Simonov suggested a play; Stalin responded, "This theme must be worked on." Simonov, of course, obeyed, turning to a leading Soviet microbiologist for assistance with the specifics.[84]

Nikolai Krementsov narrated the consequential unfolding of events in his 2002 book, *The Cure*. In Simonov's play, called *Alien Shadows* (*Chuzhaia ten'*), the protagonist is a scientist who "has discovered a way to decrease or increase the 'infectious strength' of various microbes, thus paving the way for creating effective vaccines against any infectious disease." The scientist shares his findings with colleagues in America because, as he sees it, "from the point of view of true humanitarianism, it is ridiculous to speak of secrets in producing vaccines against disease." It soon becomes evident, however, that the Americans are not interested in producing vaccines, but in using those same techniques to produce biological weapons. After rounds of edits and approvals, which went all the way up to Stalin himself, the play opened on December 15, 1949. It did not go unnoticed abroad.[85]

A June 1952 report, titled "Communist Bacteriological Warfare Propaganda," explained that "Moscow began its first serious effort to associate the US with preparations for bacteriological warfare in 1949"

---

[83] Krementsov, Nikolai. *The Cure*. Chicago: University of Chicago Press, 2002, 75–91, 112–113, 136.
[84] Ibid., 94–96, 104–105, 136–137.   [85] Ibid., 136–143.

with the premiere of *Alien Shadows*. It was, the report explained, the first of the "early charges" against the United States.[86] In the play, the USSR accused the United States of doing exactly what it was doing with rinderpest at Grosse Île: vaccine production research had given way to biological weapons research. Fear about that knowledge reaching Moscow is why the DRB tried to keep FAO in the dark about its renewed activities on the island in 1950. The kinds of manipulations of the virus that Grosse Île researchers were attempting, however, were exactly of the kind that Rosebury and Kabat had described as possible in their 1942 (published 1947) article and an expansion on the work that Shope and his team had published in 1946. It did not take a great deal of imagination on the part of the Soviets to create accusations that had a good chance of sticking. The technology was already "out there," which is what made the accusations such an easy weapon.

Alien Shadows was domestic propaganda with a decidedly limited audience, but Moscow quickly expanded its scope of attack and, it seems likely, its own BW program, as evidenced by the December 1949 trial of the Japanese army officers charged with attempting to "unleash germ warfare against the Allies during the war." The trial, the 1952 report explained, "was described as a 'blow at the American imperialists who dream of using lethal bacteria for their mad plans.'" The Soviets pointed to the continued presence of the emperor as a clear example of American "collaboration in Japanese bacteriological warfare plans." In March, Radio Moscow broadcast in Japanese that the US protection of the Japanese researchers that it had caught clearly demonstrated the United States "is actively preparing for a war utilizing" bacteriological weapons. Moscow published the transcript of the trial in English in May. In June, "the East German Government accused the US of dropping potato bugs on its fields from the air." It had been the victim, it insisted, of an "experiment in biological warfare." Moscow's propaganda machine even looked to the past for options for attack. In July, "*Pravda* and the *Literary Gazette* ... accused the US of shipping infected seed to the USSR in 1943 under the guise of UNRRA aid."[87] This was small-scale stuff, however. The accusations would need to be louder and more aggressive to have a global impact. Moscow needed something bigger. It found it in Korea.

---

[86] Communist Bacteriological Warfare Propaganda (June 16, 1952), 40. Communist Charges Bacteriological and Chemical Warfare, Box 10, State: Topical File Subseries, Papers of Harry S. Truman: SMOF: Selected Records Relating to the Korean War, HSTL.
[87] Ibid.

In the spring of 1951, China and North Korean accused the United States of deliberately spreading diseases and using poison gas. *Pravda* picked up the story that General MacArthur was "engaged in the large-scale production of bacteriological weapons for use against the Korean army and people." The North Korean minister of foreign affairs sent a cablegram (in Russian) to the UN Security Council protesting the "monstrous crime which is being committed by the American interventionists." They had, he alleged, worked with "the staff of Syngman Rhee's army" to "carry on secret biological warfare against the North." Although more aggressive than the previous accusations, this too was a "small-scale" attack, the 1952 report explained, "not intended for world-wide treatment." It was preparatory. "In contrast ... the 1952 campaign was projected for international implementation on the most extensive scale."[88] Moscow was ready to fight. It brought allies with it.

At first, it seemed that 1952 would not be significantly different from 1951. "On February 21 there was a brief Pei-p'ing Radio report that the US had been dropping bacteria on north Korea," which was not dissimilar from anything that had been said before. This time, however, the "sharply rising volume of Soviet and Chinese radio comment" indicated that the intensity of the attack had changed. Moscow seized upon the charges with evident zeal. The alleged US atrocities – now against both Korea and China – quickly became "the focal point for mass Communist agitation – protest meetings, posters, et cetera – both behind the Iron Curtain and throughout the non-Communist world." Newspapers carried photographs of an alleged "burst germ bomb" and "eyewitness reports."[89] The controversy spread around the international state system, cropping up, at times, in unexpected places.

At an OIE regional conference in Pakistan in May of 1952, when the delegates were supposed to be reporting on conditions in their home countries, "the Chinese Delegate could give no information regarding the disease situation in the People's Republic but stated that the U.S.A.

---

[88] Ibid.; Pak Hen En to the President of the Security Council (May 18, 1951), 40. Communist Charges Bacteriological and Chemical Warfare, Box 10, State: Topical File Subseries, Papers of Harry S. Truman: SMOF: Selected Records Relating to the Korean War, HSTL.

[89] Communist Bacteriological Warfare Propaganda (June 16, 1952), 40. Communist Charges Bacteriological and Chemical Warfare, Box 10, State: Topical File Subseries, Papers of Harry S. Truman: SMOF: Selected Records Relating to the Korean War, HSTL; Intelligence Report: Survey of Communist Bacteriological Warfare Campaign (March 27, 1952), 40. Communist Charges Bacteriological and Chemical Warfare, Box 10, State: Topical File Subseries, Papers of Harry S. Truman: SMOF: Selected Records Relating to the Korean War, HSTL.

was engaged in Bacterial warfare which affected both men and beasts."
He asked the conference to "take note of these artificial epizootics in
North China and Korea." The delegate from the USSR then "sprang to
his feet in support." He "insisted that the question was not a political one
and deserved scientific treatment. If the Germ-War, he continued, is
allowed to go on in Asia, all the good resolutions on eradication of
diseases from this part of the world were of no practical value whatever."
Likely flustered by the outburst, the conference chairman insisted that the
OIE was not the forum for the conversation. He "advised" that it be
"taken up by China and the U.S.S.R. at the Security Council" and tried
to quickly steer the meeting back to the subject at hand.[90]

The Chinese and Soviet delegates no doubt viewed the conference as
just another international space in which they could wage the attack on
the United States, but it is interesting that the Soviet delegate particularly
stressed the point about the "resolutions on eradication" not being of
value, because the most important of those resolutions dealt with rinderpest. The Chinese had not accused the United States of using that particular virus as a biological weapon, yet the Soviet Union was clearly
suspicious that the United States was working on being able to do so.
Perhaps Moscow had intelligence of that nature. Perhaps it was just an
accusation that the Soviets thought could cause particular trouble at an
OIE meeting, where they knew a representative of the United States would
not be in attendance. Cold War attacks sometimes took ambiguous forms.

For the US government, a March State Department intelligence report
explained, "The paramount question has been whether the campaign
represents an effort to justify new political or military action against the
US or whether it has a more limited purpose of simply providing a new
theme with which to stimulate hatred and distrust of the US."[91] Both
options were troubling. The Psychological Strategy Board, which had
been created the year before, considered the allegations a psychological
attack. "Propaganda of this type is itself a horror-weapon," they argued,
and a clever one. "The BW propaganda campaign has already provided
the Soviet Union with a means of harnessing the forces of nature to their
propaganda advantage." Any outbreaks of disease or sudden plagues of

---

[90] 1ère Conférence de la Commission d'Asia de l'OIE – Karachi, Pakistan (May 3–9, 1952);
available at http://www.oie.int/doc/en_document.php?numrec=3176803.

[91] Intelligence Report: Survey of Communist Bacteriological Warfare Campaign (March
27, 1952), 40. Communist Charges Bacteriological and Chemical Warfare, Box 10, State:
Topical File Subseries, Papers of Harry S. Truman: SMOF: Selected Records Relating to
the Korean War, HSTL.

insects can easily be labeled American attacks, putting the US on the defensive, forcing it to prove that it did not cause them. "Troubled backward areas," the Board continued, "may be constantly invited to blame diseases and plagues on U.S. imperialist machinations. Even American efforts to help the peoples of these areas may be used against the U.S." American technicians abroad could easily be accused of spreading the diseases they claimed to be fighting. The Soviets are trying to "link the United States, and particularly the U.S. armed forces, in the public mind with the perversion of science," the Board insisted, and the United States needed to fight back. Notably, it did not say "with propaganda." The United States ostensibly only had to fight with information. It is "an unusually good occasion to seek from friendly nations and from international organizations more adequate recognition that the U.S. has yet been given for its disinterested efforts to utilize our technological resources for the relief of human want and suffering throughout the world," the Board noted. American-funded disease and pest campaigns spoke for themselves, provided they were given enough press.[92]

The US government was furious that the Soviets and the Chinese were "harnessing the forces of nature to their propaganda advantage," but it was also concerned that it hinted that the Soviets were preparing to use those forces to a different advantage as well; perhaps, the Americans worried, Moscow was attempting to establish a US "first use," so that they would have a reason to retaliate.[93] Dean Acheson asked the Chinese and North Korea to allow the International Red Cross and/or the World Health Organization to investigate, but they refused, turning instead to sympathetic Western scientists, who formed an International Scientific Commission (ISC). The ISC found the Americans guilty, citing not only domestic "evidence" of the alleged crimes (including the "confessions" of captured US airmen who retracted their statements as soon as they were back on American soil), but also the documents that the United States government had published about its BW research: Merck's report and Rosebury and Kabat's article, for example. As a whole, it was less than persuasive. As Jacob Darwin Hamblin recently argued, "Examining the evidence today, it is difficult to understand how it could have appeared as

---

[92] Psychological Strategy Board, Staff Study – Preliminary Analysis of the Communist BW Propaganda Campaign, with Recommendations (August 7, 1952), File #1 – PSB Staff, 729.2, Papers of Harry S. Truman, SMOF: Psychological Strategy Board Files, Box 37, HSTL.

[93] Hamblin, *Arming Mother Nature*, 54–55.

anything but staged."[94] The ISC investigation did not prove that the United States had used biological weapons, but it suggested that it could. In fact, the Soviet and Chinese campaign of 1952 made such action more possible, as the United States responded to the attack with a ramping up of BW research activity, primarily at Camp Detrick, but also at Grosse Île.[95]

In early April 1952, at a meeting of the BW Research Panel at Grosse Île, Mitchell gave a progress report on 'R' work. He started with vaccine production, noting that they were currently working on methods of drying the lapinized vaccine so that it could be transported without dry ice or refrigeration. They were also studying altering passages between rabbits and swine and rabbits and eggs, Mitchell reported. Next, he noted, "extensive work has been done on the investigation of a method of producing 'R' in cattle." Mitchell detailed the findings:

(a) Subcutaneous injections of the Kabete strain of concentrations 10 to the -7 or greater, produce infection.
(b) Much larger amounts than this swabbed on the mucous membranes, added to the food, dropped into the eye, etc., will not produce infection.
(c) Infection is produced by animal contact in about 50% of the cases but if two calves are separated by as little as 18" there is no take. It is assumed from this that aerosol infection by the nasal route is not effective.
(d) Some samples of a Turkish strain of "R" were obtained which work better, infections being achieved by the nasal route using ground spleen. The use of infected feathers, gelatin capsules, molasses mixtures, etc., was not successful however.[96]

Future research, he continued, "in examination of the part played by insect vectors in spreading 'R,'" will begin "shortly."[97] Using rinderpest as a biological weapon required learning how to wield it. Researchers at Grosse Île, as evidenced in Mitchell's report, tested a number of options: from "infected feathers" to "subcutaneous injections." The virus proved

[94] Ibid., 51–55. See also, Furmanksi, Martin and Mark Wheelis. "Allegations of Biological Weapons Use," in *Deadly Cultures*, ed. Wheelis et al., 253–261.
[95] Moon, "The US Biological Weapons Program," in *Deadly Cultures*, ed. Wheelis et al., 20–24.
[96] Memo to File: Progress Report on "R" Work at GIES (April 10, 1952), Biological Warfare – Rinderpest Virus, RG24, Series F-1, Vol. 4224, LAC.
[97] Ibid.

more difficult to spread than an analysis of its history would have predicted. This was good news from the perspective of having to fight *against*
an attack of rinderpest, but bad news from the perspective of trying to use
it. Research continued.

In the meantime, the United States had begun researching the virus at
Britain's new Muguga laboratory in Kenya.[98] It also began biological
weapons research at Fort Terry, a deactivated base on Plum Island in
Long Island Sound. The Chemical Corps reactivated Fort Terry
on April 15, 1952, "to establish and pursue a program of research and
development of certain anti-animal (BW) agents."[99] The Chemical Corps
subsequently deposited there the strains of rinderpest that it collected from
around the world through an effort known as "Project 1001." By the time it
turned Fort Terry over to the US Department of Agriculture in the summer
of 1954, there would be forty-one strains of rinderpest on Plum Island.
It also sent strains to Grosse Île – twenty-seven cubic feet worth of frozen
samples in the spring of 1954 – to expand the work being conducted
there.[100]

Researchers remained busy on Grosse Île, eager to tease out more of
'R's' secrets. They produced reports, two of which were of particular
interest: DRB Report No. 74: "Rinderpest Virus Adapted to the
Chorioallantoic Membrane of the Chick Embryo – Its Attenuation and
Use as a Vaccine" and DRB Report No. 84: "Application of the
Complement-Fixation Test to the Demonstration of Rinderpest Virus in

---

[98] Chas. A. Mitchell to E. Ll. Davies (December 31, 1953); Biological Warfare – Rinderpest
      Virus, RG24, Series F-1, Vol. 4224, LAC. British officials built the laboratory at Muguga
      because Kabete had become "overcrowded and congested" in the aftermath of the 1948
      creation of the East African Veterinary Research Organization (E.A.V.R.O.) and a new
      Colonial Office commitment to take an "African-wide view ... in rinderpest research
      and production." For more on this see, East African Veterinary Research Organization,
      CO 927/186/5, NAUK; Ad Hoc Veterinary Panel, Review of the Present Position of the
      East African Veterinary Research Organization Vis-à-vis the Kenya Veterinary
      Department, CO 927/93/1, NAUK; P.E. Mitchell to A. Creech Jones (April 26, 1949),
      CO 927/93/1, NAUK; East Africa High Commission (October 11, 1949), CO 927/186/
      6, NAUK; Shipp to J. G. Hibbert (November 23, 1950), CO 927/186/3, NAUK; Notes
      from a meeting of the East Africa Advisory Council on Agriculture, Relation between
      E.A.A.F.R.O. and E.A.V.R.O. and Territories adjoining East Africa (January 20, 1951),
      CO 927/186/4, NAUK. See also Binns, H. R. *The East African Veterinary Research
      Organization: Its Development, Objectives and Scientific Activities*. Nairobi, Kenya:
      East African Standard Ltd. (1957).
[99] Millett, Piers. "Antianimal Biological Weapons Programs," in *Deadly Cultures*, ed.
      Wheelis et al., 225.
[100] Ibid., 226; H. E. Staples to Canadian Defense Scientific Service (March 8, 1954),
      Biological Warfare – Rinderpest Virus, RG24, Series F-1, Vol. 4224, LAC.

the Tissue of Infected Cattle Using Rabbit Antiserum."[101] The titles seem innocuous, but the reports caused a flurry of conversation about who should be able to see them. The question went to the heart of the struggle between research for the sake of science and research for the sake of national security.

Mitchell forwarded the reports to the DRB for guidance. Writing about Report No. 84, in February of 1955, he explained that "because of the extraordinary fine piece of work which has been carried out, one is tempted to publish the material in the open literature. I think, however, having regard to the surrounding circumstances at present, it would be advisable to maintain this method in our hands because of the obvious advantage which it gives in quickly identifying the nature of an out-break." If we release it too widely, he continued, "the objective at hand may be defeated."[102] The paper described a way to test for the presence of rinderpest in cattle tissue using rabbit serum instead of having to test it in live cattle. It saved money and time and would, without question, be of use to veterinarians throughout Asia and Africa who would then be able to diagnose rinderpest more quickly. It would also be of use to any nation that the US tried to attack with rinderpest in the future, which is what concerned Mitchell. His internationalism warred with his national security concerns. Sharing Report No. 84 would help everyone defensively, but it could potentially limit the damage of a Tripartite attack with rinderpest on the Soviets.

The DRB took Mitchell's concerns seriously, but it, too, struggled to weigh the obvious benefits of releasing the report with the potential costs. Mitchell was told that both Report 74 and Report 84 were "classified SECRET," but was informed that the "Director of Scientific Intelligence has asked whether the information in these reports could be released to Commonwealth and NATO countries because of the value of the information." The director asked Mitchell for advice as to "the real value of these reports" and whether or they "could be downgraded from Secret to Confidential" so that they could be shared beyond Australia and New Zealand, which were "the only Commonwealth countries to which secret information could be

---

[101] This was the second in a series, it had been preceded by "The Use of the Complement-Fixation Test for the Demonstration of Rinderpest Virus in Rabbit Tissue Using Rabbit Antisera," author Paul Boulanger, which can be found in *Canadian Journal of Comparative Medicine* 21:11 (November 1957): 363–369.

[102] Chas. A. Mitchell to G. R. Vavasour (February 7, 1955), Biological Warfare – Rinderpest Virus, RG24, Series F-1, Vol. 4224, LAC.

released."[103] Mitchell responded immediately in a "confidential" letter that revealed a great deal about the offensive research going on under his leadership.

I do not believe that there would be any advantage in distributing the present SECRET reports to NATO countries over their publication in the open literature. I think it is fair to assume that before long the information would be distributed beyond the Iron Curtain. As you know, the objective of our work has to some extent altered within these last few months and more attention is being paid to the aggressive aspects of BW in relationship to animal disease. The evidence indicates that we have two satisfactory agents available and that considerable progress has been made in developing the rinderpest virus into a suitable agent for this purpose. Since we have in advance developed a highly protective method it would seem rather ridiculous to distribute information concerning this on a wide scale and conceivably carry on the aggressive aspects of our work. Obviously, if the agent were ever employed it would be more efficient if well developed defense measures were not available, that we should keep these in our own hands would seem to me advisable.[104]

Mitchell's national security concerns outweighed his humanitarian ones. His team had created two "satisfactory" anti-animal biological agents and they were close to making rinderpest a "suitable" biological weapon. It was too risky to present the enemy with a better protective shield against it.

The DRB passed Mitchell's letter on to A. F. B. Stannard, the Director of Scientific Intelligence, who found it not "entirely satisfying." He had a number of questions that evidenced both security and international public relations concerns for the DRB.

Would it be possible to supply vaccine to NATO & Commonwealth countries who do have a ~~serious~~ Rinderpest problem without prejudicing the method of manufacture?" "Is the fact that we have complete protection really sensitive information? It would ~~prevent~~ discourage a potential enemy from considering R as a candidate agent but would not necessarily divulge our intentions." "Is the knowledge that protection can be obtained of any real help to the Russians in developing a similar material? I suppose it would give them a more definite intelligence target.[105]

---

[103] G. R. Vavasour to C. A. Mitchell (April 5, 1955), Biological Warfare – Rinderpest Virus, RG24, Series F-1, Vol. 4224, LAC; A. F. B. Stannard to G. R. Vavasour, Biological Warfare – Rinderpest Virus, RG24, Series F-1, Vol. 4224, LAC.

[104] Chas. A. Mitchell to G. R. Vavasour (April 7, 1955), Biological Warfare – Rinderpest Virus, RG24, Series F-1, Vol. 4224, LAC.

[105] A. F. B. Stannard to G. R. Vavasour (April 14, 1955), Biological Warfare – Rinderpest Virus, RG24, Series F-1, Vol. 4224, LAC.

Stannard closed the letter with an acknowledgment that it was "somewhat rambling." He was thinking aloud on paper. Discretion won out over humanitarianism. The reports would not go abroad, even to NATO and Commonwealth countries, for "we must not encourage their interest and then refuse the main information." And the main information could not be shared; it would reveal too much about what really was going on on Grosse Île.[106]

The conversation, however, did not stop there. Indeed, Stannard had insisted that it should not, because there were clearly consequences to the decision not to share the reports with others. Handwritten notes from later in April and in May reveal an internal debate that weighed the importance of sharing valuable scientific knowledge against the danger of compromising the offensive aspects of the Grosse Île research activities. One note to Stannard insisted, "this has always been primarily defensive work and on that basis I would support that we might well consider offering it to our Allies, etc." Another recommended a "situation analysis" – "What USSR defenses exist, etc." – before making a decision. Another, referencing the 1946 publication of the Grosse Île war research, warned that "release would carry with it a risk of compromising our interest in certain lines of BW and draw unnecessary attention to GIES [Grosse Île Experimental Station]."[107] The full details of those "lines" remain unknown today, but a 1954 DRB report explained that "four viruses had been developed by the Joint US–Canada anti-animal program as efficient agents 'for the destruction of ... food-bearing animals.'" The list included rinderpest, fowl plague, African warthog fever, and African swine fever.[108] The last two are actually the same disease, which might explain why Mitchell listed only two plus rinderpest in his 1955 letter.

The debate highlighted the inherent tension at the heart of the rinderpest research at Grosse Île. The scientists wanted dominion over the virus to be able to manipulate it to their purposes of both using it and destroying it at will. FAO's eradication campaign might have made that effort seem an odd choice, but the mutability of the virus that allowed it to be

---

[106] Ibid.
[107] G. R. Vavasour to A. F. B. Stannard (April 29, 1955), Biological Warfare – Rinderpest Virus, RG24, Series F-1, Vol. 4224, LAC; Burke to Unknown (undated, but note from unknown to A. F. B. Stannard on it dated April 29), Biological Warfare – Rinderpest Virus, RG24, Series F-1, Vol. 4224, LAC; Unknown to A. F. B. Stannard (May 16, 1955), Biological Warfare – Rinderpest Virus, RG24, Series F-1, Vol. 4224, LAC.
[108] Avery, *Pathogens for War*, 74.

attenuated into vaccines also made it a tempting choice for BW. Researchers had learned a great deal about the virus over the past ten years of vaccine research, much of it useful for thinking about how to weaponize it. The very success of the live vaccines drove the interest in creating a rinderpest strain that could get around them. The virus was just so delightfully mutable.

In the end, the DRB decided to keep the reports secret – shared only with the US and the UK, its Tripartite allies – for the time being. When they did finally release the reports, they released them to everyone, publishing them in the *Canadian Journal of Comparative Medicine* in November of 1957, with open acknowledgment of their origins at Grosse Île Experimental Station "under direction of C. A. Mitchell."[109] By that point, Grosse Île was on its way to its "demise," as G. R. Vavasour described it in a November 1958 note.[110] Grosse Île returned to the Department of Agriculture that year. The department accepted "the responsibility of storing and maintaining the biological properties of the mother material for rinderpest vaccine and in addition hold a small store of the vaccine itself." Notably absent was responsibility for continuing BW research. Grosse Île's days as a biological warfare research facility were over; it became a quarantine site once again, but this time for animals, not humans. Canada's Department of Agriculture shipped a large of amount of its viral stock off to the USDA's research facility on Plum Island and to the East African Veterinary Research Organization's Muguga laboratory.[111] In those places, rinderpest research continued.

Muguga soon became the center of ground-breaking vaccine research discussed in the next chapter, while Plum Island shrouded itself in a silence that continues to this day. As Piers Millett has reported, "it is unclear whether the USDA also assumed responsibility for antianimal BW

[109] McKercher, P. D. "Rinderpest Virus Adapted to the Chorioallantoic Membrane of the Chick Embryo – Its Attenuation and Use as a Vaccine." *Canadian Journal of Comparative Medicine* 21:11 (November 1957): 374–378; Boulanger, Paul. "Application of the Complement-Fixation Test to the Demonstration of Rinderpest Virus in the Tissue of Infected Cattle Using Rabbit Antiserum." *Canadian Journal of Comparative Medicine* 21:11 (November 1957): 379–388.

[110] Note handwritten on Defense Research Board, Review of "Suspended" Projects (January 27, 1958), Biological Warfare – Rinderpest Virus, RG24, Series F-1, Vol. 4224, LAC.

[111] G. R. Vavasour to E. W. Henselwood (February 5, 1959), Biological Warfare – Rinderpest Virus, RG24, Series F-1, Vol. 4224, LAC; K. F. Wells to H. M. Barrett (September 5, 1958), Biological Warfare – Rinderpest Virus, RG24, Series F-1, Vol. 4224, LAC.

research" when it assumed control of Plum Island from the Chemical Corps.[112] In its 1956 publication, *The Plum Island Animal Disease Laboratory*, the USDA asserted that "knowledge of specific animal diseases and methods of combating them developed at the Laboratory can become important protections for this country's supplies of food and other animal products." They could also become weapons, as the Grosse Île story clearly showed, but the USDA did not publish that. Unlike Grosse Île, Plum Island has kept most of its secrets secret.[113] But not all of them.

In his 2004 book, *Lab 257*, Michael Christopher Caroll quoted from Charles Mitchell's 1956 Plum Island dedication day remarks:

I often think and almost tremble at what could have taken place had our Teutonic enemies been more alive to this. It is said that some of their scientists pointed out the advantages to be obtained from the artificial sowing of disease agents that attack domestic animals. Fortunately blunders existed in the Teutonic camp as in our own. Consequently, this means of attack was looked upon as a scientific poppy dream. If [as much] time and money were invested in biologic agent dispersal as in one bomber plane, the Free World would have almost certainly gone down to defeat.

Caroll wrote that "one dedication day VIP stirred uncomfortably" at Mitchell's speech. He was Dr. Erich Traub, the director of a new virus lab in Tübingen, West Germany. He was at Plum Island that day at the express invitation of its director. He was also the man whom, in 1944, Kurt Blome had sent to Turkey to collect the rinderpest virus that Heinrich Himmler had obtained for him. "Ironically," Caroll continued, "Traub spent the prewar period of his scientific career on a fellowship at the Rockefeller Institute ... perfecting his skills in viruses and bacteria under the tutelage of American experts," including Richard E. Shope. In 1958, Caroll reported, the USDA offered Traub the position of senior scientist at Plum Island, explaining "he was the most desirable candidate from any source." Shope wrote a letter of support describing Traub as "careful, skillfull, productive, and *very original*"; "one of the world's most outstanding virologists." "During the war," Shope added, "he was in Germany serving in the German Army." Shope likely believed that Traub's wartime efforts were no different than his own. The United States government did a good job covering up the pasts of foreign

---

[112] Millett, "Antianimal Biological Weapons Programs," in *Deadly Cultures*. ed. Wheelis et al., 229.

[113] USDA, *The Plum Island Animal Disease Laboratory*, Miscellaneous Publication No. 730. Washington, DC: United States Government Printing Office, September, 1956, back cover.

scientists that it wished to employ. Traub, nevertheless, declined the job, to the USDA's disappointment. He had been its top pick for its own "peculiar purposes" on Plum Island.[114]

The USDA was not the only organization thinking along those terms in 1958. That year, the Communist Party of the Soviet Union and the Council of Ministers issued a decree calling for "strengthening work in the field of microbiology and virology." The decree reportedly "established six specialized research institutes and branches under the USSR Ministry of Agriculture ... to conduct anti-animal and anti-plant offensive BW programs."[115] At Ministry of Agriculture facilities in Russia and in Kazakhstan, scientists worked on weaponizing rinderpest, African Swine fever, Newcastle disease, and more.[116] Secret work on biological weapons did not, however, preclude public calls for eradicating diseases by the Soviet Union any more than it did by the United States. Both nations supported both ideas at the same time.

At a 1958 meeting of the World Health Assembly, the Soviet deputy minister of health, Dr. Viktor Zhdanov, presented a detailed proposal for a global campaign against smallpox. Former director of the World Health Organization, Brock Chisholm, had advocated for the same in 1953, but his proposal had not found support among the key donor countries – most importantly the United States – which had been far more interested in pursuing malaria control. Two years later, WHO approved a global malaria eradication campaign, a decision, as the historian Randall Packard has explained, that "reflected a combination of hopes and fears regarding the power of pesticides, the threat of pesticide resistance, the future of malaria control, and the role of the World Health Organization in leading international-health efforts across the globe." It had been as much a political as a scientific decision.[117]

[114] Carroll, Michael Christopher. *Lab 257*. New York: William Morrow, 2004, 6–11.
[115] Rimmington, Anthony. "The Soviet Union's Offensive Program: The Implications for Contemporary Arms Control," in *Biological Warfare and Disarmament: New Problems/New Perspectives*. ed. Susan Wright. Lanham, MD: Rowman & Littlefield, 2002: 113–115.
[116] Bozheyeva, Gulbarshyn, Yerlan Kunakbayev, and Dastan Yeleukenov, "Former Soviet Biological Weapons Facilities in Kazakhstan: Past, Present, and Future." Occasional Paper No. 1, Chemical and Biological Weapons Nonproliferation Project, Center for Nonproliferation Studies, Monterey Institute of International Studies (June 1999); Vogel, Kathleen M. "Pathogen Proliferation: Threats from the Former Soviet Bioweapons Complex." *Politics and the Life Sciences* 19:1 (March 2000): 3–16.
[117] Packard, Randall M. *A History of Global Health: Interventions in the Libes of Other Peoples*. Baltimore: The Johns Hopkins University Press, 2016, 144.

The Soviet Union had withdrawn from WHO in 1949 in protest over its policies and the influence of the United States within it and had only rejoined the organization in 1956. Zhdanov's smallpox proposal marked the USSR's attempt to begin asserting leadership in the organization. It was born of domestic concerns: although the Soviets had stopped national transmission of the virus during the 1930s, they were constantly having to deal with its reintroduction into the Central Asian Republics from neighboring countries. A new heat-stable, freeze-dried vaccine appeared to offer the necessary technological machinery for eradication and WHO could offer the necessary bureaucratic machinery. The Soviets pledged to provide large quantities of the vaccine to support the effort. In response to Zhdanov's proposal, the World Health Assembly called on the director general to prepare a report on the feasibility of a global smallpox eradication campaign, which he presented the following year. The subsequent proposal was modest: it called for national vaccination campaigns committed to vaccinating at least 80 percent of the population with WHO providing "technical assistance when asked" and assistance with the "development of vaccine production," all of which would apparently be covered by a budget allocation of only $100,000 a year. The assembly approved the plan, making smallpox eradication an official WHO goal, but one whose success was going to depend a great deal on national governments for both implementation and funding. It was not, in that way, very different from FAO's rinderpest campaign.[118]

The combined interest in the 1950s in both eradicating diseases and weaponizing them were two sides of the same coin: a recognition, as Arnold Toynbee had described it, of mankind's enhanced "ability to make nonhuman nature produce what man requires from her."[119] That power opened the door to a variety of options and humans explored them. Vaccine technologies blended with political ambitions to inspire different kinds of actions: weaponized pathogens and eradication were the boldest of them – the most *transforming* of them. The pursuit of both at the same time by the same national entities speaks to the complicated nature of international relations in the second half of the twentieth century. Within the new international system that developed around the United Nations,

[118] Henderson, D.A. *Smallpox: The Death of a Disease*. New York: Prometheus Books, 2009, 60-62; Packard, *A History of Global Health*, 137–151; Manela, Erez. "A Pox on Your Narrative: Writing Disease Control into Cold War History." *Diplomatic History* 34:2 (April 2010): 299–323.
[119] Toynbee, Arnold. "Not the Age of Atoms but of Welfare for All." *New York Times* (October 21, 1951):168.

national governments worked more closely together in pursuit of global ambitions than ever before in history. At the same time, however, nationalism and a preoccupation with national security continually threatened to pull the international system apart. That that system survived the century speaks, in part, to the power of the achievements that the global community secured by working together. Disease control was one of the most important of them. Pursuit of that goal had played a key role in bringing nations together in the first half of the twentieth century; it strengthened those bonds in the second half.

6

"Freedom from Rinderpest"

In 1964, Gordon R. Scott, formerly the director of the Muguga laboratory, published a lengthy and definitive study of rinderpest in *Advances in
Veterinary Science* that opened with the good news that the "past decade
has witnessed a surge of activity in the study of rinderpest and the
incidence of the disease has fallen spectacularly." The Office
International des Epizooties reported that there have been "around
40,000 outbreaks and just over 1 million deaths" during that time, he
continued, which was an astoundingly low number considering that "as
late as 1949 it killed over 2 million cattle each year." The disease has been
"suppressed" in many places, Scott cheerfully reported, and "eradication
programs are currently underway" in many others. Despite these successes, Scott remained wary that total eradication would happen soon.
Much had been done, but there was still much left to do. Rinderpest's
persistence in a country, he wrote, is related inversely to the veterinary
staff and financing available for eradication programs. The OIE, FAO,
and the Inter-African Bureau for Animal Health are all "woefully under-
staffed," he warned, and unable to offer all the necessary assistance
required. "The gap," he insisted, "can only be filled by aid from wealthier
nations."[1]

The many national rinderpest eradication campaigns that had been
initiated in the postwar years had, by 1964, born quite a bit of fruit. At
a joint OIE-FAO conference on epizootics in Asia that year, representatives from India reported that their country's ten-year-old rinderpest

[1] Scott, "Rinderpest," in *Advances in Veterinary Science*, Vol. 9. ed. Bandly and Jungherr,
195.

campaign had resulted in the vaccination of about 131.9 million bovines out of a population of around 140 million and a decrease in outbreaks from over 8,000 a year in 1959 to 354 in 1963.[2] China had been rinderpest-free since 1955, South Korea since 1947, and Thailand since 1959. Cambodia was in the middle of a national campaign aided by FAO, "which has sent the experts," and the member countries of the Colombo Plan, which sent "capital and materials." Japan alone had sent field veterinarians, tens of thousands of doses of Nakamura's vaccine, motorcycles, and ear tags to mark vaccinated cattle. A similar campaign was also underway in Laos. These campaigns were all separate national efforts, but, as a participant at the 1967 joint OIE-FAO meeting on Asia would insist, ultimate success required that "countries render their effort to the international campaign to eradicate Rinderpest."[3]

"The international campaign" was an admittedly amorphous concept in the 1960s, just as it had been in the 1950s. FAO had led the charge and it continued to send "the experts," but it had never been able to fund the effort. It had always relied primarily on the countries themselves to organize and run their own campaigns, with assistance, as Scott has pointed out, from wealthier nations. The "international campaign" was, in that way, really the combination of dozens of national campaigns, linked by FAO and the OIE's networks of experts and conferences and by outside support from foreign aid agencies. This bureaucratic machinery of development – international organizations and national aid programs alike – created in the late 1940s and early 1950s had been supposed to also become the bureaucratic machinery of eradication. It had not done so yet – not in the case of rinderpest, or malaria, or smallpox – but it had made significant progress and that progress had not gone unnoticed.

In response to a 1961 speech at the UN General Assembly by Prime Minister Jawaharlal Nehru, the UN decided to declare 1965 – the twentieth anniversary of the United Nations – International Cooperation Year. In anticipation, in 1964, US President Lyndon B. Johnson argued for a specific vision of what that cooperation should be for, proposing that humans "dedicate this year to finding new techniques for making man's knowledge serve man's welfare. Let this be the year of science. Let it be a

[2] "Conference Regionale O.I.E.-F.A.O. Sur Les Epizooties En Asie (Nouvelle-Dehli, 24–29 Novembre 1964)." *Bulletin of the Office of International Épizooties* 63:1–2 (1965): 49.
[3] "Conférence Régionale O.I.E.-F.A.O. Sur Les Epizooties En Asie Et En Extrême-Orient (Tokyo, Japan, October 2–9, 1967)." *Bulletin of the Office of International Épizooties* 69:1–2 (1968): 3–24.

turning point in the struggle – not of man against man, but of man against nature." Science, he insisted, gives us the power to fight poverty, disease, and "diminishing natural resources." We need only to put it to work. Johnson did not use the term "eradication" in his speech, but announced that the United States would expand its efforts "to prevent and to control disease in every continent, cooperating with other nations which seek to elevate the well-being of mankind."[4] Those efforts, it was clear, would depend upon the combined utilization of the technological *and* bureaucratic machinery of development: vaccines as well as UN agencies.

It was a welcome promise, even for veterinary officials who knew that rinderpest was not one of the diseases that Johnson was thinking about when he made it. An enhanced commitment to international disease control per se was a step in eradication's direction, helping, as it did, to create new networks of connection between nations, international agencies, non-governmental organizations, scientists, doctors, veterinarians, and more. This is exactly what Gordon Scott called for in his 1964 article, warning that eradication was impossible without it. Even with it, however, eradication was far from assured, for, as the concept's greatest advocate, Frederick Soper, admitted, "perfection is the minimum permissible standard for a successful eradication campaign."[5] It remained to be seen if the machinery of eradication could reach it; there were many political obstacles standing in the way, there were also technical ones. Eradication required overcoming all of them.

***

Efforts to perfect the machinery of rinderpest eradication took multiple forms in the 1960s: some technical and some bureaucratic. The most important technical change was the development of a new vaccine: Tissue Culture Rinderpest Vaccine or TCRV, which marked a vast improvement over the attenuated vaccines that had dominated the campaigns of the 1950s. Its creation, like those of most of the vaccines that went before it, stemmed from a combination of the skillful application of new techniques and luck. The man behind it, Walter Plowright, was a young veterinary pathologist at Muguga who, upon learning that John F. Enders had discovered a way to

---

[4] Lyndon B. Johnson, Commencement Address at Holy Cross College (June 10, 1964), available at http://www.presidency.ucsb.edu/ws/?pid=26305.

[5] Soper quoted in Packard, *A History of Global Health,* 171. For more on Soper, see Stepan, Nancy Leys. *Eradication: Ridding the World of Diseases Forever?* Ithaca, NY: Cornell University Press, 2011.

grow the measles virus (MeV) in human kidney cells, decided to see if he could do the same with the rinderpest virus (RPV) in bovine kidney cells.[6]

There were many advantages to being able to grow viruses in tissue cultures, not the least of which being that scientists no longer had to rely upon living animals or eggs for keeping the virus alive in the laboratory. Plowright and his colleague, R.D. Ferris, began trying to grow RPV in tissue cultures in the hope that they would be able to find a new method of serological testing. They were successful, eventually finding a way to grow Kabete 'O' virus in petri dishes of bovine embryonic kidney tissue cells. After five passages, they could detect the presence of the virus through visible morphological changes in the cells. Plowright and Ferris had done what they set out to do. They had found a way to test for the virus without injecting animals, but that was not all that they had accomplished. Their research had also, they explained in a 1959 article on their findings, "furnished another attenuated strain suitable for the immunization of cattle." Passaging Kabete 'O' on the kidney cells at first increased its pathogenicity for cattle from about 60 percent to 100 percent. The virus initially mutated to become deadlier, reaching its peak around the tenth passage, but thereafter it started to decline. By the twenty-first passage, injected cattle showed no sign of infection, but could resist subsequent injections of large doses of un-passaged Kabete 'O.'[7]

---

[6] In 1947, John F. Enders, who had become interested in virology after meeting Hans Zinsser years earlier, decided to tackle the frustrating problem of getting viruses to grow in tissue cultures. The technique had been first invented in 1907 and scientists had, in the decades since, demonstrated that it was possible to get several viruses to grow in various attempted mediums, but success had been sporadic. Enders and his collaborator, Thomas Weller, created a new technique that ultimately proved favorable to the growth of a variety of viruses and transformed the field of virology, earning the team the Nobel Prize in Medicine in 1954. That same year, Enders adapted measles virus to grow in human kidney cells. He then gradually attenuated that strain "by successive passages in primary cell cultures of human kidney, human amnion, embryonated eggs, and finally in chick embryo tissue cultures." The vaccine that he created in the process, Edmonston B., was released for public use in 1963, but was already well-known among virologists before then. For more on this, see Weller, Thomas H. *Growing Pathogens in Tissue Cultures*. Canton, MI: Science History Publications, 2004, 41–73; Oshinsky, David M. *Polio: An American Story*. New York: Oxford, 2006, 121–130; "Today's Facts about Tomorrow's Vaccine for Measles." *The American Journal of Nursing* 62:6 (June 1962): 68–69; "Vaccination against Measles." *The British Medical Journal* 2:5195 (July 30, 1960): 368; Enders, John F., et al., "Measles Virus: A Summary of Experiments Concerned with Isolation, Properties and Behavior." *American Journal of Public Health* 47:3 (March 1957): 275–282.For more on Plowright, see Roeder, Peter. "Remembering Walter Plowright." (March 2010), available at http://www.fao.org/ag/againfo/programmes/en/empres/plowright_080410.htm.

[7] Plowright, W. and R. D. Ferris, "Cytopathogenicity of Rinderpest Virus in Tissue Culture." *Nature* 179 (February 9, 1957): 316; Plowright, W. and R. D. Ferris. "Studies with Rinderpest Virus in Tissue Culture, I. Growth and Cytopathogenicity." *Journal of*

Plowright and Ferris had created a new vaccine. How they had done so remained unclear. Why had passage in bovine kidney tissues led to such a startling reduction in pathogencity? Attenuation via passaging in goats, rabbits, and eggs came from the dramatic change in environment (i.e., foreign cells). Passaging in particular animals created rinderpest sub-populations that no longer caused serious harm, but created immunity in cattle. But how could passaging in cattle tissue do the same thing? It appears to have been a bit of luck. A 2006 essay explained, "it may be supposed that the process of attenuation described by Plowright repre-sented the overgrowth of either one, or a succession of avirulent variants at the expense of the more virulent sub-populations present in the original virus pool."[8] The variants that won out in the end happened to be the best ones from the researchers' perspective, not so from rinderpest's. Those mutations were costly for the virus. After ninety serial passages the strain officially became TCRV. It was the best vaccine yet. With continued work on it in the early 1960s, Plowright created a vaccine that was safe for all kinds of cattle of any age, that conferred lifetime immunity, and that was cheap and easy to produce. It had only one significant weakness: thermol-ability. It broke down in the heat and therefore required refrigeration. It was not perfect, but it solved some of the key technical problems that had plagued rinderpest campaigns during the 1950s. The international com-munity still needed to solve some of the key structural ones, however.[9]

TCRV, though critically important, was still just a tool that needed to be wielded effectively in order to work. The central question of the 1960s

Comparative Pathology 69 (1959): 152–172; Plowright, W. and R.D. Ferris, "Studies with Rinderpest Virus in Tissue Culture, II. Pathogenicity for Cattle of Culture-Passaged Virus." *Journal of Comparative Pathology* 69 (1959): 173–184.

[8] Taylor, William P., et al. "History of Vaccines and Vaccination," in *Rinderpest and Peste des Petits Ruminants.* ed. Thomas Barrett, Paul-Pierre Pastoret, and William P. Taylor. Amsterdam: Elsevier, 2006, 233. M.D. Baron, et al., argue that "the high attenuation and stability of the current vaccine are due to the accumulation of a number of separate mutations" (Baron, M.D., et al. "The Plowright Vaccine Strain of Rinderpest Virus Has Attenuating Mutations in Most Genes." *Journal of General Virology* 86 [2005]: 1093–1101).

[9] Plowright, W. "The Application of Monolayer Tissue Culture Techniques in Rinderpest Research. II. The Use of Attenuated Culture Virus as a Vaccine for Cattle." *Bulletin de l'OIE.* 57 (1962c): 253–276; Plowright, W. and R. D. Ferris. "Studies with Rinderpest Virus in Tissue Culture: The Use of Attenuated Culture Virus as a Vaccine for Cattle." *Research in Veterinary Science* 3 (1962): 172–182; Taylor, W. P. and W. Plowright. "Studies on the Pathogenesis of Rinderpest in Experimental Cattle. III. Proliferation of an Attenuated Strain in Various Tissues Following Subcutaneous Inoculation." *Journal of Hygiene* 63:2 (June 1965): 263–275.

FIGURE 6.1 Two important figures in rinderpest vaccine development, Walter Plowright and Kazuya Yamanouchi, in Tokyo in 1994.
Photograph courtesy of Kazuya Yamanouchi.

was figuring out the whos and hows of that wielding. The 1950s had laid a strong groundwork with its national campaigns supported by international "expert advice" and limited funding, but it seemed clear that the final push was going to require a more sophisticated structure of coordination. The effort to create one began in Africa in 1961.

That year, the Commission for Technical Cooperation in Africa South of the Sahara (the CCTA – the organization that the French and British had founded in 1950 in response to the sudden burst of new technical cooperation initiatives) launched a new, large-scale rinderpest campaign: Joint Project 15 (JP15). JP15 was designed to overcome the limitations of previous efforts on the continent which had, a 1965 report explained, "for the most part been carried out without co-ordination between neighboring territories in regard to timing, methods, and other factors," largely because those territories often "belong[ed] to different colonial powers."[10] European colonial officials had tried to overcome those difficulties, as the existence of the CCTA itself testified, but they had not yet

[10] STRC Join Project N° 15: Rinderpest Immunization Campaign – Proposed Extension to Eastern Africa (July 2, 1965), OD 31/064 NAUK.

managed to put into place the strong inter-imperial *and* international bureaucratic framework that continental eradication required. Cooperation was not enough; the effort demanded structured coordination.

JP15 used an imperial framework to construct an international campaign. H.E. Lépisier of France and I.M. Macfarlane of the United Kingdom were appointed Coordinator and Deputy Coordinator. They organized their efforts through the Inter-African Bureau for Animal Health (IBAH), which the CCTA had created in 1951.[11] The CCTA had been criticized during the 1950s for being "a ploy by the colonial powers to keep the United Nations out and to reduce the publicity given to African problems." There is no question that politics had been involved in its creation, but the organization had "made an effort to bring together the numerous fieldworkers and scientists in Africa into a common communications network." It had encouraged cooperation between individuals as well as between the various colonial offices involved.[12] With JP15, it took that encouragement to a new level.

Though JP15's initial structure was imperial, its funding was not. This was a critical difference from previous African programs. Most of the funding came from the European Development Fund (EDF) and the United States Agency for International Development (USAID), not from colonial offices in London and Paris, which were rapidly losing power in the face of independence movements across the continent. They would lose power in the rinderpest campaign as well. Africa was changing. The bureaucratic structure of the human fight against rinderpest changed with it.[13]

The CCTA would disappear in the middle of JP15. It initially tried to survive by becoming less imperial. CCTA's head, Claude Cheysson, told the American consul general in Tangiers that he was "moving his agency from the colonial orbit to a position from where it was clearly serving Africa." Part of this move consisted in petitioning the United States for assistance, despite concern from some member nations ("especially in the Union of South Africa, Portugal, and to a lesser extent in Belgian circles") that American assistance would be "a prelude to an aggressive policy

[11] Ibid. The Bureau had been created by the CCTA in 1951 as the Inter-African Bureau of Epizootic Disease, but the name was changed to the IBAH in 1956 to encompass a broader perspective on animal health. Its headquarters remained in Nairobi.

[12] Gruhn, "The Commission for Technical Co-Operation in Africa, 1950–1965," 459–460.

[13] Lepissier, H. E. *OAU/STRC Joint Campaign Against Rinderpest in Central and West Africa (1961-1969).* Lagos: OAU/STRC, 1971, 38–58.

which would undermine their position in the continent."[14] The CCTA further tried to disassociate itself from its imperial foundations by moving its headquarters from London to Lagos in 1959 and opening membership to the independent states of Ghana, Liberia, Guinea, and Cameroon.[15] It was not enough. The new Organization of African Unity, created in 1963, announced at it first Council of Ministers session that it wanted "to give the CCTA an African character and to integrate it in the OAU."[16] It subsequently created the Scientific Technical and Research Commission (STRC) in 1965 to take over the CCTA's functions, including the running of JP15.[17] JP15 began as an imperial program, but it spent most of its existence as an international one.

JP15 initially involved twenty-two countries (seventeen of which were currently struggling with rinderpest).[18] Vaccinations began in the fall of 1962 in the Lake Chad Basin "not only because it was continually troubled by numerous outbreaks, but also because it is a point of junction of several independent African states." To work, JP15 had to be international in action as well as in structure. Phase I moved out from Chad to include vaccinations in Cameroon, Niger, and Nigeria. Vaccination teams administered about nine million vaccines the first year and eleven million the second. By the end of the third year, they could report that they had reached "well above 80% of the entire cattle population in the area." JP15 initiated Phase II in 1964, moving on to Dahomey (now Benin), Togo, Ghana, Upper Volta (now Burkina Faso), Côte d'Ivoire, and Mali. It also continued vaccinations in Niger and Nigeria. That same year, requests

---

[14] Leo G. Cyr, Foreign Service Dispatch (February 4, 1960), CCTA/FAMA; Box 07, Subject Files, 1957–1961, Office of Africa & European Operations: West Africa Division; RG 469, NARA; Fred L. Hadsel, Foreign Service Dispatch (September 26, 1958), CCTA/FAMA, Box 07, Subject Files, 1957–1961, Office of Africa & European Operations: West Africa Division; RG 469, NARA; H. Z. Jaffels, Commission for Technical Cooperation in Africa South of the Sahara (March 23, 1961), CCTA/FAMA, Box 07, Subject Files, 1957–1961, Office of Africa & European Operations: West Africa Division; RG 469, NARA.

[15] H. Z. Jaffels, Commission for Technical Cooperation in Africa South of the Sahara (March 23, 1961), CCTA/FAMA; Box 07, Subject Files, 1957–1961, Office of Africa & European Operations: West Africa Division; RG 469, NARA.

[16] Resolutions and Recommendations of the Ordinary Session of the Council of Ministers Held in Dakar, Senegal, from August 2 to 11, 1963, available at https://au.int/sites/default/files/decisions/9565-council_en_2_11_august_1963_council_ministers_first_ordinary_session.pdf.

[17] Hailemariam, Solomon, Rene Besin, and Datsun Kariuki. *Wiping the Tears of African Cattle Owners.* Bloomington, IN: Xlbris, 2010, 2.

[18] Roeder, Peter, et al. "Rinderpest: The Veterinary Perspective on Eradication." *Philosophical Transactions of the Royal Society* 368 (2013): 2.

came in to the JP15 from an FAO meeting in Addis Ababa "to facilitate, in cooperation with the international agencies, the extension of this project to the whole region of Africa affected with Rinderpest." East Africa wanted in. JP15 obliged, but East Africa would have to wait for Phase III, which included Mauritania, Senegal, Gambia, Sierra Leone, Guinea, Liberia, and "the rest of the Ivory Coast, Mali and Chad," to end before Phase IV could begin in 1969. Phases IV-VI focused on Sudan, Uganda, Kenya, Tanzania, Somalia, and Ethiopia.[19]

JP15's plan centered on mass vaccination: "all cattle every year for three successive years."[20] Technicians used the Dried Goat Virus (DGV) "in adult Zebu type cattle," for whom it was still very safe, and TCRV "in all young animals," who could not be safely vaccinated with the cheaper caprinized strain. Officials used approximately 100 million doses of TCRV during the campaign. The new vaccine proved critical to securing the cooperation of cattle owners because it did not kill any calves, making owners justifiably more eager to bring them in for vaccination.[21] Solomon Hailemariam, who was JP15's Deputy Coordinator in southern Ethiopia, described efforts there: "During some days we used to vaccinate up to seven thousand cattle per day. During the whole operation we vaccinated in Ethiopia about forty-five million cattle." Vaccinated cattle were marked with a "clover leaf shaped cutting" in their ears to keep track of progress. "In five years time," Solomon wrote, "the vaccination campaign reached to the whole country except in some remote parts in Tigray and Eritrea," which were "no-go areas ... partly controlled by rebel movements."[22] Such unavoidable weak spots in the campaign would eventually prove costly, but, at the time, it seemed that they would not. Rinderpest appeared beaten in Africa by herd immunity levels of over 80 percent.

[19] STRC Join Project N° 15: Rinderpest Immunization Campaign – Proposed Extension to Eastern Africa (July 2, 1965), OD 31/064, NAUK. See also Lepissier, H. E. *OAU/STRC Joint Campaign Against Rinderpest in Central and West Africa (1961–1969)*. Lagos: OAU/STRC, 1971.

[20] Roeder et al., "Rinderpest: The Veterinary Perspective on Eradication," 2.

[21] STRC Join Project N° 15: Rinderpest Immunization Campaign – Proposed Extension to Eastern Africa (July 2, 1965), OD 31/064; NAUK; Roeder, Peter and Karl Rich. "The Global Effort to Eradicate Rinderpest." IFPRI Discussion Paper 00923 (November 2009): 19.

[22] Solomon Hailemariam. *The Diary of an African Veterinary Doctor*. Bloomington, IN: Xlibris, 2010, 5-6. The Ethiopian Ministry of Agriculture and Rural Development puts the number of cattle vaccinated in the country during JP15 at 59,000,000. This was 78.67 percent of the total cattle population (Ministry of Agriculture and Rural Development Animal and Plant Health Regulatory Directorate, *Ethiopia Freed from the Most Dangerous Cattle Disease*. [Addis Ababa, June 2009], 21).

Celebration ensued, and not just for the victory over the virus, but for the success of the project itself.

In a 1971 report about JP15's initial phases, its coordinator, H. E. Lepissier, wrote, "in a word, a 'global strategy' has thus been defined and implemented for the first time." This concept, he insisted, "must be maintained in the future for sanitary matters, inside which 'particular policies and projects must only be integrated elements' according to a recent expression of Mr Robert McNamara."[23] The president of the World Bank, and former US Secretary of Defense, had brought the Kennedy and Johnson administrations' "global liberalism" with him to his new job, expanding the bank's mission and its framework for action.[24] Lepissier was particularly focused on the point that the program had demonstrated that a "combined operation, multi-financed, as vast as the Joint Campaign has been, could be drawn up, then carried out across the centre and west of the African continent without major difficulties." The campaign provided a successful example of truly international coopera-tion. That cooperation had not, in actual fact, been global, but regional, but Lepissier did not see that as an issue. The larger point was that the campaign had shown that international organization around one targeted development goal could work. It was now simply a question of scale, not structure.[25]

Lepissier's insistence might have had something to do with JP15's competition for the right to claim that "a 'global strategy' has thus been defined and implemented for the first time." In 1971, the international community was supporting two human-focused disease-eradication cam-paigns. By 1970, twenty-six of the fifty countries that had initiated malaria eradication campaigns as part of the WHO-led Malaria Eradication Programme (MEP) had eliminated the disease within their borders and many others had drastically reduced the disease's ability to move through their population. This was, as Randall Packard has argued, a "significant achievement," but it was not global eradication and the international community was, by 1971, largely disenchanted with the MEP.[26] It was not, however, disenchanted with the idea of eradication. It had simply

---

[23] Lepissier, *OAU/STRC Joint Campaign Against Rinderpest in Central and West Africa*, 58.
[24] Milobsky, David and Louis Galambos. "The McNamara Bank and its Legacy, 1968–1987." *Business and Economic History* 24:2 (Winter 1995): 167–195.
[25] Lepissier, *OAU/STRC Joint Campaign Against Rinderpest in Central and West Africa*, 57–58.
[26] Packard, *A History of Global Health*, 152.

shifted its attention to another target. With key support from the United States, WHO had launched the Intensified Smallpox Eradication Program (SEP) in 1967.

Like JP15, the SEP was a multination effort coordinated by an international organization. Unlike JP15, it was not confined to one continent; it vaccinated targets in many places, though not, it is fair to say, everywhere on earth. It focused on the countries that "either had endemic smallpox or were adjacent to countries where smallpox was endemic." About fifty countries met that criteria in 1967 and they were responsible for implementing and running their own eradication programs. WHO took a decidedly more hands-off approach with SEP than it had with the malaria program, demanding only that participant nations vaccinate and create surveillance systems to find every case of the disease. WHO advisers worked with participant governments to design and coordinate programs. At its headquarters in Geneva, officials allocated the limited funds that they had and monitored the campaign's progress.[27]

SPE received a great deal more attention than JP15; it was, after all, fighting a virus that attacked humans. The stakes were different, though many of the issues of structure and organization were not. International cooperation for disease eradication was international, regardless of the targeted disease; the larger point was the establishment of bureaucratic machinery that could adequately tackle "sanitary matters" on a global scale. In 1971, both JP15 and the SEP were still working out the process.

In the end, JP15 proved an excellent example of political, but not scientific, success. It "ran out of steam in about 1975 having come close to eliminating rinderpest from Africa," though it officially lasted until 1976. The effort's "major deficit," Peter Roeder later concluded, "was that it failed to recognize continuing covert circulation in domestic and wild ungulates in West and eastern Africa and, most importantly, persistent reservoirs of infection in the extensive pastoral communities of the Senegal River basin of West Africa and in eastern Africa."[28] Rinderpest had been run underground, but it had not been vanquished, as would soon become apparent. Countries were supposed to continue vaccination campaigns on their own, but they were not told when they should cease, and they needed to cease in order for researchers to start testing to make sure

[27] Ibid., 165; Henderson, *Smallpox*, 79–105, 108, 129–156; Reinhardt, *The End of a Global Pox*, 124–158. See also Manela, "Globalizing the Great Society" and "A Pox on Your Narrative."

[28] Roeder et al., "Rinderpest: The Veterinary Perspective on Eradication," 2.

rinderpest was no longer present. There was "no exit strategy." "Ultimately," researchers later explained, "international leadership ... failed to recognize that mass vaccination was only the beginning of the eradication, and that an accreditation process was also required."[29] That lesson would come from the SEP, but it had to await that program's own conclusion, which came after JP15 wound down.[30]

JP15 did not eradicate rinderpest in Africa, but it did play a critical role in the evolution of the bureaucratic machinery for eradication. The program helped the Organization of African Unity co-opt the CCTA, turning an imperial commission into an international one by giving states a reason to want to keep cooperating and by ensuring that there was financial support available from abroad for that cooperation. JP15 offered a tangible example of the benefits of multination action in Africa and provided a framework for making it happen. It proved that international cooperation of the kind that required states to give up some sovereignty in the name of the larger mission for disease eradication was possible on a continental level. That level of cooperation went far beyond anything FAO had coordinated during the 1950s.

JP15 had not just required sharing information and vaccines, it had required national government ministries to commit to working with the program's designated coordinators who were "entirely responsible for the direction of the campaign as far as 'time, place, and method'" and who were "authorized to go at any time to any part of the zone to see the progress of operations." The countries involved also had to make commitments about financing, "making available to the Project all the material means at their disposal." Money secured through foreign assistance for JP15 could not be diverted to other projects and participant states were required to undertake and maintain specific sanitary prophylaxis measures. The point, Lepissier explained, was that an "operation of such scope needs not only to be *co-ordinated* but also *directed* in the proper meaning of the term" (emphasis original). JP15 emphasized the need for direction, as opposed to just cooperation.[31]

[29] Taylor, William P., Peter L. Roeder, and Mark M. Rweyemamu. "Use of Rinderpest Vaccine in International Programmes for the Control and Eradication of Rinderpest," in *Rinderpest and Peste des Petits Ruminants*. ed. Barrett et al., 266; Njeumi, F, et al. "The Long Journey: A Brief Review of the Eradication of Rinderpest." *Revue Scientifique et Technique (International Office of Epizootics)* 31:3 (2012): 735.
[30] Reinhardt, *The End of a Global Pox*, 150–154.
[31] Lepissier, *OAU/STRC Joint Campaign against Rinderpest in Central and West Africa*, 26–27, 51, 60.

Lepissier argued that such direction made JP15 more appealing to outside donors and would make future "global strategies" that did the same more appealing as well. It is clear, he wrote that many "Foreign Aids" (listing "USAID, World Bank, French, British, German Aids, etc., the list is a long one") are "undergoing noticeable reductions in the amount of credit at their disposal" and are becoming increasingly reluctant to take on new projects "whose profitability is debatable precisely because they do not form part of a whole." A large-scale project, in contrast, whose potential profitability was not in doubt was a safer use of limited funds. JP15 had relied primarily on assistance from USAID and the European Development Fund, but it had also gotten limited assistance from the British Ministry for Overseas Development, Canada, and Germany. JP15's leaders were in contact with FAO and the OIE throughout the campaign, but those organizations had not provided funding or direction.[32] A more expansive eradication campaign would necessarily demand a more expansive support network, but it also might be more likely to get it, because it would be more likely to work.

Lepissier noted that aid agencies were, in the early 1970s, under intense pressure to show demonstrable results. Early supporters of development had become disillusioned over time with a perceived lack of results. The new director of USAID, John H. Hannah, who had just overseen a task force on international development, wrote in 1970 that he had found in the process "that the underdeveloped world had not changed much" since his first exposure to it via the Marshall Plan and Point Four two decades earlier. Such grim findings made it difficult to sell new programs. "Much," however, was not "at all." Hannah noted that he recently participated in a program "marking the vaccination of the one hundredth millionth person ... in a campaign to eliminate smallpox and control measles as a cause of death." That program, which predated the larger SEP program, was working. In just over two years, he noted, "smallpox as a cause of death has practically disappeared" in the region. It was, he readily admitted, "a great accomplishment."[33]

An even greater accomplishment awaited the world at the end of the decade. At a special meeting in Geneva in 1980, WHO announced global

---

[32] Ibid., 68–69.

[33] Hannah, John A. "New Directions in Foreign Aid for the 1970's." *American Journal of Agricultural Economics* 52:2 (May 1970): 302-307. For more on the program Hannah referenced (the CDC's West and Central African Smallpox Eradication and Measles Control Program), see Reinhardt, *The End of a Global Pox*, 86–123.

"freedom from smallpox," noting that the victory "demonstrated how nations working together in a common cause may further human progress."[34] WHO had done via smallpox eradication what FAO had been hoping to do via rinderpest eradication in the 1950s: provide the ultimate proof of utility of internationalism. WHO's victory must have stung a bit in Rome, particularly since FAO was at a low point in its history.

FAO struggled throughout the 1970s, even "floundered," "demonstrating little initiative and undermining much of the faith of the international community."[35] It had continued to provide some support to national rinderpest campaigns throughout the 1960s and 1970s, but its attention – and those of most of the countries themselves – had moved elsewhere in light of the largely successful campaigns of the 1950s. During the 1960s and into the 1970s, agricultural ministries around the world put expanding crop production ahead of livestock development. FAO followed suit.[36] The goal of rinderpest eradication gave way to satisfaction with rinderpest control, because the disease had been so well contained via extensive national campaigns in Asia and JP15 was producing great results in Africa. Expanded human dominion over rinderpest made eradication a far less pressing concern than it had been in the immediate postwar period. It also made it a less desirable target in terms of public relations. Despite Lepissier's assertions that rinderpest eradication could unite the global community, the world – and FAO – spent the 1970s focused on a global food crisis that it defined primarily as a problem of grain production and distribution. Controlled rinderpest was not threatening enough to warrant a global campaign. But rinderpest did not stay controlled, which, alongside the SEP's victory over smallpox, opened the door to a revitalized conversation about eradication.

***

By the late twentieth century, the world had been fighting rinderpest in some degree as an international community for over a hundred years. Concerned officials had organized the first international veterinary conference in Hamburg in 1863 to address the threat and had come back together again in Vienna in 1871 to better coordinate their efforts. Then, as would

---

[34] Henderson, D. A. *Smallpox: The Death of a Disease.* New York: Prometheus Books, 2009, 248–249.

[35] Staples, *The Birth of Development,* 182-183; Talbot, Ross B. "The International Fund for Agricultural Development." *Political Science Quarterly* 95:2 (Summer 1980): 261–276.

[36] FAO Office for Corporate Communication. *70 Years of FAO: 1945–2015* (2015), 33.

also be the case in the late twentieth century, humans tended to act when the virus was on the move and to get complacent when it retreated to small spaces – pockets of endemic infection that were easy to tolerate. In the aftermath of JP15, residual reservoirs of infection remained in cattle in the Senegal River basin in West Africa and in the Greater Horn in East Africa. In Ethiopia, awareness of the danger came slowly, because it began with giraffes and lesser kudus before it jumped into cattle. By 1976, as JP15 officially ended, rinderpest was in the herds. It traveled with them, entering Sudan around 1978. From there, it continued west, following the growing cattle trade toward Nigeria, whose booming economy was funding a booming beef trade. At the same time, that trade was also carrying rinderpest eastward out of the Senegal River basin. The two separate strains of rinderpest – east and west – moving toward Central Africa in the early 1980s, created the Second Great African Rinderpest Panzootic.[37]

As had been the case throughout history, rinderpest moved along the networks of human commerce and conflict. It was less lethal than it had been a hundred years earlier, now killing only around 50 percent of its victims, but that was still far too many for farmers and pastoralists who depended on their herds. The problem was worst in Africa, but it was not confined there. The Near East Panzootic of 1969–1973, which had started in Afghanistan, reached virtually every country in the Middle East and left small pockets of endemic infection in its wake. A decade later, the Syrian and Israeli armies carried the virus out of Lebanon in looted cattle, creating fresh outbreaks at home. Iraq, Saudi Arabia, Oman, and the United Arab Emirates regularly suffered infections from cattle imported from India and Pakistan, where rinderpest remained stubbornly endemic despite decades of vaccination efforts. In 1978, Indian peacekeepers brought rinderpest to Sri Lanka via infected goats. The Asian cattle trade led to outbreaks in Turkey, Iran, Yemen, and Nepal during the 1970s and 1980s as well.[38]

[37] Roeder et al., "Rinderpest in the Twentieth and Twenty-first Centuries," in *Rinderpest and Peste des Petits Ruminants*. ed. Barrett et al., 105–142; Conversation with Peter Roeder (August 26, 2017).

[38] Ibid.; Roeder and Rich, "The Global Effort to Eradicate Rinderpest," 7–11; Taylor, William P. "Epidemiology and Control of Rinderpest." *Revue Scientifique et Technique (International Office of Epizootics)* 5:2 (1986): 407–410; Friedgut, O. "A Brief History of Rinderpest in Palestine-Israel." *Israel Journal of Veterinary Medicine* 66:3 (September 2011): 65–68; Mathur, Satish Chandra. "The West Asian Rinderpest Eradication Campaign." *EMPRESS: Freedom From Rinderpest, Bulletin* 38 (2011): 26–31; Sasaki, Masao et al. "Global Rinderpest Eradication and the South Asia Rinderpest Eradication Campaign." *EMPRES: Transboundary Animal Disease Bulletin* 38 (2011): 32–40; Scott, "Rinderpest," in *Advances in Veterinary Science*, Vol. 9. ed. Bandly and Jungherr, 195.

The outbreaks were costly. There were estimates of 100 million cattle deaths in Africa alone and economic losses ran into the billions. Rinderpest exacerbated food insecurity and social strife. It was – as it had been in the 1950s – described as a development problem. Observers placed blame for its resurgence on "budgetary problems" which had resulted in the "breakdown of state veterinary services" in many of the affected nations.[39] Those nations had, of course, had many other things to worry about and could ill afford to prioritize rinderpest vaccination and surveillance. The outbreaks of the early 1980s raised the economic stakes of rinderpest's continued circulation and brought the disease back into the spotlight as a development problem that both needed to be solved and ostensibly could be solved – just as smallpox had been. Cognizant that rinderpest eradication could give it the victory that WHO had gotten via the SEP, FAO decided once again to prioritize the fight against the virus.

The first step was finding partners for the effort, which FAO could not undertake on its own. It turned first to the OIE. WHO credited its own smallpox victory to a combination of both mass vaccination *and* a reporting-surveillance system at the national level that allowed it to quickly investigate and contain outbreaks.[40] The OIE was the obvious choice to organize such a surveillance system for rinderpest, as it was already the official repository for national veterinary reports from its member countries. Expanded OIE cooperation was critical to moving forward, but it was just the beginning. Neither FAO nor the OIE had the funds available to pay for the campaigns that they were hoping to develop and support. They turned to the international development community for help. That community was far larger in the 1980s than it had been in the 1950s; it also commanded far more resources. FAO set out to tap them.[41]

FAO and the OIE focused first on Africa, where rinderpest was doing the greatest damage. In 1982, they sent two veterinarians – Solomon Hailemariam of Ethiopia and Samba Sidibe of Mali – to Europe to begin fundraising efforts for a new pan-African campaign. Their efforts, Solomon reported, "helped to raise the first fifty million Euro for supporting the program."[42] Meanwhile, back on the ground in Africa, FAO initiated a number of emergency vaccination campaigns to try to shut

[39] Walsh, John. "War on Cattle Disease Divides the Tropics." *Science* 237:4820 (September 11, 1987): 1289–1291.

[40] Henderson, *Smallpox*, 89–92.

[41] Jolly, Richard, et al. *UN Contributions to Development Thinking and Practice*. Bloomington: Indiana University Press, 2004, 224.

[42] Solomon, Hailemariam. *Diary of an African Veterinary Doctor*, 245–246.

down the current panzootic. These campaigns would bring the number of African nations reporting active outbreaks from eighteen down to four by 1986, demonstrating the possibilities of a revitalized international campaign. The effort swayed some members of the international development community, but not all. With the assistance of the Organization for African Unity and the Inter-African Bureau of Animal Resources, FAO and the OIE used the funding from the European Development Fund to launch the Pan-African Rinderpest Campaign (PARC) in 1986. They had hoped to have more support, but met with resistance from the World Bank and USAID – two of the most powerful institutions in the international development community. The Bank insisted that it could not fund PARC, which did not meet its requirement that its loans only go to specific development projects, but it also expressed "firm reservations" about the program from the beginning. Concern centered on the current state of veterinary services in Africa and on conflicts that would make several key locations virtually inaccessible to vaccinators. USAID was also skeptical, citing JP15's failure as a reason to be wary of investing in another massive vaccination campaign. USAID's reluctance, however, was not just linked to JP15. The agency was simply not very interested in livestock development, preferring, instead, to put its money toward crop development, which allowed for a quicker return on its investments. Returns mattered to an agency that was constantly trying to defend its budget. Instead of putting money directly into PARC, then, American aid authorities announced that they would contribute to the larger struggle against the virus by helping fund research into the development of new vaccines that could overcome TCRV's cold-chain limitations which kept vaccinators frustratingly dependent upon refrigeration in order to ensure the vaccine remained potent.[43]

Despite the lack of World Bank and USAID support, PARC began operations in Burkina Faso, Ethiopia, Mali, Nigeria, and Sudan in September of 1987. An article in *Science* that month observed that "the next decade will show whether FAO, the European Community, and their African partners can muster the will and resources to eradicate rinderpest in Africa." There was a lot riding on the program, which would eventually encompass thirty-five countries.[44] As USAID's response had

---

[43] Walsh, "War on Cattle Disease Divides the Tropics," 1289–1291. Additional information from author interview with Jeffrey Mariner (August 14, 2017).
[44] Ibid.; Chibeu, Dickens M. and Ahmed El-Sawalhy. "Rinderpest Eradication in Africa." *EMPRES: Transboundary Animal Disease Bulletin* 38 (2011): 21–25.

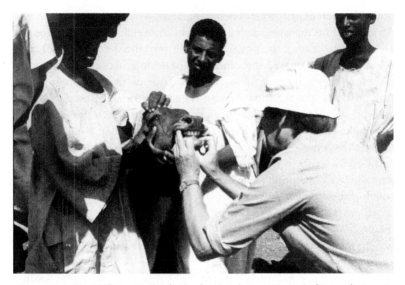

FIGURE 6.2 Dr. William Taylor looks for the characteristic rinderpest lesions in the mouth of a cow in Sudan in 1987.
©*FAO/F. Paladini.*

demonstrated, JP15's failure had cast a long shadow. Aid organizations who had to answer for their budgets were leery of investing in another expensive campaign, because eradication set such a high bar for success, but there were reasons for the hope that inspired PARC. Humans had achieved major victories over rinderpest since the end of World War II. The virus claimed far fewer victims a year – even in the height of the 1980s resurgence – than it had before the international community had started working together to try to control it. The many campaigns that humans had waged against the virus over the decades had accomplished a great deal and there was reason to believe that they could accomplish even more. The SEP's victory over smallpox had proven that eradication was possible. That achievement reverberated throughout the development community.

The most obvious immediate consequence of the SEP's success was a shift in the international public health community toward focusing the limited development funds available for the expansion of public health in the Global South on targeted, top-down interventions via biomedical technologies. This shift displaced efforts to try to improve health from below by fighting poverty and a general lack of access to basic public health services. It was not that the people involved did not want people to have better public health services. They did, but they did not think that they could build them with the resources at their command, so they turned to what they thought they could do: targeting diseases, prioritizing vaccination for both control and for eradication. In 1977, the final year of the SEP, WHO announced that its Expanded Programme on Immunization would vaccinate all of the world's children against diphtheria, pertussis, tetanus, polio, measles, and tuberculosis by 1990. UNICEF and the UN signed on to help in 1983 and 1985 respectively. In 1988, WHO launched the Global Polio Eradication Initiative with a target date of 2000. The international community united around immunization, making vaccination one of the important acts of development in the final decades of the twentieth century.[45]

Those involved turned to the SEP for more than just inspiration. WHO published *Smallpox and Its Eradication* in 1988. The massive book, written by five central figures involved in the SEP, was created at the request of the World Health Assembly "in order to preserve the unique historical experience of eradication and thereby contribute to the development of other health programmes." To make sure that that contribution reached as many people as possible, WHO took the additional step of publishing a condensed version of its most important chapter in the *Bulletin of the World Health Organization* in 1987. Written by Donald Henderson, who had been the chief of the Smallpox Eradication Unit in WHO from 1966 to 1977, "Principles and Lessons from the Smallpox Eradication Programme" explained why the SEP had worked. Henderson cited several key factors: some technical, some bureaucratic. The "provision of adequate quantities of heat-stable freeze-dried vaccine of assured potency" and "improved vaccination instruments" had been critical, but they were far from enough to secure eradication. Factors in the structure of the program itself had proven just as important. Henderson particularly stressed a 1967 shift from measuring success by the number of vaccinations administered to instead focusing on the ultimate goal of "a zero incidence of smallpox." This shift made surveillance as important a part of the program as vaccination. In response, "new methods were devised for discovering cases and containing outbreaks." The SEP's structure, he noted, had also played a central role in its success. It had been decentralized, because WHO "provided only a small proportion of the resources and had no authority over national programmes other than

[45] Packard, *A History of Global Health*, 249–266.

moral persuasion." Instead of being a weakness, however, decentralization had helped, giving local programs the freedom to adapt to the needs of their region. The final key, Henderson wrote, was creating an independent international assessment system that could officially award nations certificates of smallpox eradication. Certification based on extensive surveillance had provided the certainty necessary to declare victory and end vaccination.[46]

Eradication had not been easy, Henderson explained, but it had happened and there was reason to believe that it could happen again. Some of the experiences of the SEP could help in the struggle against other diseases, but not all of it. Every disease was different; every relationship between humans and their microbial enemies was unique. But there were also always similarities and much of the machinery of smallpox eradication could be utilized in campaigns against other pathogens – even the ones that did not attack humans.

Key figures in the international veterinary community used the "principles and lessons" of the SEP to rethink their approach to rinderpest. In 1987, FAO hosted an Expert Consultation on rinderpest to evaluate the international situation. Everyone involved agreed that countries needed a prescribed exit strategy for ending their vaccination programs, something that had been sorely lacking in JP15. In response, the OIE began constructing a three-stage pathway into eradication, which it formally adopted in 1989. The pathway began with a shift from vaccination to surveillance. After two years of no sign of the virus, a country was ordered to cease vaccination. This would prove the hardest step, a researcher recalled later, because farmers did not want to stop vaccination, but there was no other way to determine if the virus was really gone, because antibody tests could not distinguish between an animal that had been vaccinated and one that has survived a natural infection. At that point, a country was "provisionally free from rinderpest." After three more years of continued surveillance without any evidence of the virus in unvaccinated stock, it could apply to the OIE to gain official designation of "freedom from rinderpest disease." Following external verification of its results, it could, after an additional two years of successful surveillance, apply for the status of "freedom from rinderpest infection," which would be granted following external verification. Global eradication

---

[46] Henderson, D. A. "Principles and Lessons from the Smallpox Eradication Programme." *Bulletin of the World Health Organization* 65:4 (1987): 535–546. See also, Fenner, F., et. al. *Smallpox and Its Eradication.* Geneva: World Health Organization, 1988.

could be declared only after every infected nation had earned the final status.[47]

The OIE Rinderpest Pathway was a clear example of the evolution of the machinery of eradication in response to lessons learned in the past – via both JP15 and the SEP – and it would prove critically important. It was, however, only one of many changes in the human struggle against rinderpest that took place during the 1980s. The growing popularity of vaccination campaigns for humans spilled over into the veterinary community as well, opening the door to a renewed focus on rinderpest in nations that were, in the aftermath of the early 1980s outbreaks, eager to banish the disease once and for all. Following the PARC model, the West Asia Rinderpest Eradication Campaign (WAREC) began its operations in 1989 in eleven countries with funding from the United Nations Development Programme. Consequent FAO efforts to launch the South Asia Rinderpest Eradication Campaign (SAREC) ultimately did not come to fruition, but the EU agreed to begin supporting national eradication programs in India, Nepal, Bhutan, and Pakistan the same year. FAO did not run any of these campaigns. Most of the aid was administered bilaterally and national veterinary services implemented both vaccination and surveillance efforts, but FAO and the OIE both played critical roles in keeping the separate efforts connected and abreast of the latest developments in the technologies of eradication. That work fit very well with the model that Henderson had described as critical to the SEP's success: bureaucratic decentralization combined with widely shared technological innovation.[48]

Much of that innovation was itself reminiscent of what had happened with smallpox, including the creation of a "heat-stable freeze-dried vaccine of assured potency."[49] Walter Plowright's tissue culture rinderpest vaccine (TCRV) had been an enormous step forward in the struggle against rinderpest. The vaccine would earn Plowright the 1999 World Food Prize, with the explanation that TCRV "has helped save countless

---

[47] Rweyemamu et al. "Towards a Global Eradication of Rinderpest," in *Rinderpest and Peste des Petits Ruminants*. ed. Barrett et al., 308–309; Normile, Dennis. "Driven to Extinction." *Science* 319:5870 (March 21, 2008): 1608–1609.

[48] Njeumi, F., et al. "The Long Journey," 735–736; Taylor et al. "Use of Rinderpest Vaccine in International Programmes," in *Rinderpest and Peste des Petits Ruminants*. ed. Barrett et al., 271; Rweyemamu, Mark M., Peter L. Roeder and William P. Taylor, "Towards a Global Eradication of Rinderpest," in *Rinderpest and Peste des Petits Ruminants*. ed. Thomas Barrett, Paul-Pierre Pastoret, and William P. Taylor. Amsterdam: Elsevier, 2006), 306–309.

[49] Henderson, "Principles and Lessons from the Smallpox Eradication Programme," 536.

lives, while ensuring that our global food supply remains abundant and safe for future generations."[50] TCRV had been the most important vaccine in the struggle against rinderpest for decades, and it continued to be widely used, but it had one critical weakness: even after having been freeze-dried it still required a "cold chain" from the laboratory to the field. Researchers, including Plowright himself, had been trying to find a way around the problem for decades.[51] In the late 1980s, Jeffrey C. Mariner of Tufts University, in collaboration with scientists at the USDA's Plum Island Animal Disease Research Center solved the problem. The Department of Agriculture wanted an improved vaccine to have ready to go if ever needed in the United States. The scientists from Tufts wanted an improved vaccine that could be immediately employed on the ground in the global struggle against rinderpest. Thermostability was not a critical concern of the USDA, but it was for the global effort. With some prodding, Mariner and his team were able to get about a million dollars from USAID to help support their efforts to create a better vaccine. They did so by growing ninety-second passage Kabete 'O' on Vero cells and then freeze-drying them via a more complicated process than what had been used in the past to lyophilize, or freeze-dry, Plowright's vaccine.[52]

The switch to Vero cells was key. Vero cells are an established line of African green monkey (*Cercopithecus Aethiops* or grivet) kidney cells that were created in Japan in 1962. Telomerase activity is much higher in Vero cells than in normal (diploid) cells, which means that Vero cells can overcome the Hayflick limit (as can HeLa cells, another famous continuous cell line). Vero cells can divide indefinitely; they are immortal. They are also highly susceptible to a number of viruses, but it took additional research to figure that out.[53] In 1965, officials at Japan's National

[50] "1999: Plowright," The World Food Prize website, available at https://www.worldfood prize.org/en/laureates/19871999_laureates/1999_plowright/.

[51] Plowright, W., C. S. Rampton, W. P. Taylor, and K. A. J. Herniman. "Studies on Rinderpest Culture Vaccine, III. Stability of the Lyophilised Product." *Research in Veterinary Science* 11:1 (1970): 71–81.

[52] Author interview with Jeffrey Mariner (July 14, 2017). The Expert Consultation referred to the process as "soft lyophilization," which required "extended (70 hours) lyophilization with high final shelf temperature and the use of vacuum regulation which accelerates the rate of drying in the final stages" (FAO. *FAO Expert Consultation on the Strategy for Global Rinderpest Eradication.* Rome: FAO, 1993: 27).

[53] There are now several lineages of Vero cells used in research. All descend from the original strain. Yasumara, Y. and Y. Kawatika. "Studies on SV40 Virus in Tissue Cultures." *Nihon Rinsho* 21 (1963): 1201–1215 (in Japanese). For the fascinating history of HeLa cells, see Skloot, Rebecca. *The Immortal Life of Henrietta Lacks.* New York: Broadway Books, 2010.

Institute of Health discovered that they could grow a strain of measles virus (MeV) in Vero cells. This finding convinced Kazuya Yamanouchi to try also growing rinderpest (RPV) and canine distemper (CDV) in the cells. He was, he recalls, surprised by how susceptible the Vero cells proved to be to rinderpest. Yamanouchi and his fellow researchers were able to make a "clear-cut demonstration" of the close relationship between measles, rinderpest, and canine distemper by showing that the three viruses produced the same kind of cytopathic effect in the same cell culture. They also paved the way for the first Vero cell rinderpest vaccine (based on Nakamura III), which was created in Japan in 1976. The Japanese government stockpiled this virus for its own emergency supply, just in case.[54]

Vero cells proved a boon to virologists working on rinderpest in several ways. First, they no longer had to rely on bovine kidney cell tissue for growing the virus, which saved trips to local slaughterhouses and removed the danger of cross-contamination with undetected bovine pathogens. Additionally, it resulted in greater consistency in results, because researchers were always working with cells from the same line. Immortal Vero cells also produced high yields of the virus – approximately 100 times higher than bovine kidney cells. This was critical for vaccine production, as Mariner and his team noted in their own research, praising Vero cells for their advantages in yields, "quality control, reproducibility, and supply." They also proved better candidates for lyophilization, allowing researchers to reduce the residual moisture level to about 1 percent as opposed to the previous 3 percent. Using Vero cells, Mariner and his team created a thermostable rinderpest vaccine (TRV) called ThermoVax that could survive thirty days (and sometimes longer) without refrigeration. Tufts immediately began working with PARC to transfer the technology to African vaccine manufactures.[55]

---

[54] Shishido, A., K. Yamanouchi, M. Hikita, T. Sako, A. Fukuda, and F. Kobune. "Development of a Cell Culture System Susceptible to Measles, Canine Distemper, and Rinderpest Viruses." *Archiv für die gesamte Virusforschung* 22:3–4 (1967): 364–380 (Shishido is listed as the lead author because he was the director of the department, but it was Yamanouchi's research project.); Sonoda, Akiro. "Production of Rinderpest Tissue Culture Live Vaccine." *Japan Agricultural Research Quarterly* 17:3 (1983): 191–198.

[55] Mariner, Jeffrey C., et al. "Comparison of Effect of Various Chemical Stabilizers and Lyophilization Cycles on the Theromostability of a Vero Cell-Adapted Rinderpest Vaccine." *Veterinary Microbiology* 21:3 (January 1990): 195–209; *FAO Expert Consultation on the Strategy for Global Rinderpest Eradication*, 24; Sheets, Rebecca. "History and Characterization of the Vero Cell Line." Report for the Vaccines and Related Biological Products Advisory Committee Meeting (May 12, 2000), available at http://www.fda.gov/ohrms/dockets/ac/00/backgrd/3616b1a.pdf; Taylor, William P., P. N. Bhat, and Y. P. Nanda. "The Principles and Practice of Rinderpest Eradication."

At the time, there were about eleven labs in Africa producing rinderpest vaccine. It was, Mariner explained, too many. These national labs were a source of pride for their home governments. The ability to produce one's own vaccines had, as was discussed earlier in this book, been an important part of many national development efforts, but many of these labs produced so few rinderpest vaccines each year that their product suffered from quality control issues. A 1983 investigation by FAO had found that only 20 percent of the rinderpest vaccines produced in Africa were of good quality. To avoid that problem with the new vaccine, Mariner and his team began working with only three labs: one in Ethiopia, one in Cameroon, and one in Mali. Within three years, Ethiopia and Cameroon were producing almost all of the ThermoVax being used in Africa. They were soon joined by the Botswana Vaccine Institute, which had initially refused the offer to produce ThermoVax, but changed its mind when its best customers (countries, NGOs, and international agencies) started shopping at the other labs. High-quality control standards meant that vaccinators in the field could be confident that their injections were potent and eradication depended upon it.[56]

ThermoVax was not the only advance happening in rinderpest vaccine technology by the early 1990s. Researchers were also creating recombinant rinderpest vaccines by injecting some of the genetic material responsible for triggering an immune response into different, harmless viruses, turning them into vectors that would, after being injected into a host, stimulate its immune system into producing antigens against the ferried rinderpest genes.[57] It was an exciting moment in the history of the human struggle against rinderpest – just as the late 1940s had been.

Veterinary Microbiology 44 (1995): 362–363; Roeder et. al, "Rinderpest: The Veterinary Perspective on Eradication," 4; Henderson, "Principles and Lessons from the Smallpox Eradication Programme," 536. Figure of "approximately 100 times higher than bovine kidney cells" told to author by Kazuya Yamanouchi.

[56] Author interview with Jeffrey Mariner (July 14, 2017); Nwankpa, Nick and Charles Bodjo, "(AU-PANVAC): Ensuring Vaccine Quality, Producing Basic Diagnostic Reagents and Maintaining Africa Free from Rinderpest." Power Point presented at 21st Conference of OIE Regional Commission for Africa in Rabat (February 16–20, 2015), available at http://www.rr-africa.oie.int/docspdf/fr/2015/RC21/PANVAC.pdf.

[57] Kazuya Yamanouchi, now at the Nippon Institute for Biological Science, and Thomas Barrett, at Pirbright Laboratory, who had been working separately on the attenuated vaccinia virus vector, joined forces to create the most promising of these recombinant vaccines, which was made by inserting the H gene from a lapinized rinderpest strain into the haemagglutinin region of the vaccinia virus. The researchers tested it first in rabbits, then in cattle in a high containment laboratory at Pirbright. It worked perfectly. It had several advantages. First, as the vaccinia virus is one of the most heat-stable viruses, the

FAO seized on that excitement by hosting, in the fall of 1992, an Expert Consultation on the Strategy for Global Rinderpest Eradication in Rome "to examine the justification for and the feasibility of global rinderpest eradication by the year 2010, having considered past and present campaigns against rinderpest as well as the experience of the WHO, particularly the global smallpox eradication programme."[58] There were many reasons to try. As had been the case in the 1940s, there were new, better vaccines.[59] There were also new innovations in the international bureaucracy. FAO and the OIE had learned from their own previous failures and from WHO's SEP success and developed the Pathway in response. Innovations in molecular epidemiology also promised to be of vital assistance in helping scientists narrow in on the virus, which had, as a direct result of the 1980s campaigns, been pushed back into a few key geographic zones of infection. In this respect, 1992 was very different from the late 1940s; the virus was far more controlled and the international community far more connected. Despite those advantages, however, there was still much to do and many reasons why a global eradication campaign could fail. Participants at the 1992 meeting considered all of them, starting with an analysis of how the existing regional campaigns were going. The situation was not ideal.

recombinant vaccine remained "almost unchanged for one month at both 37°C and 45° C." It, too, broke the "cold chain." It actually broke it more decisively than did Thermovax, but the vaccine had the additional advantage of producing a different kind of immune response in vaccinated cattle than any other kind of available vaccine. Cattle only developed antibodies to the H protein, not the NP protein, which they always would in a natural infection. Therefore, the recombinant vaccine would make it much easier during a country's post-vaccination period to determine that tested cattle had developed antibodies from a vaccine, rather than from a natural, undetected infection. See Yamanouchi, K. and T. Barrett. "Progress in the Development of a Heat-Stable Recombinant Rinderpest Vaccine Using an Attenuated Vaccinia Virus Vector." *Scientific and Technical Review of the Office International des Epizooties* 13:3 (1994): 721–735; Yamanouchi, K., T. Barrett and C. Kai. "New Approaches to the Development of Virus Vaccines for Veterinary Use." *Scientific and Technical Review of the Office International des Epizooties* 17:3 (1998): 641–653.

[58] *FAO Expert Consultation on the Strategy for Global Rinderpest Eradication.* Rome: FAO, 1993, 1.

[59] The recombinant vaccines ended up – for largely bureaucratic reasons – not playing a role in eradication. The researchers, Yamanouchi recalls, were stymied in their efforts to initiate the necessary field tests by a lack of funding. The Indian Veterinary Research Institute and the Kenya Agricultural Research Institute both proposed hosting field tests of the new vaccine, but the proposals never made it from their governments to Japan's foreign ministry, which was necessary in order to secure funding from the Japan International Cooperation Agency (Author correspondence with Kazuya Yamanouchi [July 2015]).

PARC had made great progress, but it had yet to tackle its hardest job: East Africa. Ethiopia, the official report noted, "with 35 million head, has more cattle than the whole of West Africa." Sudan had an additional 21 million, Kenya an additional 14 million, and Uganda an additional 4.2 million. "In terms of PARC funding per bovine," it continued, "West Africa has received four times as much assistance as East Africa." And vaccination in East Africa was troubled by "conflict and instability," dotted with "no-go areas ... where programme activities are limited or curtailed owing to security problems." In southern Sudan, Red Cross and UNICEF teams had vaccinated only 1.6 million cows in the past three years, "which reduced disease and generated disease intelligence but is not an eradication programme." Difficulties were not limited to Africa.[60]

WAREC was also struggling with financing and security problems. Although the program had secured four-and-a-half years of support from the United Nations Development Program, individual countries still needed to finance their own control programs and they had widely disparate abilities of doing so. Yemen had endemic rinderpest and almost no funding, while Egypt received "generous assistance from USAID," for reasons, of course, that had nothing to do with rinderpest. These discrepancies mattered, because rinderpest was still a problem in the region. The Gulf War, the report gloomily noted, had "disrupted WAREC," leading to the relocation of its coordination office from Baghdad to Amman. The "illegal road movement of animals from the Iraq/Iran border" had also led to an outbreak in Turkey "which threatened Europe" and which cost Turkey 6,000 animals and 32 million dollars in control efforts. Meanwhile, SAREC remained an aspiration, not a program, although India was heavily invested in its own national program.[61]

Obstacles clearly remained, yet it seemed that the time was right for a final, coordinated push that would make the most of the work that had already been done and finish the job. The Expert Consultation "concluded that global eradication of rinderpest was justified and feasible and that a Global Rinderpest Eradication Programme (GREP) should be launched."[62] With the strong support of its new Director General, Jacques Diouf, FAO did so in 1994 as part of the larger Emergency Prevention System for Transboundary Animal and Plant Pests and Diseases (EMPRES), which it created at the same time. EMPRES, implementing goals stressed at the 1992 meeting, linked laboratories around the

---

[60] FAO Expert Consultation on the Strategy for Global Rinderpest Eradication, 10–11.
[61] Ibid., 11–15.   [62] Ibid., 2.

world in order to improve "information gathering," "disease monitoring," and "vaccination efficacy and coverage." The GREP Secretariat, headquartered in Rome, was designed "to link the activities of the regional programmes." GREP, an FAO pamphlet later explained, was "conceived as an international coordination mechanism to promote the global eradication of rinderpest and the verification of freedom from rinderpest, while providing technical guidance to achieve these goals."[63] It set itself a 2010 deadline for doing so.

The first step, Peter Roeder, who was in charge of the effort from 1995 to 2007, explained, was to figure out where rinderpest was. No one was certain, both because some outbreaks went unreported locally and because some went unreported internationally by countries who did not want the world to know that they had rinderpest within their borders. Roeder and his team began traveling the world searching for it and organizing regular meetings to convince national veterinary leaders to be more transparent. They knew they needed both cooperation and commitment to have a chance at eradication; FAO and the OIE could not do it by themselves.[64]

GREP was a program, not a campaign. FAO provided global coordination; regional organizations (or, in the case of South Asia, the individual countries) ran the actual campaigns. Those organizations changed over time, acronym followed acronym, as the 1990s gave way to the 2000s. Most notably, in Africa, the Pan African Programme for the Control of Epizootics (PACE) replaced PARC in 1999. PACE was truly international, as opposed to multi-national: "In contrast to PARC, PACE was managed and coordinated by the African Union-Interafrican Bureau for Animal Resources (AU-IBAR), with 32 participating countries each allocated a portion of the total budget."[65] The campaigns were independent in funding and action, but they were united by shared technology, information, and purpose.

FAO helped to coordinate some of the research behind that technology and that information. It worked with the International Atomic Energy

---

[63] Conversation with Peter Roeder (July 26, 2017). Rweyemamu et al., "Towards a Global Eradication of Rinderpest," in *Rinderpest and Peste des Petits Ruminants.* ed. Barrett et al., 308, 318; FAO, "The Global Rinderpest Eradication Programme: Progress Report on Rinderpest Eradication: Success Stories and Actions Leading to the June 2011 Global Declaration," available at http://www.fao.org/ag/againfo/resources/documents/AH/GR EP_flyer.pdf

[64] Conversation with Peter Roeder (July 26, 2017).

[65] Chibeu and El-Sawalhy. "Rinderpest Eradication in Africa." *EMPRES: Transboundary Animal Disease Bulletin* 38 (2011): 23.

Agency (IAEA) to get new field serosurveillance technology (using serum samples to test for the presence of the virus) into the hands of testers on the ground so that they could confirm an absence of rinderpest. In 1994, FAO designated the UK's Institute for Animal Health, Pirbright Laboratory (where Edwards had conducted his secret rinderpest research during WWII), as the World Reference Laboratory for rinderpest. Molecular epidemiology research there revealed that there were three existing, distinct, viral lineages of rinderpest responsible for all of the outbreaks of the past fifty years: the Asian lineage and African lineages I and II. This information proved critical, because it allowed researchers to trace outbreaks back to their source and discover previously unknown reservoirs of the virus. Researchers discovered that the Somali ecosystem (which spread across parts of Kenya and Ethiopia and all of Somalia and Djibouti) not only had African lineage I but also had a strain of African lineage II that had become so low in virulence that researchers had not realized that it existed. Additional research used modeling data to determine the basic reproductive number (R_o) of each lineage, which provided the minimum herd immunity requirement necessary in a given population. With lineage II in the Somali ecosystem, scientists found that herd immunity only needed to reach only around 33 percent to be effective. The number was shockingly low, but "rinderpest was eradicated from the Somali ecosystem with herd immunity levels that probably never exceeded 50 per cent." This saved time and vaccines.[66]

By 1996, FAO identified just seven areas of the world as possible reservoirs of rinderpest: two in Asia, two in the Middle East, and three in East Africa. Eradication required getting rid of the "possible" part through vaccination and/or intense surveillance. The ultimate goal was closer than ever and, in order to help keep up the momentum, FAO launched Intensified GREP in 1999 under the slogan "Seek, Contain, Eliminate." The money to pay for it came from a variety of sources. The countries involved were primarily responsible, but they reached out for help to the European Commission, USAID, the UN Office for the

---

[66] Roeder and Rich, "The Global Effort to Eradicate Rinderpest," 22–23; Roeder et al. "Rinderpest: The Veterinary Perspective on Eradication," 3, 7; Roeder et al. "Rinderpest in the Twentieth and Twenty-first Centuries," in *Rinderpest and Peste des Petits Ruminants*. ed. Barrett et al.,124–132; Baron, Michael D. "The Institute for Animal Health's Contribution to the Eradication of Rinderpest." *EMPRES: Transboundary Animal Disease Bulletin*, 38 (2011): 51–53; Mariner, J. C. and P. L. Roeder. "Use of Participatory Epidemiology in Studies of the Persistence of Lineage 2 Rinderpest Virus in East Africa." *The Veterinary Record* 152 (2003): 641–647.

Coordination of Humanitarian Affairs, and more, arguing that eradication would aid food security and economic development. Critically, countries needed funding not only for vaccination, but for surveillance, which had to continue long after vaccination had ended. It was time-consuming work, but eradication depended upon it.[67]

At the center of the effort, GREP became widely understood to mean more than the specific office at FAO, but, in fact, the sum total of all of the individual campaigns. It was the "international coordination mechanism" that was bringing eradication ever closer. The OIE received the dossiers to prove it: more than 260 of them between 1999 and 2011 from countries eager to move forward along its Pathway. In 2009, it established, with FAO, the Joint FAO/OIE Committee for Global Rinderpest Eradication "to determine whether the world could be declared free of rinderpest and/ or recommend the actions to be taken for this achievement to be confirmed."[68] They had reason to be hopeful that they could announce the former. Rinderpest had not been detected since 2001, when it was found in wild African buffalos near Mount Meru National Park in Kenya. The last vaccination had taken place in 2006. Targeted surveillance teams continued to be unable to find the virus anywhere. It was an excellent sign.

Veterinarians on the ground attributed their success to a number of factors, but stressed the importance of technologies that enabled them to bring local communities into the global campaign in a meaningful way. The thermostable vaccine was clearly critical, in part because it opened the door to community-based vaccination and surveillance campaigns that placed the power in the hands of the livestock owners themselves. Roeder explained the change to a reporter.

"The way we previously did it was really mindless ... We'd get up before dawn to drive long distances. We'd be wrestling the animals to the ground, it'd get stinking hot, and pretty soon the locals would get fed up and walk away." The cattle were nervous and hard to handle, and no wonder, he said: They lived day and night with their owners and now were being roped and tackled by white men wearing khaki and reeking of unfamiliar soaps and deodorants. "But some local, dressed as a

---

[67] Roeder et al., "Rinderpest: The Veterinary Perspective on Eradication," 8–9. As an example of the funding structure, between 2000 and 2004, the budget for the PACE/ Ethiopia program was 7.2 million EURO. 2.9 million of that came from the Ethiopian government and 4.9 million came from outside assistance (Ministry of Agriculture and Rural Development Animal and Plant Health Regulatory Directorate, *Ethiopia Freed from the Most Dangerous Cattle Disease.* [Addis Ababa, June 2009], 45). For more information about the technologies of surveillance, see Crowther, J. R. "Rinderpest: At War with the Disease of War." *Science Progress* 80:1 (1997): 21–43.
[68] Njeumi, F., et al., "The Long Journey," 741.

local, with mutton fat rubbed in his hair, could walk among them and stick in a needle and barely be noticed." ... "We'd be lucky to get 20 percent immunity in a herd; our local guys could get 90, 95 percent."[69]

Locals were not only better vaccinators, but also vital sources of information about the movements of herds and the presence or absence of disease. Diagnostic kits created at Pirbright helped to further expand their role by decentralizing surveillance. Requiring only an eye swab, the new kits could alert a user to the presence or absence of rinderpest antibodies in ten minutes. The technology brought the laboratory to the field and it did so in a way that was accessible to non-specialists.[70] The new technologies made it far easier to pursue eradication work in isolated locations, which is exactly where rinderpest made its final stands. The technology linked the local to the global. The work of eradication was transnational work. The politics of it, however, remained decidedly international.[71]

The final struggles against rinderpest took place in spaces that did not line up with political borders, such as the "Somali Ecosystem, an area covering southern Somalia and the adjoining parts of Ethiopia and Kenya," which highlighted the necessity of an international approach. GREP and the OIE, however, plotted eradication along firm national lines. Maps charting the "Global Rinderpest Situation" were delineated by political borders and entire countries either were or were not "infected." "Freedom from rinderpest infection" was an official status granted only to nations: India received it in 2004; Nepal in 2002; Pakistan in 2007; Bangladesh in 2010; Bhutan in 2005; Afghanistan in 2007; Tajikistan in 2007; Uzbekistan in 2008; Kazakhstan in 2011; Laos in 2011; Myanmar in 2006; Sri Lanka in 2011; Tukey in 2005; Iraq in 2009; Iran in 2008; Oman in 2009; Jordan in 2008; Lebanon in 2008;

---

[69] McNeil, Donald G., Jr. "Rinderpest, Scourge of Cattle, Is Vanquished." *New York Times* (June 27, 2011).

[70] Mariner, Jeffrey, Peter Roeder, and Berhanu Admassu. "Community Participation and the Global Eradication of Rinderpest," in *PLA Notes 45: Community Based Animal Health Care.* ed. Andy Catley and Tim Leyland (October 2002): 29–33; Mariner, Jeffrey C., et al., "Rinderpest Eradication: Appropriate Technology and Social Innovations." *Science* 337:6100 (September 14, 2012): 1309–1312.

[71] According to Akira Iriye, "Internationalism may be seen as an idea of fostering cooperation among nations through inter-state cooperation" while "Transnationalism as an ideology, in contrast, underlies the efforts by private individuals and non-state actors in various countries to establish bridges toward one another and to engage in common activities" (Iriye, Akira. "The Making of a Transnational World," in *Global Interdependence: The World after 1945.* ed. Akira Iriye. Cambridge: Harvard University Press, 2014, 692).

FIGURE 6.3 President Mwai Kibaki of Kenya and other dignitaries at the unveiling of a buffalo statue at Meru National Park in 2010. Meru was the site of the last known outbreak of rinderpest in 2001. ©*FAO/Tony Karumba.*

Yemen in 2010; Saudi Arabia in 2011; UAE in 2011; Armenia in 2009; Belarus in 2008; Brunei in 2009; Serbia in 2008; Azerbaijan in 2011; Kosovo in 2011; Comoros in 2011; Liberia in 2011; Sao Tome and Principe in 2011; Sierra Leone in 2011; Ethiopia in 2008; Kenya in 2009; and Somalia in 2010. The "Global Rinderpest Situation" map gradually became all one color as surveillance teams kept failing to find the virus. "By the end of 2010, FAO was confident that all rinderpest viral lineages (except for those conserved in laboratories) had been proven to be extinct." It had finally achieved the goal it had first imagined possible in 1946.[72]

On June 28, 2011, the thirty-seventh session of the Food and Agriculture Organization of the United Nations Conference endorsed the Declaration of Global Freedom from Rinderpest that the World Organization for Animal Health (the new name of the OIE) had adopted

---

[72] FAO, "The Global Rinderpest Eradication Programme," available at http://www.fao.or g/ag/againfo/resources/documents/AH/GREP_flyer.pdf; Sasaki, Masao, et al. "Global Rinderpest Eradication and the South Asia Rinderpest Eradication Campaign." *EMPRES: Transboundary Animal Disease Bulletin* 38 (2011): 32–40.

the previous month.[73] "Mindful of the devastation caused by rinderpest, a viral disease of cattle, buffalo and many wildlife species that led to famines, demise of livelihoods in Africa, Asia and Europe, and loss of animal genetic resources over centuries," the conference "solemnly" declared "that the world has achieved freedom from rinderpest in its natural setting."[74] The campaigns were over.

***

Global "freedom from rinderpest" had many consequences. The most quickly celebrated of them were economic. FAO estimated that eradication only cost about $610 million, yet was "worth billions of dollars" in Africa alone, making it a very efficient use of resources. The numbers are, of course, imprecise. The question of the economic benefits of eradication is not just one of cattle and buffalo saved, but the consequences of that survival in terms of trade, downstream sectors, the health of owners and their families who depend on the animals for food and/or labor, and so on.[75]

Acknowledging the necessary imprecision, researchers have performed cost-benefit analysis of eradication in Pakistan, Kenya, Ethiopia, Chad, and India, and found eradication to have been economically beneficial in every case. Pakistan's beef exports rose from 1,000 tons a year in 2003 to 3,000 tons a year in 2006 following its status as provisionally free of rinderpest in 2003. Increased milk production in Kenya following eradication was worth an estimated $951 million and increased production of beef in Ethiopia was worth an estimated $434 million. Researchers estimated that eradication increased the GDP of the former by 0.5 percent and the latter by 2.4 percent. In Chad, researchers found that rural households "would have had incomes 2.6 percent lower in the absence of rinderpest control" and the country's GDP would have been 1 percent lower. The numbers matter: "for low-income countries, a few percentage points of GDP can make the difference between meeting basic needs and large-scale human misery." Saving livestock saves lives and bolsters economies. In India, even though actual deaths to livestock were "usually under 5,000 animals per year," rinderpest's eradication in 1990 allowed

---

[73] Its full title was The Declaration of Global Freedom from Rinderpest and on the Implementation of Follow-up Measures to Maintain World Freedom from Rinderpest.

[74] FAO. *Declaration of Global Freedom from Rinderpest – Thirty-seventh Session of the FAO Conference, Rome 25 June–2 July 2011 FAO Animal Production and Health Proceedings.* 17. Rome: FAO, 2013, 34.

[75] FAO, *70 Years of FAO*, 114.

for "a massive increase in market access for buffalo meat in particular, as trading partners have accepted India's rinderpest-free status." That market continues to grow. A 2016 *New York Times* article reported, "Over the last five years, with the market for flash-frozen buffalo meat booming in Saudi Arabia, Egypt and China, India had quietly become the world's largest meat exporter." Eradication created new economic opportunities. Freeing countries from the label of "infected" made domestic cattle and buffalo far more valuable commodities than they had been before.[76]

The economic benefits of eradication went beyond expanding national economies. Livestock provide "food and income for one billion of the world's poor," particularly those who live in arid regions that cannot support significant crop production. In such places, livestock are an essential part of food security and outbreaks of animal diseases threaten human lives. The successful eradication of rinderpest highlighted the necessity of taking that threat more seriously and improving both national and regional veterinary services. Rinderpest was far from the only disease to plague livestock and its removal from circulation only emphasized the importance of combating others.[77]

The world was now better prepared to do so; the struggle for eradication had lasting benefits. For example, the emphasis on improving vaccine efficacy led to the creation in Africa in 1986 of two regional vaccine quality control and training centers. In 1993, the two were united into the Pan African Veterinary Vaccine Centre (PANVAC), which worked to ensure quality control across the continent. Between 1985 and 1997, the proportion of rinderpest vaccine lots produced in African labs that met international quality standards rose from 33 percent to over 90 percent. PARC insisted that only PANVAC-certified vaccines be used in national rinderpest campaigns in Africa and PANVAC-certified vaccines were sold

---

[76] Rich, Karl M., David Roland-Holst, and Joachim Otte. *"An Assessment of the Socio-Economic Impacts of Global Rinderpest Eradication."* FAO Animal Production and Health Working Paper (Rome 2012), vi–vii, 1–5, 51–53; Barry, Ellen. "7 Indian Women Wage War with Their Village over Jobs." *New York Times* (January 30, 2016). There are environmental consequences from the expansion of the cattle industry in these countries, but people are working to mitigate the effects. For an example, see Community Development Research, *Ethiopia: Methane Emissions from Agriculture Waste,* Country Resource Assessment for Global Methane Initiative (October 2011) available at https://www.global methane.org/documents/ag_ethiopia_res_assessment.pdf and Thornton, Philip K. "Livestock Production: Recent Trends, Future Prospects." *Philosophical Transactions of the Royal Society* 365 (2010): 2653–2867.

[77] Roeder, Peter. "Making a Global Impact: Challenges for the Future." *Veterinary Record* 169 (2011): 672.

to Asian countries battling the disease as well. When the struggle against rinderpest was over, the institution now known as AU-PANVAC – and the "network of vaccine production laboratories throughout Africa and the Near East" that it had created – remained, free now to turn their attention to other deadly animal pathogens.[78]

They did so with new models of vaccine creation and vaccine delivery that were also products of the final years of the human struggle against rinderpest. ThermoVax had demonstrated the critical importance of thermostability. Breaking the cold chain broke more than just technical restraints; it also broke government-imposed professional ones. ThermoVax's lack of dependence on refrigeration allowed for the democratization of the vaccination process by opening the door to greater participation by community-based animal health workers (CBAHWs) who were "livestock owners selected by their communities to be trained and equipped for treating priority animal diseases." In most cases, compared to government officials, CBAHWs had better access to the animals who needed vaccination, better knowledge about those animals, better relations with those animals' owners, and better incentive to make sure that vaccination with high-quality vaccines occurred. CBAHW participation played a vital role in eradication and helped lead to the creation of a new branch of veterinary epidemiology known as participatory epidemiology (PE). Jeffrey Mariner created a manual for participatory epidemiology for FAO in 2000, describing it as "the use of participatory techniques for the harvesting of qualitative epidemiological intelligence contained within community observations, existing veterinary knowledge and traditional oral history." This shift in focus and action toward the local within the international veterinary community mirrored a turn within the international development community as a whole. The use of CBAHWs in the final years of the rinderpest campaigns demonstrated how the popular idea of local empowerment could work very well in practice.[79]

---

[78] Tounkara, Karim. Nick Nwankpa, and Charles Bodjo. "The Role of the African Union Pan African Veterinary Vaccine Centre (AU-PANVAC) in Rinderpest Eradication." *EMPRES: Transboundary Animal Disease Bulletin* 38 (2011): 43–45.

[79] Mariner, Jeffrey C. with additions by Roger Paskin, FAO Animal Health Manual 10 – Manual on Participatory Epidemiology – Method for the Collection of Action-Oriented Epidemiological Intelligence (FAO 2000), available at http://www.fao.org/docrep/003 /X8833E/X8833E00.HTM; Roeder et al., "Rinderpest: The Veterinary Perspective on Eradication," 4–5; Catley, Andrew, et al. "Participatory Epidemiology: Approaches, Methods, Experiences." *The Veterinary Journal* 191 (2012): 151–160. For more on the history of the idea of community development, see Immerwahr, *Thinking Small*.

FIGURE 6.4 Tom Olaka, a community animal worker in Karamoja, Uganda, using thermostable rinderpest vaccine to vaccinate local cattle in 1994. Olaka identified and reported the last outbreak of rinderpest in north-eastern Uganda. *Photograph courtesy of Dr. Christine C. Jost.*

In response, in 2007, the OIE, FAO, AU-IBAR, the International Livestock Research Institute (ILRI), CDC, Royal Veterinary College of London University, Vétérinaires Sans Frontières-Belgium, Veterinarians Without Borders-Canada, the African Field Epidemiology and Laboratory Network, and Tufts Cummings School of Veterinary Medicine joined together to create the Participatory Epidemiology Network for Animal and Public Health (PENAPH) to connect people interested in using PE methods to control both existing and emerging diseases. The partnership described its overall goal as being to "enhance epidemiological services in the developing world by making them more representative of and responsive to the needs of beneficiaries." A grant from the Rockefeller Foundation to ILRI allowed PENAPH to host its first technical workshop in Thailand in 2012. Additional assistance to support participant travel came from the Centre de Coopération Internationale en Recherche Agronomique pour le Développement, the South East Asian One Health University Network, and FAO. PENAPH's very existence testified to the ways in which the international development community had changed in the decades since World War II. Most obviously, there had been a massive expansion in the number of stakeholders in that community: newer acronyms sit next to a few familiar old ones. That growth allowed for new pairings in the pursuit of development in all its many forms. Rinderpest eradication had depended on such pairings – aid and action had come from many places. PENAPH was, in part, a recognition that such pairings had not only been necessary, but that they had been useful. Eradication had depended upon both centralization *and* decentralization. It was an important insight.[80]

There was, in the aftermath of eradication, a great deal of new machinery – both technical and bureaucratic – for eradication. This was a vital part of the campaigns' legacy, because there were many more diseases that threatened human well-being by threatening livestock, including one that became more obvious after rinderpest's eradication: its morbillivirus relative, peste des petits ruminants (PPR).

PPR was first described in 1942 in Côte d'Ivoire as a disease of sheep and goats that looked just like rinderpest in its clinical symptoms, but which did not infect cattle or buffalo. It was classified as the fourth morbillivirus (along with measles, rinderpest, and distemper) in 1979. Like rinderpest, PPR is highly contagious and kills anywhere from 30 percent to 100 percent of infected animals. It is a serious development problem, because "small ruminants are often the gateway to a bigger farm, and women are more likely to use small ruminant farming than cattle as a means to self-sufficiency." An outbreak of PPR can take all of those opportunities away. It seems likely that PPR was under-diagnosed

---

[80] ILRI, *Participatory Epidemiology Network for Animal and Public Health: Proceedings of the First Technical workshop, Chiang Mai, Thailand, 11–13 December 2012* (2013); available online at https://cgspace.cgiar.org/bitstream/handle/10568/32781/PenaphWor kshopReport.pdf?sequence=4&isAllowed=y. See also the PENAPH website at https:// penaph.net/. Also based in part on author interview with Jeffrey Mariner (July 14, 2017). For more on the expansion of the international development community, see Iriye, *Global Community*, 96–193; Frey, Marc, Sönke Kunkel, and Corrina R. Unger. ed., *International Organizations and Development, 1945–1990*. New York: Palgrave Macmillan, 2014; Adler, Paul. *For People and Planet: U.S. NGOs and Global Inequalities*. Philadelphia: University of Pennsylvania Press (in preparation).

for decades, often being mistaken for rinderpest. In the aftermath of eradication, PPR was both more noticeable and perhaps also a bit more of a problem as it possibly took advantage of formerly closed livestock trade routes that had been opened by eradication.[81]

In 2011, the OIE and FAO started officially talking about PPR eradication. In 2012, the Bill and Melinda Gates Foundation gave the OIE a grant of $3,618,386 "to establish a PPR vaccine bank, increase the supply of PPR vaccines, and develop a pilot strategy to progressively control/eradicate PPR in Africa." The OIE and FAO set up a working group to develop that strategy, announcing it to the international development community at the International Conference for the Control and Eradication of *peste des petits ruminants* in Abidjan, Côte d'Ivoire, in 2015. The OIE/FAO plan pointed out many reasons why conditions were favorable for PPR's eradication, starting with "the experience gained from eradicating rinderpest." They went on to note the favorable technical aspects of the task: "a battery of diagnostic and surveillance tools, effective and inexpensive vaccines that covers all known strains/lineages of the virus, no long-term virus carriers and no significant role of wildlife." They finished by citing "a growing political commitment from various decision-makers at national, regional and global levels to invest in a control and eradication strategy for PPR." Rinderpest's eradication made the international donor community – most importantly the Gates Foundation – interested in driving another animal disease out of circulation. PPR was the best candidate for the job, so that is where FAO and the OIE had turned their attention. Eradication, they announced, would take fifteen years (divided into three five-year phases) and would cost between 7.6 billion and 9.1 billion dollars. It was an expensive price tag and it reflected the nature of their chosen strategy: mass

---

[81] There is a bit of a debate in the international veterinary community about what role, if any, rinderpest eradication had in what may or may not have been an increase in PPR since 2000. Jeff Mariner told me that in his experience "PPR was very much under diagnosed or under reported prior to the eradication of RP for a variety of reasons" and that an analysis of its "'rate of spread' to new countries is about the same before as after RP eradication" (correspondence with Jeffrey Mariner [July 27, 2017]); Jeffrey Mariner quoted in Rajewski, Genevieve. "The Fight Against Goat Plague." *Tufts Now* (June 21, 2017), available at http://now.tufts.edu/articles/fight-against-goat-plague; Banyard, Ashley C., et al. "Global Distribution of Peste des Petits Ruminants Virus and Prospects for Improved Diagnosis and Control." *Journal of General Virology* 91 (2010): 2885–2897.

vaccination campaigns in seventy infected and fifty at-risk countries. But it was not the only option.[82]

In 2013, the Gates Foundation sponsored a working group of nine members (including Jeffrey Mariner and Peter Roeder) of the international animal health community to come up with a PPR eradication plan. The group agreed with the OIE and FAO that now was the time to act, for many of the same reasons. "The conditions that favored the eradication of RP," they argued, "are also largely present for PPR." The existing PPR vaccines "are among the most effective vaccines available for any disease." But, even more than that, humans can benefit from the many lessons that they learned while combating rinderpest, including the vital importance of targeted "smart vaccination" based on knowledge gained through "vigorous" surveillance and communication with local farmers and herders. GREP, they argued, "taught us that there should be far more to disease control and eradication than institutionalized, pulsed vaccination." Targeted vaccination was not only necessary, they argued, but also faster and cheaper. The working group estimated that their eradication program would take twelve years and cost 3.1 billion dollars. Decentralization would make it possible. "None of the expertise, whether technical, managerial, or political," to do it, they concluded, "is to be found in a single organization or group of people, nor should any one organization seek to exert ownership of the global effort." Instead, leadership should be located in a "multipartite oversight body representative of the range of stakeholders participating in the program" along the lines of the Global Fund to Fight AIDS, Tuberculosis and Malaria.[83]

The Global Fund describes itself as "a 21st-century partnership organization": "a partnership between governments, civil society, the private sector and people affected by the diseases." It is "a financing institution, providing support to countries in the response to the three diseases." It encourages the countries themselves "to take the lead." Its own funding comes from wealthy nations (including the United States, Norway,

[82] Bill and Melinda Gates Foundation, Grants Database, available at https://www.gatesfoundation.org/How-We-Work/Quick-Links/Grants-Database/Grants/2012/10/OPP1057689; FAO and OIE. *Global Strategy for the Control and Eradication of PPR* (OIE and FAO, 2015); available online at http://www.fao.org/3/a-i4460e.pdf.

[83] Mariner, Jeffrey C., et al., "The Opportunity to Eradicate Peste des Petits Ruminants." *The Journal of Immunology* 196 (2016): 3499–3506; Jones, Bryony A., et al., "The Economic Impact of Eradicating Peste des Petits Ruminants: A Benefit-Cost Analysis." *PLOS ONE* (February 22, 2016). Also based on conversations with Jeffrey Mariner (August 14, 2017) and Peter Roeder (July 26, 2017).

Sweden, France, the UK, Japan, and more) and private foundations/institutions (most notably the Gates Foundation).[84] The Fund's structure and mission testify to the changing shape of the international development community in the twenty-first century. Frustration with the bureaucracy and the politics of the UN institutions and the bilateral aid agencies has encouraged the search for new frameworks of action that are more flexible, more efficient, and more attuned to local needs.[85] It is no coincidence that the independent PPR eradication plan argues that success will depend upon creating an operational structure that prioritizes those very things. Ideas about what effective international cooperation looks like are changing – in part because of what that cooperation looked like on the ground in the final years of the rinderpest eradication effort: it was flexible, efficient, and attuned to local needs, and it succeeded.[86]

The extensive international interest in PPR eradication and the competing visions of how it can be done are all part of the rinderpest campaigns' legacy. Eradication has spurred further interest *in* eradication. This was true of smallpox and it was true of rinderpest, and not only in terms of PPR.[87] WHO hosted a Global Technical Consultation to Assess the Feasibility of Measles Eradication in June of 2010. Participants were told that rinderpest eradication was expected in October, an announcement, they agreed, that "provides further support for the biological feasibility of eradication of measles-like virus." In partnership with the American Red Cross, the CDC, UNICEF, and the United Nations Foundation, WHO launched the Global Measles and Rubella Strategic Plan in 2012, calling for the elimination of the viruses in five WHO Regions by 2020. They planned to establish target dates for eradication down the road. Although measles is rinderpest's closest relative, the two

---

[84] The Global Fund, https://www.theglobalfund.org/en/.

[85] Wall, Imogen. "'Outdated and Resistant to Change': How Can We Fix the Humanitarian System?" *The Guardian* (February 10, 2016). available at https://www.theguardian.com /global-development-professionals-network/2016/feb/10/outdated-and-resistant-to-change-how-can-we-fix-the-humanitarian-system.

[86] Something similar has happened with the international effort to eradicate Guinea worm, which has reduced the incidence of infection more than 99.99%. There were only twenty-five recorded cases in 2016, down from 3.5 million in 1986 (The Carter Center, Guinea Worm Eradication Program, available at https://www.cartercenter.org/health/guinea _worm/). Jeffrey Mariner is currently leading a 2.5 million dollar project funded by USAID that is testing these strategies for PPR eradication in Uganda (Rajewski, Genevieve. "The Fight Against Goat Plague." *Tufts Now* (June 21, 2017), available at http://now.tufts.edu/articles/fight-against-goat-plague).

[87] For more on the debate about eradication in the international public health community since smallpox, see Stepan, *Eradication*, 225–261.

viruses differ in some important respects beyond their differing hosts. Measles is one of the most infectious pathogens on the planet: its basic reproductive number (the number of people infected by each host) is estimated to range between 12 and 18. Rinderpest's basic reproductive number ranges between 1.5 and 4, depending on the strain. Before the vaccine became widely available, over 90 percent of children contracted measles by age fifteen. Its high infectivity rate means that eliminating the virus in a population requires a herd immunity of around 95 percent and immunity requires at least two doses of non-thermostable vaccine per person. WHO initially said that it would set an official target date for measles eradication in 2015; it still has not yet done so.[88]

Eradication is not easy. That point, too, is one of the rinderpest campaigns' legacies, and part of the struggle for ownership of the achievement. In its 2015 retrospective, *70 Years of FAO*, FAO placed the eradication of rinderpest first in its list of ten "greatest achievements."[89] Representatives of the OIE wrote that the "main contribution" had come "from the countries themselves and an uncountable number of highly dedicated individuals, be they farmers, veterinarians, scientists or local community workers," who had conducted the vaccinations and the monitoring on the ground. They were also quick to point out, however, that eradication "would not have been possible without international solidarity across continents, as well as firm commitment of international and regional organizations, without encouraging countries' transparency in reporting the disease situation, without OIE's efforts in disseminating new scientific information, and without the continued support from donors such as the EU."[90] Rinderpest eradication was one of the greatest achievements of development, of international cooperation, of the ever-lasting human struggle with microbes. It was fitting that FAO and the OIE were the ones who announced the victory, an act which in itself claimed ownership of the deed. The human struggle against rinderpest had played an

---

[88] WHO, "Proceedings of the Global Technical Consultation to Assess the Feasibility of Measles Eradication, 28–30 June 2010." *The Journal of Infectious Diseases* 204: Supplement 1 (July 15, 2011): S4–S13; Plemper, Richard K. and Anthea L. Hammond. "Will Synergizing Vaccination with Therapeutics Boost Measles Virus Eradication?" *Expert Opinion on Drug Discovery* 9:2 (February 2014): 201–214. For more information, and access to official documents from the campaign, visit the WHO web page on measles, available at http://www.who.int/immunization/diseases/measles/en/.

[89] FAO, *70 Years of FAO*, 114.

[90] Knopf, Lea. Kazuaki Miyagishima, and Bernard Vallat. "OIE's Contribution to the Eradication of Rinderpest." *EMPRES: Transboundary Animal Diseases Bulletin* 38 (2011): 18.

important role in the history of both institutions: the OIE had been created in response to rinderpest and rinderpest had been the reason FAO decided to make fighting animal diseases part of its mission. FAO and the OIE had been there at the beginning and they were there at the end. They had both played vital roles in eradication, but they had been far from alone.

Getting the history right matters, because, as this chapter has shown, people are drawing on it in the construction of new campaigns against other foes. The details matter, but so too does the larger story, which speaks not so much to the specifics of what made eradication possible, but to how the struggle to achieve it influenced the international development community – a community that rinderpest inadvertently helped build by encouraging human cooperation to combat it.

When Peter Roeder and Karl Rich reflected on eradication, they argued that it "was neither the outcome of a single project or program nor due to the efforts of a single agency, but was rather the result of a series of periodic, concerted, and coordinated international efforts built upon the ongoing national programs of many affected countries." Those efforts stretched far into the previous century, they noted. "If one views the start as the point when goat-adapted attenuated vaccines were readily available, then the process had taken some 70 years."[91] That dating takes us back to 1940, but, as this book has shown, international cooperation to combat rinderpest dates even earlier. Rinderpest inspired the British veterinarian John Gamgee to call for an International Conference of Veterinary Surgeons in 1863 to try to establish better regulation of cattle bodies across borders. It would be the first of many such conferences and European governments began working together in a more systematic way to keep the disease outside their borders with strict cull and kill policies and quarantines, effectively policing the virus out of the region. They did so out of a recognition that disease control was a national issue with international political and commercial implications.[92]

That cooperation originally centered on regulation. It manifested itself in quarantines and inspections and in laws that controlled the movement of cattle bodies at the borders of nations and empires. This kind of international cooperation was designed to further the needs of the nation and/or the empire. It was still the ruling order of the day when representatives came together in Paris in 1921 at the International Conference for the Study of Epizootics. They created the OIE "to assist governments in

---

[91] Roeder and Rich, "The Global Effort to Eradicate Rinderpest," 4.
[92] Harrison, *Contagion*, 139–173, 215; Spinage, *Cattle Plague*, 228.

the pursuit and enforcement" of agreed-upon sanitary regulations. But that was not all they created it to do.[93] They also charged it to promote research about epizootic diseases and to collect and publicize new information discovered about them.[94] The possibilities of disease control had changed in the early twentieth century. The OIE's mission reflected the political repercussions of the biological revolution. Participants at the conference were acutely aware that they were living in a world where nations were connected not just by politics and economics, but also environmentally.

World War II strengthened that sense. The war killed the League of Nations as a political body, but it did not kill the idea of an international community united by shared hopes and shared fears. In fact, the war strengthened that idea. The extent of its horrors only emphasized the necessity of building a better world order when it was over – one that led to better lives for everyone. This was the internationalism of John Boyd Orr, who, in his 1949 Nobel Lecture, called for the application of "science to develop the resources of the earth for the benefit of all."[95] Cattle and buffalo were two of those resources. The rinderpest vaccines created in the war – most importantly the avianized vaccine from Grosse Île, but also Junji Nakamura's lapinized vaccine – helped convince people, including Orr, to believe that the development of the world itself – not just individual nations or empires – was possible and that it should be the goal of the United Nations to secure it. The vaccines gave tangible form to the ideology. Vaccination turned international development into a physical act that could be repeated all over the world, and internationalists embraced it.

FAO initially pursued development-via-vaccination in the postwar world in the name of "a common humanity" – the most profound form of internationalism. As this book has shown, it did not go unchallenged. The United States tried to manipulate international development to its advantage, trying to create a "free world" that benefited from its political

---

[93] FAO Standing Advisory Committee on Agriculture, Minutes of Meeting of Subcommittee on Animal Health (March 31–April 4, 1947), Sub-Committee on Animal Health, 1946–1947, Animal Production and Health Division, 10AGA407, FAO; International Agreement for the Creation of an Office International Des Epizooties in Paris (January 25, 1921); available at http://www.oie.int/en/about-us/key-texts/basic-texts/international-agreement-for-the-creation-of-an-office-international-des-epizooties/.

[94] "International Conference on Epizootic Diseases in Domestic Animals." *Journal of the American Veterinary Medical Association* 60, 13:1 (October 1921): 124-138.

[95] Orr, John Boyd. "Nobel Lecture" (December 12, 1949), available at http://www.nobelprize.org/nobel_prizes/peace/laureates/1949/orr-lecture.html

and economic vision. FOA's development efforts did not survive the Cold War unscathed. Meanwhile, in Africa, the European imperial powers, who were pursuing colonial development-via-vaccination, resented FAO's presence as well as its mission. They sometimes actively frustrated FAO's efforts and sometimes grudgingly acquiesced to them, but their power did not last. Imperialism ended, as, eventually, did the Cold War. The vision of "developing the vast potential wealth of the earth for the benefit of all" survived.[96]

By that point, international development and internationalism were far greater than FAO and the other UN specialized agencies who had championed them in the immediate postwar world.[97] But FAO and the other UN specialized agencies still mattered; they still had a unique part to play. Orr had said of them back in 1949, "here at last mankind has the machinery through which governments can join in eliminating hunger, poverty, and disease."[98] WHO helped proved it with smallpox eradication. FAO helped prove it with rinderpest eradication. Neither did so alone, but that in itself only demonstrated how thoroughly the commitment to international development in the name of internationalism had engaged the international community and helped to expand it.

The eradication of rinderpest was a victory of both technology and ideology. Internationalism won. *Homo Sapiens* won. The virus lost, but it deserved their thanks for having helped make possible its defeat via both the technology that had come out of its genetic code *and* the ideology that it had encouraged and strengthened. Rinderpest had helped "the peoples of the world to a mutual understanding and to a realization of the common humanity and common tasks which they share."[99] And that had helped them eradicate it.

[96] Ibid.
[97] For more on this, see Iriye, *Global Community*, 157–209 and Iriye, "The Making of a Transnational World," 681–847.
[98] Orr, John Boyd. "Nobel Lecture" (December 12, 1949), available at http://www.nobel prize.org/nobel_prizes/peace/laureates/1949/orr-lecture.html.
[99] Huxley, *UNESCO*, 13.

Conclusion

"Typhus is not dead. . . . But its freedom of action is being restricted, and more and more it will be confined, like other savage creatures, in the zoölogical gardens of controlled disease."

Hans Zinsser, *Rats, Lice and History*, 1935

Rinderpest is not extinct. It survives today in confinement in several places. Grosse Île is not one of them. The island is now a national park. Tour boats head there and back twice a day during the warmer months, packed primarily with descendants of the immigrants (mostly Irish) who were detained there during the 105 years in which it was a quarantine station. They come to visit the buildings: a decontamination hall where all luggage was sorted for steam cleaning, topped by a room lined with metal shower stalls were humans underwent similar treatment; first, second, and third class dormitories; a church; an isolation building for the sick on the other side of the island, with a smallpox room bathed in red light from the paper pasted over the windows. They come to visit the monuments to the dead. There are many more buildings that they can walk by – but not enter – including a cattle stable built during World War II. Few linger there. It is not part of the history that they have come to relive. Neither is it part of the park's narrative. There is a placard outside, but the young Park Canada guides do not send visitors to look at it. The island's biological warfare heritage is not denied, but it is not featured. People do not come looking for that; they come looking for their roots, for the seeds of their families' North American strain, for the ones who made it off the island and for the ones who did not.

Rinderpest made it off the island, but the avianized strain created there has no descendants today. All its lines died out. Ironically, the attenuated vaccine that played the most important role in stimulating the postwar campaigns played the smallest role in them. It was the vaccine that captured imaginations. Other, easier-to-produce vaccines, did most of the actual work. Yet, capturing imaginations was important in its own right. The vaccine encouraged action in the pursuit of many goals: war relief and rehabilitation, internationalism, development, and biological warfare. The vaccine was full of possibility. Eradication was the boldest of all of them. It required international cooperation to work. It still requires it today.

In 2011, FAO and OIE member countries passed two resolutions agreeing to destroy any remaining rinderpest stocks or "to safely store them in a limited number of relevant high containment laboratories approved by FAO and OIE."[1] Member countries additionally agreed to ban any research that used the live virus without FAO and OIE approval. As of 2012, more than forty laboratories around the world still had stocks of rinderpest. That year, FAO's Chief Veterinary Officer warned, "We must remain vigilant so that rinderpest remains a disease of the past, consigned to history and the textbooks of veterinarians to benefit from the lessons we've learned." [2] FAO and the OIE continue to work together to help make that happen, publishing pamphlets such as *Ten Reasons for NOT Maintaining or Storing Rinderpest Virus* that warn countries that the best way to prevent an outbreak "is not to have the virus in the country in the first place." A single release of the virus, an FAO/OIE poster warns, could "kill millions of cattle, endanger valuable wildlife, undermine international veterinary biosecurity, threaten animal welfare, decrease food security, reduce rural livelihoods, restrict local and international commerce, cost millions of dollars to re-eradicate, delay other development goals, and reverse a historical achievement." There is a great deal at stake in keeping rinderpest in confinement.[3]

---

[1] FAO, "Call for Countries to Comply with Moratorium on Research Using Living Rinderpest Virus" (July 23, 2012), available at http://www.fao.org/news/story/en/item/152953/icode/.

[2] Ibid.

[3] Hamilton, Keith, et al., "Identifying and Reducing Remaining Stock of Rinderpest Virus." *Emerging Infectious Diseases* 21:12 (December 2015): 2117–2121; FAO and OIE, *Ten Reasons for NOT Maintaining or Storing Rinderpest Virus*, available at http://www.oie.int/fileadmin/Home/eng/Media_Center/docs/pdf/Ten_reasons_leaflet_bandejaune.pdf; FAO and OIE. *10 Reasons for Not Keeping Rinderpest Virus in Laboratories*, available at http://www.fao.org/resources/infographics/infographics-details/en/c/345558/.

To help ensure that the virus remains eradicated, in 2015 the OIE World Assembly designated five facilities (in Ethiopia, Japan, the United States, and the United Kingdom) as the approved holders of RVCM (rinderpest virus-containing materials) for the global community. That same year, twenty-four countries responded to an OIE survey that they still had rinderpest materials in labs within their borders under varying biosecurity levels.[4] To address that issue, in 2016, FAO, in collaboration with the OIE, hosted its second "Maintaining Global Freedom from Rinderpest" meeting with representatives from Bangladesh, Canada, China, Iran, Kazakhstan, the Netherlands, the European Union, Pakistan, Russia, Saudi Arabia, South Africa, Switzerland, Turkey, Uzbekistan, Ethiopia, Japan, the United States, the United Kingdom, and international and regional organizations. Participants reviewed "the progress of each country towards their obligations to destroy or safely relocate (sequester) their stocks of rinderpest virus" in one of the five approved facilities and discussed the risks of failing to do so. The official summary of the meeting reported that the "conclusion from these discussions was that there was no justifiable reason to keep RVCM," but it seems clear that not all participants agreed.[5]

While Switzerland declared at the meeting that it had already destroyed all its RVCM, South Africa announced that it was transferring all its material to AU-PANVAC in Ethiopia, and Canada reported that it was going to either destroy or sequester its remaining material, other countries were not so eager to go along with the FAO/OIE mandate. The Netherlands and Russia "reported to store RVCM in their laboratories." So, too, did Iran, Turkey, and China, who said that they were "storing vaccine and seed virus to enable future vaccine production." China announced that it was applying to become a designated holder of RVCM "while Russia and Turkey remained undecided." The politics of eradication continue past eradication.[6]

The OIE and FAO have repeatedly insisted that there is no reason for countries to maintain their own stock of rinderpest viruses and vaccines because the international community is storing them in the five designated

---

[4] Hamilton, Keith, et al. "Identifying and Reducing Remaining Stock of Rinderpest Virus." *Emerging Infectious Diseases* 21:12 (December 2015): 2117–2121; Fournié, Guillaume, et al. "Rinderpest Virus Sequestration and Use in Posteradication Era." *Emerging Infectious Diseases* 19:1 (January 2013): 151–153.
[5] FAO. *Summary Report of the Meeting: Maintaining Global Freedom from Rinderpest, 20–22 January 2016*. Rome, Italy: FAO, 2016, available at http://www.fao.org/3/a-b c963e.pdf.
[6] Ibid.

laboratories. In addition, scientists do not need living virus any more to have access to the virus in an emergency; it would be possible to reassemble the virus from its genetic information, which is stored in public databases. The complete genomes of Kabete O and Nakamura III, along with many other strains, are available today on GenBank. The strains live on as data, ensuring that they are not lost to history, while the risk of infection is removed.[7] The ongoing sequence and destroy effort reminds us that eradication is not just something that the international community did, but something that it continues to do. Eradicated, but not extinct, RPV lives on "in the zoölogical gardens of controlled disease." Keeping it there demands continued international cooperation and continued international oversight. FAO and the OIE are eager to provide it. They do not want to lose their greatest victory.

***

FAO's 2017 report, *The Future of Food and Agriculture: Trends and Challenges,* opened declaring the organization's vision "of a 'world free from hunger and malnutrition, where food and agriculture contribute to improving the living standards of all, especially the poorest, in an economically, socially and environmentally sustainable manner.'"[8] With the exception of the "sustainable" part, FAO's vision remains today what it was at its founding.[9] "Freedom from Want" still remains elusive for many: the UN reports that one in nine people (795 million total) today are undernourished. Ending hunger is the second of the 17 Sustainable Development Goals announced by the UN in 2015. "The food and agriculture sector," it explained, "offers key solutions for development, and is central for hunger and poverty eradication." The target date for that eradication is 2030. The world has not yet achieved freedom from hunger and poverty for all, but freedom from rinderpest helped it get one step closer.[10]

---

[7] National Center for Biotechnology Information, GenBank X98291.3, Rinderpest virus (strain Kabete O) complete genome, genomic RNA, https://www.ncbi.nlm.nih.gov/nuccore/X98291; National Center for Biotechnology Information, GenBank AB547190.1, Rinderpest virus genomic RNA, complete genome, strain: lapinized Nakamura III, https://www.ncbi.nlm.nih.gov/nuccore/AB547190. Also based on conversation with Michael Baron on July 27, 2017.

[8] FAO. The Future of Food and Agriculture: Trends and Challenges. [Rome, Italy]: FAO, 2017, 4; available at http://www.fao.org/3/a-i6881e.pdf.

[9] For the history of how "sustainable" was included, see Macekura, *Of Limits and Growth.*

[10] UN, Goal 2, Sustainable Development Goals, available at http://www.un.org/sustainabledevelopment/hunger/; FAO, Sustainable Development Goals Overview, available at http://www.fao.org/sustainable-development-goals/overview/en/.

Select Bibliography

ARCHIVE GUIDE FOR FOOTNOTES

FAO   Food and Agriculture Organization of the United Nations, Rome
HHPL   Herbert Hoover Presidential Library, West Branch
HIA   Hoover Institution Archives, Stanford
HSTL   Harry S. Truman Presidential Library, Independence
LAC   Library and Archives Canada, Ottawa
NARA   National Archives and Records Administration, College Park
NAUK   The National Archives, Kew
OU   Oklahoma State University Library Special Collections and University
           Archives, Stillwater
RAC   Rockefeller Archive Center, Sleepy Hollow
UN   United Nations Archives, New York
SL   Schlesinger Library, Radcliffe Institute, Cambridge

PUBLISHED MATERIAL

Acheson, Dean. *Present at the Creation*. New York: W.W. Norton & Company, 1969.
    "The Requirements of Reconstruction." *The Department of State Bulletin* 16:411 (May 18, 1947): 991–994.
Adas, Michael. *Dominance by Design: Technological Imperatives and America's Civilizing Mission*. Cambridge: Harvard University Press, 2006.
    Machines as the Measure of Men: Science, Technology, and Ideologies of Western Dominance. Ithaca: Cornell University Press, 1990.
Akami, Tomoko. "Beyond Empires' Science: Inter-Imperial Pacific Science Networks in the 1920s," in *Networking the International System: Global Histories of International Organizations*. ed. Madeleine Herren. Switzerland: Springer International Publishing, 2014, 107–132.

"A Quest to Be Global: The League of Nations Health Organization and Inter-Colonial Regional Governing Agendas of the Far Eastern Association of Tropical Medicine 1910–25." *The International History Review* 38:1 (2016): 1–23.

Internationalizing the Pacific: The United States, Japan and the Institute of Pacific Relations in War and Peace, 1919–1945. London: Routledge, 2002.

Alacevich, Michele. "The World Bank and the Politics of Productivity: The Debate on Economic Growth, Poverty, and Living Standards in the 1950s." *Journal of Global History* 6:1 (March 2011): 53–74.

Allen, Arthur. *The Fantastic Laboratory of Dr. Weigl.* New York: Norton, 2014.

Allen, Charles E. "World Health and World Politics." *International Organizations* 4:1 (February 1950): 31.

Amrith, Sunil. *Decolonizing International Health: India and Southeast Asia, 1930–1965.* London: Palgrave MacMillan, 2006.

Amrith, Sunil and Patricia Clavin. "Feeding the World: Connecting Europe and Asia, 1930-1945." *Past & Present,* 218: Supplement 8 (January 1, 2013): 29-50.

Amrith, Sunil and Glenda Sluga. "New Histories of the United Nations." *Journal of World History* 19:3 (2008): 251–274.

Appuhn, Karl. "Ecologies of Beef: Eighteenth-Century Epizootics and the Environmental History of Early Modern Europe." *Environmental History* 15:2 (April 2010): 268–287.

Arndt, H. W. *Economic Development.* Chicago: University of Chicago Press, 1987.

Avery, Donald. *Pathogens for War.* Toronto: University of Toronto Press, 2013.

Avery, Donald H. and Mark Eaton, ed. *The Meaning of Life: The Scientific and Social Experiences of Everitt and Robert Murray, 1930–1964.* Toronto: The Champlain Society, 2008.

Baker, James A. "VIII. Rinderpest Infection in Rabbits." *American Journal of Veterinary Research* 6:23 (April 1946): 179–182.

Baker, James A. and A. S. Greig. "XII. The Successful Use of Young Chicks to Measure the Concentration of Rinderpest Virus Propagated in Eggs." *American Journal of Veterinary Research* 6:23 (April 1946): 196–198.

Ballard, Charles. "The Repercussions of Rinderpest: Cattle Plague and Peasant Decline in Colonial Natal." *The International Journal of African Historical Studies* 19:3 (1986): 421–450.

Balmer, Brian. *Britain and Biological Warfare.* London: Palgrave, 2001.

Banyard, Ashley C., et al. "Global Distribution of Peste des Petits Ruminants Virus and Prospects for Improved Diagnosis and Control." *Journal of General Virology* 91 (2010): 2885–2897.

Baranowski, Shelley. *Nazi Empire: German Colonialism and Imperialism from Bismarck to Hitler.* Cambridge: Cambridge University Press, 2010.

Baron, M.D., et al., "The Plowright Vaccine Strain of Rinderpest Virus Has Attenuating Mutations in Most Genes." *Journal of General Virology* 86 [2005]: 1093–1101.

Baron, Michael D. "The Institute for Animal Health's Contribution to the Eradication of Rinderpest." *EMPRES: Transboundary Animal Disease Bulletin*, 38 (2011).

Barrett, Thomas, Paul-Pierre Pastoret, and William P. Taylor, ed. *Rinderpest and Peste des Petits Ruminants: Virus Plagues of large and Small Ruminants*. Amsterdam: Elsevier, 2006.

Barry, Ellen. "7 Indian Women Wage War with Their Village over Jobs." *New York Times* (January 30, 2016).

Barry, John M. *The Great Influenza: The Epic Story of the Deadliest Plague in History*. New York: Penguin Books, 2004.

Bennett, Brett M. and Joseph M. Hodge, ed. *Science and Empire: Knowledge and Networks of Science across the British Empire, 1800–1970*. New York: Palgrave MacMillan, 2011.

Beveridge, W. I. B. *Influenza: The Last Great Plague*. London: Heinemann, 1977.

Bhattacharya, Sanjoy and Sharon Messenger, ed. *The Global Eradication of Smallpox*. New Delhi: Orient BlackSwan, 2010.

Biggs, David. *Quagmire: Nation-Building and Nature in the Mekong Delta*. Seattle: University Of Washington Press, 2010.

Binns, H. R. *The East African Veterinary Research Organization: Its Development, Objectives and Scientific Activities*. Nairobi, Kenya: East African Standard Ltd., 1957.

Black, Megan. "Interior's Exterior: The State, Mining Companies, and Resource Ideologies in the Point Four Program." *Diplomatic History* 40:1 (January 2016): 81–110.

Blancou, Jean. *History of the Surveillance and Control of Transmissible Animal Diseases*. Paris: OIE, 2003.

Bok, Bart J. "The United Nations Expanded Program for Technical Assistance." *Science* 117:3030 (January 23, 1953): 67–70.

Borgwardt, Elizabeth. *A New Deal for the World*. Cambridge: Harvard University Press, 2005.

Boroway, Iris. *Coming to Terms with World Health: The League of Nations Health Organization 1921–1946*. Frankfurt am Main: Peter Lang, 2009.

Boulanger, Paul. "Application of the Complement-Fixation Test to the Demonstration of Rinderpest Virus in the Tissue of Infected Cattle Using Rabbit Antiserum." *Canadian Journal of Comparative Medicine* 21:11 (November 1957): 379–388.

Boynton, William H. "Rinderpest, with Special Reference to Its Control by a New Method of Prophylactic Treatment." *The Philippine Journal of Science* 36 (May 1928): 1–35.

Bozheyeva, Gulvarshyn, Yerlan Kunakbayev, and Dastan Yeleukenov. "Former Soviet Biological Weapons Facilities in Kazakhstan: Past, Present, and Future." Occasional Paper No. 1, Chemical and Biological Weapons Nonproliferation Project, Center for Nonproliferation Studies, Monterey Institute of International Studies (June 1999).

Branagan, D. and J. A. Hammond. "Rinderpest in Tanganyika: A Review." *Bulletin of Epizootic Diseases of Africa* 13:3 (September 1965): 225–245.

Brandly, C. A., et al., "Newcastle Disease and Fowl Plague Investigations in the War Research Program." *Journal of the American Veterinary Medical Association* 108:831 (June 1946): 369–371.

Brantz, Dorothee. "'Risky Business': Disease, Disaster and the Unintended Consequences of Epizootics in Eighteenth- and Nineteenth-Century France and Germany." *Environment and History* 17 (2011): 35–51.

Brewer, Anthony. "Adam Ferguson, Adam Smith, and the Concept of Economic Growth." *History of Political Economy* 31:2 (1999): 237–254.

"The Concept of Growth in Eighteenth-Century Economics." *History of Political Economy* 27:4 (1995): 609–638.

Brophy, Leo P., Wyndham D. Miles, and Rexmond C. Cochrane. *The Chemical Warfare Service: From Laboratory to Field.* Washington, DC: Center of Military History, United States Army, 1988.

Brotherston, J. G. "Lapinised Rinderpest Virus and Vaccine: Some Observations in East Africa." *Journal of Comparative Pathology* 61 (1951): 263–288.

"Rinderpest: Some Notes on Control by Modified Virus Vaccines, II." *Veterinary Reviews and Annotations* 3 (1957): 45–56.

Brown, Karen. "Tropical Medicine and Animal Diseases: Onderstepoort and the Development of Veterinary Science in South Africa 1908–1950." *Journal of South African Studies* 31:3 (September 2005): 513–529.

Brown, Karen and Daniel Gilfoyle, ed. *Healing the Herds: Disease, Livestock Economies, and the Globalization of Veterinary Medicine.* Athens: Ohio University Press, 2010.

Brown, Tad. "Await the *Jarga*: Cattle, Disease, and Livestock Development in Colonial Gambia." *Agricultural History* 90:2 (Spring 2016): 230–246.

Brown, Jr. William Adams, and Redvers Opie. *American Foreign Assistance.* Washington, DC: The Brookings Institution, 1953.

Buesekom, Monica M. van. *Negotiating Development: African Farmers and Colonial Experts at the Office du Niger, 1920–1960.* Oxford: James Currey, 2002.

Burnet, F. M. "Inapparent Virus Infections: With Special Reference to Australian Examples." *The British Medical Journal* 1:3915 (January 18, 1936): 99–103.

"Influenza Virus on the Developing Egg. IV. The Pathogenicity and Immunizing Power of Egg Virus for Ferrets and Mice." *British Journal of Experimental Pathology* 18:1 (1937): 37–43.

Burton, W. E. F. "The Rinderpest Outbreak in Western Australia in 1923." *Australian Veterinary Journal* 56 (April 1980): 200–201.

Bynum, W. F. "Policing the Heart of Darkness: Aspects of the International Sanitary Conferences." *History and Philosophy of the Life Sciences* 15:3 (1993): 421–434.

Campbell, Bruce M.S. "Nature as Historical Protagonist: Environment and Society in Pre-Industrial Britain." *The Economic History Review* 63:2 (2010): 281–314.

Campbell, Gwyn. "Disease, Cattle, and Slaves: The Development of Trade between Natal and Madagascar, 1875–1904." *African Economic History* 19 (1990–1991): 105–133.

Carroll, Michael Christopher. *Lab 257.* New York: William Morrow, 2004.

Catley, Andrew, et al. "Participatory Epidemiology: Approaches, Methods, Experiences." *The Veterinary Journal* 191 (2012): 151–160.

Chibeu, Dickens M., and Ahmed El-Sawalhy. "Rinderpest Eradication in Africa." *EMPRES: Transboundary Animal Disease Bulletin* 38 (2011): 21–25.

Clavin, Patricia. *Securing the World Economy: The Reinvention of the League of Nations, 1920–1946*. Oxford: Oxford University Press, 2013.

Clavin, Patricia and Jens-Wilhelm Wessels. "Transnationalism and the League of Nations: Understanding the Work of Its Economic and Financial Organization." *Contemporary European History* 14:4 (November 2005): 465–492.

Cleveland, Harlan. "Economic Aid to China." *Institute of Pacific Relations* 18:1 (January 12, 1949): 1–6.

Coen, Ross. *Fu-Go: The Curious History of Japan's Balloon Bomb Attack on America*. Lincoln: University of Nebraska Press, 2014.

Cohen, Ed. "The Paradoxical Politics of Viral Containment; or, How Scale Undoes Us One and All." *Social Text* 106 29:1 (Spring 2011): 15–35.

Collingham, Lizzie. *The Taste of War: World War II and the Battle for Food*. New York: Penguin, 2013.

Conant, Jr., Michael. "JCRR: An Object Lesson." *Far Eastern Survey* 20:9 (May 2, 1951): 89.

Constantine, Stephen. *The Making of British Colonial Development Policy*. London: Frank Cass, 1984.

Cooper, Frederick. "Development, Modernization, and the Social Sciences in the Era of Decolonization: The Examples of British and French Africa." *Revue d'Histoire des Sciences Humaines* 10 (2004): 9–38.

"Reconstructing Empire in British and French Africa." *Past and Present*, Supplement 6 (2011): 196–210.

Cooper, Frederick and Randall Packard, ed. *International Development and the Social Sciences: Essays on the History and Politics of Knowledge*. Berkeley: University of California Press, 1997.

Crawford, Dorothy H. *Viruses: A Very Short Introduction*. Oxford: Oxford University Press, 2011.

Creager, Angela N.H. *The Life of a Virus: Tobacco Mosaic Virus as an Experimental Model, 1930–1965*. Chicago: University of Chicago Press, 2002.

Cribelli, Teresa. "'These Industrial Forests': Economic Nationalism and the Search for Agro-Industrial Commodities in Nineteenth-Century Brazil." *Journal of Latin American Studies* 45:3 (August 2013): 545–579.

Crowther, J. R. "Rinderpest: At War with the Disease of War." *Science Progress* 80:1 (1997): 21–43.

Cueto, Marcos. *Cold War, Deadly Fevers: Malaria Eradication in Mexico, 1955–1975*. Washington, DC: Woodrow Wilson Center Press, 2007.

Cullather, Nick. "The Foreign Policy of the Calorie." *American Historical Review* 112:2 (April 2007): 336–364.

The Hungry World. Cambridge: Harvard, 2010.

Cunningham, Andrew, and Perry Williams, ed. *The Laboratory Revolution in Medicine*. Cambridge: Cambridge University Press, 1992.

Curti, Merle, and Kendall Birr. *Prelude to Point Four: American Technical Missions Overseas, 1838–1938*. Madison: University of Wisconsin Press, 1954.

Daubney, R. "Récentes Acquisitions dans la Lutte contre la Peste Bovine." *Bulletin – Office International des Épizooties* 28 (1947): 36–45.

"Rinderpest: A Résumé of Recent Progress in Africa." *The Journal of Comparative Pathology and Therapeutics* 50 (1937): 405–409.

Davis, Mike. *Late Victorian Holocausts*. London: Verso, 2002.

De Bevoise, Ken. *Agents of Apocalypse*. Princeton: Princeton University Press, 1995.

De Kruif, Paul. *The Microbe Hunters*, Introduction by F. Gonzalez-Crussi. 1926, San Diego, CA: Harcourt, Inc., 1996.

We are Amphibians: Julian and Aldous Huxley on the Future of Our Species. Oakland: University of California Press, 2015.

Deichmann, Ute. *Biologists under Hitler*. Trans. Thomas Dunlap. Cambridge, MA: Harvard University Press, 1996.

DeWitte, Sharon and Philip Slavin. "Between Famine and Death: England on the Eve of the Black Death – Evidence from Paleoepidemiology and Manorial Accounts." *Journal of Interdisciplinary History* 44:1 (Summer 2013): 37–60.

Dodd, Norris E. "A Summary of Activities of the Food and Agriculture Organization in the Middle East." *Middle East Journal* 4:3 (July 1950): 352–355.

Dorsey, Kurk. "Bernath Lecture: Dealing with the Dinosaur (and Its Swamp): Putting the Environment in Diplomatic History." *Diplomatic History* 29:4 (September 2005): 573–587.

Dorsey, Kurkpatrick. *Whales and Nations: Environmental Diplomacy on the High Seas*. Seattle: University of Washington Press, 2013.

Dorwart, Reinhold A. "Cattle Disease (Rinderpest?): Prevention and Cure in Brandenburg, 1665–1732." *Agricultural History* 33:2 (April 1959): 79–85.

Easterly, William. *The Tyranny of Experts: Economists, Dictators, and the Forgotten Rights of the Poor*. New York: Basic Books, 2013.

Edwards, J. T. "The Problem of Rinderpest in India." Bulletin No. 199, Imperial Institute of Agricultural Research, Pusa (Calcutta Government of India, 1930): 1–16.

"Rinderpest: Some Properties of the Virus and Further Indications for Its Employment in the Serum-Simultaneous Method of Protective Inoculation." *Transactions of the Congress – Far Eastern Association of Tropical Medicine* 3 (1927): 699–706.

"Rinderpest: Active Immunization by Means of the Serum Simultaneous Method; Goat Virus." *Agricultural Journal of India* 23 (1928): 185–189.

"Rinderpest: Some Points on Immunity." *Transactions of the Far-Eastern Association of Tropical Medicine* 3 (1927): 707–717.

"The Uses and Limitations of the Caprinized Virus in the Control of Rinderpest (Cattle Plague) Among British and Near-Eastern Cattle." *The British Veterinary Journal* 105:7 (July 1949): 209–253.

Ekbladh, David. *The Great American Mission: Modernization and the Construction of an American World Order*. Princeton: Princeton University Press, 2010.

Enders, John F., et al., "Measles Virus: A Summary of Experiments Concerned with Isolation, Properties and Behavior." *American Journal of Public Health* 47:3 (March 1957): 275–282.

Engerman, David C. "Development Politics and the Cold War." *Diplomatic History* 41:1 (January 2017): 1–19.

Modernization from the Other Shore: American Intellectuals and the Romance of Russian Development. Cambridge: Harvard University Press, 2003.

Engerman, David C., Nils Gilman, Mark H. Haefele, and Michael E. Latham. *Staging Growth: Modernization, Development, and the Global Cold War*. Amherst: University of Massachusetts Press, 2003.

Escobar, Arturo. *Encountering Development: The Making and Unmaking of the Third World*. Princeton: Princeton University Press, 1995.

FAO Office for Corporate Communication. *70 Years of FAO: 1945-2015*. Rome: FAO, 2015.

Farley, John. *To Cast Out Disease: A History of the International Health Division of the Rockefeller Foundation (1913–1951)*. Oxford: Oxford University Press, 2002.

Fenner, F., et al. *Smallpox and Its Eradication*. Geneva: World Health Organization, 1988.

Fidler, David P. "Germs, Governance, and Global Public Health in the Wake of SARS." Journal of Clinical Investigation 113:6 (March 2004): 799–804.

Fitzgerald, Deborah. "Exporting American Agriculture: The Rockefeller Foundation in Mexico, 1943–53." *Social Studies of Science* 16:3 (August 1986): 457–483.

Fitzgerald, J. G. "An International Health Organization and the League of Nations." *The Canadian Medical Association Journal* 14:6 (June 1924): 532.

Food and Agriculture Organization. "The Global Rinderpest Eradication Programme: Progress Report on Rinderpest Eradication: Success Stories and Actions Leading to the June 2011 Global Declaration" available at http://www.fao.org/ag/againfo/resources/documents/AH/GREP_flyer.pdf.

Declaration of Global Freedom from Rinderpest – Thirty-seventh Session of the FAO Conference, Rome 25 June-2 July 2011, FAO Animal Production and Health Proceedings 17. Rome: FAO, 2013.

FAO Expert Consultation on the Strategy for Global Rinderpest Eradication. Rome: FAO, 1993.

Report to the Government of Ethiopia on the Control of Diseases of Livestock, FAO Report No. 497. Rome: FAO (May 1956): 4–7.

Report to the Government of Afghanistan on the Control of Animal Diseases, FAO Report No. 204. Rome: FAO (1953):5–8.

Report on the International Training Centre on Living Virus Vaccines (Veterinary), FAO Report No. 149. Rome: FAO (August 1953): 1.

Report to the Government of Pakistan on Control of Animal Diseases, FAO
    Report No. 103. Rome: FAO (February 1953): 2–7.
International Organization 7:1 (February 1953): 131.
Foucault, Michel. *Security, Territory, Population: Lectures at the Collège de
    France, 1977–1978,* ed. Michel Senellart, trans. Graham Burchell.
    New York: Picador, 2007.
Fournié, Guillaume, et al. "Rinderpest Virus Sequestration and Use in
    Posteradication Era." *Emerging Infectious Diseases* 19:1 (January 2013):
    151–153.
Frey, Marc, Sönke Kunkel, and Corrina R. Unger, ed. *International Organizations
    and Development, 1945–1990.* New York: Palgrave Macmillan, 2014.
Friedgut, O. "A Brief History of Rinderpest in Palestine-Israel." *Israel Journal of
    Veterinary Medicine* 66:3 (September 2011): 65–68.
Furuse, Yuki, Akira Suzuki, and Hitoshi Oshitani. "Origin of Measles Virus:
    Divergence from Rinderpest Virus between the 11th and 12th Centuries."
    Short Report, *Virology Journal* 7:52 (2010).
Geissler, Erhard, and John Ellis van Courtland Moon, ed. *Biological and Toxin
    Weapons.* Oxford: Oxford University Press, 1999.
Gilfoyle, Daniel. *The Many Plagues of Beasts.* Saarbrüken, Germany: VDM
    Verlag Dr. Müller, 2009.
    "Veterinary Research and the African Rinderpest Epizootic: The Cape Colony,
    1896–1898." *Journal of Southern African Studies* 29:1 (March 2003):
    133–154.
Gilman, Nils. *Mandarins of the Future: Modernization Theory in Cold War
    America.* Baltimore: Johns Hopkins Press, 2003.
Gonzalo, Javier. *The Idea of Third World Development: Emerging Perspectives in
    the United States and Britain, 1900–1950.* New York: University Press of
    America, 1987.
Gorman, Daniel. *The Emergence of International Society in the 1920s.*
    Cambridge: Cambridge University Press, 2012.
Graham, S. E. "The (Real)politiks of Culture: U.S. Cultural Diplomacy in Unesco,
    1946–1954." *Diplomatic History* 30:2 (April 2006): 231–251.
Gray, David F. "The Ethiopian Farmer." *Western Mail* (2 February 1950): 8–9.
Groves, Leslie R. *Now It Can Be Told.* New York: Harper & Row, 1962.
Gruhn, Isebill V. "The Commission for Technical Co-Operation in Africa."
    The Journal of Modern African Studies 9:3 (October 1971): 459–469.
Guillemin, Jeanne. *Biological Weapons.* New York: Columbia University Press,
    2005.
Hale, M. W., et al. "XIV. Immunization Experiments with Attenuated Rinderpest
    Vaccine Including Some Observations on the Keeping Qualities and Potency
    Tests." *American Journal of Veterinary Research* 6:23 (April 1946):
    212–221.
    "Rinderpest XIII. The Production of Rinderpest Vaccine from an Attenuated
    Strain of Virus." *American Journal of Veterinary Research* 6:23 (April 1946):
    199–211.
Hale, M. W., R. V. L. Walker, Fred D. Maurer, James A. Baker, and Dubois
    L. Jenkins. "XIV. Immunization Experiments with Attenuated Rinderpest

Vaccine Including Some Observations on the Keeping Qualities and Potency Tests." *American Journal of Veterinary Research* 6:23 (April 1946): 212–221.

Hambidge, Gove. *The Story of FAO*. Toronto: D. Van Nostrand Company, Inc., 1955.

Hamblin, Jacob Darwin. *Arming Mother Nature: The Birth of Catastrophic Environmentalism*. New York: Oxford, 2013.

Hamilton, Keith, et al. "Identifying and Reducing Remaining Stock of Rinderpest Virus." *Emerging Infectious Diseases* 21:12 (December 2015): 2117–2121.

Hannah, John A. "New Directions in Foreign Aid for the 1970's." *American Journal of Agricultural Economics* 52:2 (May 1970): 302–307.

Harder, Andrew. "The Politics of Impartiality: The United Nations Relief and Rehabilitation Administration in the Soviet Union, 1946–47." *Journal of Contemporary History* 47:2 (April 2012): 347–369.

Hardy, Anne. "Animals, Disease, and Man: Making Connections." *Perspectives in Biology and Medicine* 46:2 (Spring 2003): 200–215.

Harris, Sheldon H. *Factories of Death*. Revised Edition. New York: Routledge, 2002.

Harrison, Mark. *Contagion: How Commerce Has Spread Disease*. New Haven: Yale University Press, 2013.

   "Disease, Diplomacy and International Commerce: The Origins of International Sanitary Regulation in the Nineteenth Century." *Journal of Global History* 1:2 (2006): 197–217.

Havinden, Michael and David Meredith, *Colonialism and Development: Britain and Its Tropical Colonies, 1850–1960*. London: Routledge, 1993.

Headrick, Daniel R. *Power over Peoples: Technology, Environments, and Western Imperialism, 1400 to the Present*. Princeton: Princeton University Press, 2010.

Hecht, Gabrielle, ed. *Entangled Geographies: Empire and Technopolitics in the Global Cold War*. Cambridge, MA: MIT Press, 2011.

Hell, Stefan. "The Role of European Technology, Expertise and Early Development Aid in the Modernization of Thailand before the Second World War." *Journal of the Asia Pacific Economy* 6:2 (2001): 158–178.

Helleiner, Eric. *Forgotten Foundations of Bretton Woods: International Development and the Making of the Postwar Order*. Ithaca: Cornell University Press, 2014.

Henderson, D. A. "Principles and Lessons from the Smallpox Eradication Programme." *Bulletin of the World Health Organization* 65:4 (1987): 535–546.

   Smallpox: The Death of a Disease. New York: Prometheus Books, 2009.

Hodge, Joseph Morgan. "On the Historiography of Development (Part 1: The First Wave)." *Humanity* (Winter 2015): 429–460.

   Triumph of the Expert: Agrarian Doctrines of Development and the Legacies of British Colonialism. Athens: Ohio University Press, 2007.

   "Writing the History of Development (Part 2: Longer, Deeper, Wider)." *Humanity* (Spring 2016): 125–174.

Holmes, Edward C. *The Evolution and Emergence of RNA Viruses*. Oxford: Oxford University Press 2009.

Homewood, Katherine. *Ecology of African Pastoralist Societies*. Oxford: James Currey, 2008.

Hopper, Paul F. "The Institute of Pacific Relations and Origins of Asian and Pacific Studies." *Pacific Affairs* 61:1 (Spring 1988): 98–121.

Howard-Jones, Norman. "Origins of International Health Work." *British Medical Journal* 1 (May 6, 1950): 1032–1046.

Huber, Valeska. "The Unification of the Globe by Disease? The International Sanitary Conferences on Cholera, 1851–1894." *The Historical Journal* 49:2 (June 2006): 453–476.

Hudson, J. R. "Rinderpest Virus Attenuated in Eggs." *The Veterinary Record* 59:25 (5 July 1947): 331.

Hunt, Linda. "U.S. Coverup of Nazi Scientists." *Bulletin of the Atomic Scientists* 2:8 (April 1985): 16–24.

Hunter, James A. "The Rinderpest Epidemic of 1949–50 Taiwan (Formosa)." *Chinese-American Joint Commission on Rural Reconstruction, Animal Industries Series*, No. 1. Taipei: February 1951.

Hutcheon, Duncan. "Rinderpest in South Africa." *Journal of Comparative Pathology and Therapeutics* 15:4 (December 31, 1902): 300–324.

Huygelen, C. "The Immunization of Cattle against Rinderpest in Eighteenth-Century Europe." *Medical History* 41 (1997): 182–196.

Huxley, Julian, ed. *Reshaping Man's Heritage*. London: George Allen & Unwin Ltd, 1947.

UNESCO: Its Purpose and Its Philosophy. 1947, London: Euston Grove Press, 2010.

Huxley, J., et al. *When Hostilities Cease: Papers on Relief and Reconstruction Prepared for the Fabian Society*. London: Victor Gollancz Ltd, 1944.

Immerwahr, Daniel. "Modernization and Development in U.S. Foreign Relations," *Passport* (September 2012): 25.

Thinking Small: The United States and the Lure of Community Development. Cambridge: Harvard University Press, 2015.

"International Conference on Epizootic Diseases in Domestic Animals." *Journal of the American Veterinary Medical Association* 60, NS13:1 (October 1921): 124–138.

Iriye, Akira. *Cultural Internationalism and World Order*. Baltimore: The Johns Hopkins University Press, 1997.

Global Community. Berkeley: University of California Press, 2002.

"The Making of a Transnational World," in *Global Interdependence: The World after 1945*. ed. Akira Iriye. Cambridge: Harvard University Press, 2014, 681–847.

Irwin, Julia F. "The Great White Train: Typhus, Sanitation, and U.S. International Development during the Russian Civil War." *Endeavor* 36:3 (2012): 89–96.

Making the World Safe: The American Red Cross and a Nation's Humanitarian Awakening. New York: Oxford University Press, 2013.

"Taming Total War: Great War-Era American Humanitarianism and Its Legacies." *Diplomatic History* 38:4 (2014): 763–775.

Isogai, S. "On the Rabbit Virus Inoculation as an Active Immunization Method against Rinderpest for Mongolian Cattle." *The Japanese Journal of Veterinary Science* 6:5 (1944) 371–390.

Iyer, Samantha. "Colonial Population and the Idea of Development." *Comparative Studies in Society and History* 55:1 (2013): 65–91.

Jachertz, Ruth. "Coping with Hunger? Visions of a Global Food System, 1930–1960." *Journal of Global Health* 6 (2011): 99–119.

Jacotot, H. "Sur la Sensibilité du Lapin au Virus de la Peste Bovine." *Bulletin de la Société de Pathologie Exotique* 23 (November 12, 1930): 904–909.

Javaloyes, Pedro, et al. *70 Years of FAO: 1945–2015*. FAO Office for Corporate Communication. 2015, http://www.fao.org/3/a-i5142e.pdf.

Jenkins, Dubois, and Richard E. Shope. "VII. The Attenuation of Rinderpest Virus for Cattle by Cultivation in Embryonating Eggs." *American Journal of Veterinary Research* 6:23 (April 1946): 174–178.

Jolly, Richard, Louis Emmerij, Dharam Ghai, and Frédéric Lapeyre. *UN Contributions to Development Thinking and Practice*. Bloomington, IN: Indiana University Press, 2004.

Jones, Bryony A., et al. "The Economic Impact of Eradicating Peste des Petits Ruminants: A Benefit-Cost Analysis." *PLOS ONE* (February 22, 2016) available at https://doi.org/10.1371/journal.pone.0149982.

Jones, Joseph M. *The United Nations at Work: Developing Land, Forests, Oceans ... and People*. Oxford: Pergamon Press, 1965.

Kakizaki, Chiharu, Shunzo Nakanishi, and Junji Nakamur. "Experimental Studies on the Economical Rinderpest Vaccine." *Journal of the Japanese Society of Veterinary Science* 6:2 (1927): 107–120.

Kakizaki, Chiharu, Shunzo Nakanishi, and Takashi Oizumi. "Experimental Studies on Prophylactic Inoculation against Rinderpest, Report III." *Journal of the Japanese Society of Veterinary Science* 5:4 (1926): 221–280.

Kakizaki, Chiharu. "Study on the Glycerinated Rinderpest Vaccine." *Kitasato Archives of Experimental Medicine* 2 (1918): 59–66.

Kelser, R. A., S. Youngberg, and T. Topacio. "An Improved Vaccine for Immunization against Rinderpest." *Journal of the American Veterinary Medicine Association* 74 (1929): 28–41.

Kent, John. *The Internationalization of Colonialism*. Oxford: Clarendon Press, 1992.

Kesteven, K. L, ed. *Rinderpest Vaccines: Their Production and Use in the Field*, 1st Ed. Washington, DC: FAO, 1949.

Kishi, Hiroshi. "A Historical Study on Outbreaks of Rinderpest during the Yedo Era in Japan." *The Yamaguchi Journal of Veterinary Medicine* 3 (1976): 33–40.

Kjekshus, Helge. *Ecology Control and Economic Development in East African History*. 1977, London: James Currey, 1996 impression.

Knab, Cornelia. "Infectious Rats and Dangerous Cows: Transnational Perspectives on Animal Disease in the First Half of the Twentieth Century." *Contemporary European History* 20:3 (August 2011): 281–306.

Knab, Cornelia and Amalia Ribi Forclaz. "Transnational Co-Operation in Food, Agriculture, Environment and Health in Historical Perspective:

Introduction." *Contemporary European History* 20:3 (August 2011): 247–255.

Knight, R. F. and C. G. Thomson. "Brief Report on the Veterinary Institutes of Japan." *Philippine Agricultural Review* 4 (March 1911): 111–118.

Knopf, Lea, Kazuaki Miyagishima, and Bernard Vallat. "OIE's Contribution to the Eradication of Rinderpest," in *EMPRES: Transboundary Animal Diseases Bulletin* 38 (2011): 18–20.

Koch, Robert. "Prof. Robert Koch's Berichte über seine in Kimberley ausgeführten Experimentalstudien zur Bekämpfung der Rinderpest." Special reprint of *Deutschen Medicinischen Wochenschrift* numbers 15 and 16 (1897): 1–15.

"Special Report to the 'British Medical Journal' by Professor R. Koch on His Research into the Cause of Cattle Plague." *The British Medical Journal* 1:1898 (May 15, 1897): 1245–1246.

Kott, Sandrine. "International Organizations – A Field of Research for a Global History." *Zeithistorische Forschungen* 8 (2011): 446–450.

Krementsov, Nikolai. *The Cure.* Chicago: University of Chicago Press, 2002.

Kruszweski, Charles. "International Affairs: Germany's Lebensraum." *The American Political Science Review* 34:5 (October 1940): 964–975.

Latham, Michael E. *Modernization as Ideology: American Social Science and "Nation Building" in the Kennedy Era.* Chapel Hill: University of North Carolina Press, 2000.

The Right Kind of Revolution: Modernization, Development, and U.S. Foreign Policy from the Cold War to the Present. Ithaca: Cornell University Press, 2010.

Latour, Bruno. *The Pasteurization of France,* trans. Alan Sheridan and John Law. Cambridge: Harvard University Press, 1988.

League of Nations. Report of the Delegation on Economic Depressions, Part I, *The Transition from War to Peace Economy.* Geneva: League of Nations, 1943.

Leffler, Melvyn P. *A Preponderance of Power.* Stanford, CA: Stanford University Press, 1992.

For the Soul of Mankind. New York: Hill and Wang, 2007.

Leffler, Melvyn P. and David S. Painter, ed. *Origins of the Cold War.* 2nd Ed. New York: Routledge, 2005.

Leitenberg, Milton and Raymond A. Zilinskas with Jens H. Kuhn. *The Soviet Biological Weapons Program.* Cambridge, MA: Harvard University Press, 2012.

Lepissier, H.E. *OAU/STRC Joint Campaign against Rinderpest in Central and West Africa (1961–1969).* Lagos: OAU/STRC, 1971.

Lewis, Joanna. *Empire State-Building: War & Welfare in Kenya, 1925–52.* Oxford: James Currey, 2000.

Lie, Trygve. *In the Cause of Peace: Seven Years with the United Nations.* New York, 1954.

Lindenmann, Jean. "Typhus Vaccine Developments from the First to the Second World War (On Paul Weindling's 'Between Bacteriology and Virology ...')." *History and Philosophy of the Life Sciences* 24:3/4 (2002): 467–485.

Livingston, Julie and Jasbir K. Puar. "Interspecies." *Social Text 106* 29:1 (Spring 2011): 3–14.

Loth, Wilfried. "States and the Changing Equations of Power," in *Global Interdependence*. ed. Akira Iriye. Cambridge, MA: Harvard University Press, 2014, 11–199.

Lowe, Keith. *Savage Continent: Europe in the Aftermath of World War II*. New York: Picador, 2012.

Lurtz, Casey Marina. "Developing the Mexican Countryside: The Department of Fomento's Social Project of Modernization." *Business History Review* 90:3 (Autumn 2016): 431–455.

Macekura, Stephen J. *Of Limits and Growth: The Rise of Global Sustainable Development*. Cambridge: Cambridge University Press, 2015.

"The Point Four Program and U.S. International Development Policy." *Political Science* Quarterly 128:1 (Spring 2013): 127–160.

MacFarlane, David. L "The UNRRA Experience in Relation to Developments in Food and Agriculture." *Journal of Farm Economics* 30:1 (February 1948): 69–77.

Mack, Roy. "The Great African Cattle Plague Epidemic of the 1890's." *Tropical Animal Health and Production* 2:4 (December 1970): 210–219.

MacOwen, K. D. S. "Virulent Rinderpest Virus (R.B.K.)." *Annual Report Veterinary Service Department Kenya* (1955), 29.

Madsen, Thorvald. "The Scientific Work of the Health Organization of the League of Nations." *Bulletin of the New York Academy of Medicine* 13:8 (August 1937): 439–465.

Manela, Erez. "A Pox on Your Narrative: Writing Disease Control into Cold War History." *Diplomatic History* 34:2 (April 2010): 299–323.

"Globalizing the Great Society," in *Beyond the Cold War: Lyndon Johnson and the New Global Challenges of the 1960s*, ed. Francis J. Gavin and Mark Atwood Lawrence. New York: Oxford University Press, 2014, 165–181.

The Wilsonian Moment: Self-Determination and the International Origins of Anticolonial Nationalism. New York: Oxford University Press, 2007.

Marcus, Harold. *A History of Ethiopia*, updated ed. Berkeley: University of California Press, 2002.

Marcus, Harold G. *The Life and Times of Menelik II*. Oxford: Clarendon Press, 1975.

Mariner, Jeffrey C., with additions by Roger Paskin. *FAO Animal Health Manual 10 – Manual on Participatory Epidemiology – Method for the Collection of Action-Oriented Epidemiological Intelligence*. FAO, 2000.

Mariner, Jeffrey C., et al. "Comparison of Effect of Various Chemical Stabilizers and Lyophilization Cycles on the Theromostability of a Vero Cell-Adapted Rinderpest Vaccine." *Veterinary Microbiology* 21:3 (January 1990): 195–209.

"The Opportunity to Eradicate Peste des Petits Ruminants." *The Journal of Immunology* 196 (2016): 3499–3506.

"Rinderpest Eradication: Appropriate Technology and Social Innovations." *Science* 337:6100 (September 14, 2012): 1309–1312.

Mariner, J.C., and P.L. Roeder. "Use of Participatory Epidemiology in Studies of the Persistence of Lineage 2 Rinderpest Virus in East Africa." *The Veterinary Record* 152 (2003): 641–647.

Mariner, Jeffrey, Peter Roeder, and Berhanu Admassu. "Community Participation and the Global Eradication of Rinderpest," in *PLA Notes 45: Community Based Animal Health Care*. ed. Andy Catley and Tim Leyland (October 2002): 29–33. http://pubs.iied.org/pdfs/G02019.pdf.

Materials on the Trial of Former Servicemen of the Japanese Army Charged with Manufacturing and Employing Bacteriological Weapons. Moscow: Foreign Languages Publishing House, 1950.

Mathur, Satish Chandra. "The West Asian Rinderpest Eradication Campaign." *EMPRESS: Freedom from Rinderpest, Bulletin* 38 (2011): 26–31.

Maul, Daniel. "'Help Them Move the ILO Way': The International Labor Organization and the Modernization Discourse in the Era of Decolonization and the Cold War." *Diplomatic History* 33:3 (June 2009): 387–404.

Maurer, Fred D., R.V.L. Walker, Richard E. Shope, Henry J. Griffiths, and Dubois L. Jenkins. "V. Attempts to Prepare an Effective Rinderpest Vaccine from Inactivated Egg-Cultured Virus." *American Journal of Veterinary Research* 6:23 (April 1946): 164–169.

Mazower, Mark. *Hitler's Empire: How the Nazis Ruled Europe*. New York: Penguin, 2009.

Governing the World: The History of an Idea. New York: Penguin Press, 2012.

No Enchanted Palace: The End of Empire and the Ideological Origins of the United Nations. Princeton: Oxford University Press, 2009.

"Reconstruction: The Historiographical Issues." *Past and Present*, Supplement 6 (2011): 17–28.

McKercher, P.D. "Rinderpest Virus Adapted to the Chorioallantoic Membrane of the Chick Embryo – Its Attenuation and Use as a Vaccine." *Canadian Journal of Comparative Medicine* 21:11 (November 1957): 374–378.

McNeill, J. R. *Mosquito Empires*. Cambridge: Cambridge, 2010.

McNeill, J. R. and Corinna R. Unger, ed. *Environmental Histories of the Cold War*. Cambridge: Cambridge University Press, 2010.

McVety, Amanda. *Enlightened Aid: U.S. Development as Foreign Policy in Ethiopia*. New York: Oxford University Press, 2012.

"Pursuing Progress: Point Four in Ethiopia." *Diplomatic History* 32:3 (June 2008): 371–403.

Mehos, Donna C. and Suzanne M. Moon. "The Uses of Potability: Circulating Experts in the Technopolitics of Cold War and Decolonization," in *Entangled Geographies*. ed. Gabrielle Hecht. Cambridge: MIT Press, 2011, 43–74.

Mercado, Stephen C. "The Japanese Army's Noborito Institute." *International Journal of Intelligence and Counterintelligence* 17 (2004): 286–299.

Miescher, Giorgio. *Namibia's Red Line: The History of a Veterinary and Settlement Border*. New York: Palgrave Macmillan, 2012.

Milobsky, David, and Louis Galambos. "The McNamara Bank and Its Legacy, 1968–1987." *Business and Economic History* 24:2 (Winter 1995): 167–195.

Ministère de L'Agriculture, Républic Francaise. *Conférence Internationale pour L'Étude des Épizooties*, Paris, May 25–28, 1921. Paris: Imprimerie Nationale, 1921.

Ministry of Agriculture and Rural Development Animal and Plant Health Regulatory Directorate. *Ethiopia Freed from the Most Dangerous Cattle Disease*. Addis Ababa, June 2009.

Mishra, Saurabh. *Beastly Encounters of the Raj*. Manchester: Manchester University Press, 2015.

"Beasts, Murrains, and the British Raj: Reassessing Colonial Medicine in India from the Veterinary Perspective, 1860–1900." *Bulletin of the History of Medicine* 85 (2011): 587–619.

Mitchell, Timothy. "Economentality: How the Future Entered Government." *Critical Inquiry* 40 (Summer 2014): 479–507.

Rule of Experts: Egypt, Techno-Politics, Modernity. Berkeley: University of California Press, 2002.

Mitter, Rana. *Forgotten Ally*. Boston: Houghton Mifflin Harcourt, 2013.

"Imperialism, Transnationalism, and the Reconstruction of Post-War China: UNRRA in China, 1944–7." *Past and Present*, Supplement 8 (2013): 51–69.

Mitter, Rana, and Helen M. Schneider. "Introduction: Relief and Reconstruction in Wartime China." *European Journal of East Asian Studies* 11 (2012): 179–186.

Moon, Suzanne. "Empirical Knowledge, Scientific Authority, and Native Development: The Controversy over Sugar/Rice Ecology in the Netherlands East Indies, 1905–1914." *Environment and History* 10:1 (2004): 59–81.

Morens, David M., et al. "Global Rinderpest Eradication: Lessons Learned and Why Humans Should Celebrate Too." *The Journal of Infectious Diseases* 204:4 (August 15, 2011): 502–505.

Nagata, Naomi. "International Control of Epidemic Diseases from a Historical and Cultural Perspective," in *Networking the International System*, ed. M. Herren. Switzerland: Springer International Publishing, 2014, 73–88.

Nakamura, Junji and Sadao Kuroda. "Rinderpest: On the Virulence of the Attenuated Rabbit Virus for Cattle." *The Japanese Journal of Veterinary Science* 4:2 (1942): 75–102.

Nakamura, Junji, Shosaburo Wagatuma, and Kanemato Fukusho. "On the Experimental Infection with Rinderpest Virus in the Rabbits." *The Japanese Journal of Veterinary Science* 17 (1938): 185–204.

Nakamura, Junji and Takeshi Miyamoto. "Avianization of Lapinized Rinderpest Virus." *American Journal of Veterinary Research* 14 (1953): 307–317.

Nash, Linda. *Inescapable Ecologies: A History of Environment, Disease, and Knowledge*. Berkeley: University of California Press, 2006.

Neill, Deborah J. *Networks in Tropical Medicine: Internationalism, Colonialism, and the Rise of a Medical Specialty, 1890–1930*. Stanford: Stanford University Press, 2012.

Newfield, Timothy P. "A Cattle Panzootic in Early Fourteenth-Century Europe." *Agricultural History Review* 57 (2009): 155–190.

"A Great Carolingian Panzootic: The Probable Extent, Diagnosis and Impact of an Early Ninth-Century Cattle Pestilence." *Argos* 46 (2012): 200–210.

"Early Medieval Epizootics and Landscapes of Disease: The Origins and Triggers of European Livestock Pestilences, 400–1000 CE," in *Landscapes and Societies in Medieval Europe East of the Elbe*, ed. Sunhild Kleingärtner, Sébastian Rossignol and Donat Wehner, Papers in Mediaeval Studies 23. Toronto: Pontifical Institute of Mediaeval Studies, 2013: 73–113.

"Human-Bovine Plagues in Early Middle Ages." *Journal of Interdisciplinary History* 45:1 (Summer 2015): 1–38.

Nicolle, M. and Adil Mustafa. "Etudes sur la Peste Bovine. Troisième Mémoire. Expériences sur la Filtration du Virus." *Annales de l'Institut Pasteur* 16 (1902): 56–64.

Njeumi, F., et al. "The Long Journey: A Brief Review of the Eradication of Rinderpest." *Revue Scientifique et Technique (International Office of Epizootics)* 31:3 (2012): 729–746.

Normile, Dennis. "Driven to Extinction." *Science* 319:5870 (21 March 2008): 1608–1609.

Nunan, Timothy. *Humanitarian Invasion: Global Development in Cold War Afghanistan*. Cambridge: Cambridge University Press, 2016.

OFRRO Division of Public Information. *The Office of Foreign Relief and Rehabilitation Operations, Department of State*. Washington, DC, 1943.

O'Gallagher, Marianna. *Grosse Ile: Gateway to Canada, 1832–1937*. Quebec City: Livres Carraig Books, 1984.

Olitsky, Peter K. "Hans Zinsser and His Studies of Typhus Fever." *The Journal of the American Medical Association* 115:10 (March 8, 1941): 907–912.

Olmstead, Alan L. "The First Line of Defense: Inventing the Infrastructure to Combat Animal Diseases." *The Journal of Economic History* 69:2 (June 2009): 327–357.

Olmstead, Alan L. and Paul W. Rhode. *Arresting Contagion: Science, Policy, and Conflicts over Animal Disease Control*. Cambridge, MA: Harvard University Press, 2015.

Onselen, C. van. "Reactions to Rinderpest in Southern Africa 1896–97." *The Journal of African History* 13:3 (1972): 473–488.

Orr, John Boyd. *The White Man's Dilemma*. 1953, New York: British Book Centre, Inc., 1954.

Oshinsky, David M. *Polio: An American Story*. New York: Oxford, 2006.

Osterhammel, Jürgen. "'Technical Co-Operation' between the League of Nations and China." *Modern Asian Studies* 13:4 (1979): 661–680.

Owen, David. "The United Nations Expanded Program of Technical Assistance – A Multilateral Approach." *Annals of the American Academy of Political and Social Science* 323 (May 1959): 25–32.

Özkul, Türel, and R. Tamay Başagac Gül. "The Collaboration of Maurice Nicolle et Adil Mustafa: The Discovery of Rinderpest Agent." *Revue de Médicine Vétérinaire* 159 (2008): 243–246.

Packard, Randall M. *A History of Global Health: Interventions in the Lives of Other Peoples*. Baltimore: The Johns Hopkins University Press, 2016.

"Malaria Dreams: Postwar Visions of Health and Development in the Third World." *Medical Anthropology* 17:3 (May 1997): 279–296.

"'Roll Back Malaria, Roll in Development': Reassessing the Economic Burden of Malaria." *Population and Development Review* 35:1 (March 2009): 53–87.

Packenham, Robert A. *Liberal America and the Third World: Political Development Ideas in Foreign Aid.* Princeton: Princeton University Press, 1973.

Pankhurst, Richard. *Economic History of Ethiopia, 1800–1935.* Addis Ababa: Haile Sellassie I University Press, 1968.

"*The Great Ethiopian Famine of 1888–1892: A New Assessment.*" *The Journal of the History of Medicine and Applied Science* (April 1966): 95–124.

Parmar, Inderjeet. *Foundations in the American Century: The Ford, Carnegie, and Rockefeller Foundations in the Rise of American Power.* New York: Columbia University Press, 2012.

Patenaude, Bertrand M. *The Big Show in Bololand: The American Relief Expedition to Soviet Russia in the Famine of 1921.* Stanford: Stanford University Press, 2002.

Patterson, K. David. "Typhus and Its Control in Russia, 1870–1940." *Medical History* 37 (1993): 361–381.

Pedersen, Susan. "Back to the League of Nations." *The American Historical Review* 112:4 (October 2007): 1091–1117.

The Guardians: The League of Nations and the Crisis of Empire. New York: Oxford University Press, 2015.

Pemberton, JoAnne. "New Worlds for Old: The League of Nations in the Age of Electricity." *Review of International Studies* 28 (2002) 311–336.

Perren, Richard. "Filth and Profit, Disease and Health: Public and Private Impediments to Slaughterhouse Reform in Victorian Britain," in *Meat, Modernity, and the Rise of the Slaughter House.* ed. Paula Young Lee. Durham, NH: University of New Hampshire Press, 2008, 127–150.

Phillips, Ralph W. "International Cooperation to Improve World Agriculture." *The Scientific Monthly* 79:3 (September 1954): 154–164.

Phoofolo, Pule. "Epidemics and Revolutions: The Rinderpest Epidemic in Late Nineteenth-Century Southern Africa." *Past & Present* 138 (February 1993): 112–143.

"Face to Face with Famine: the BaSotho and the Rinderpest, 1897–1899." *Journal of Southern African Studies* 29:2 (June 2003): 503–527.

Plemper, Richard K., and Anthea L. Hammond. "Will Synergizing Vaccination with Therapeutics Boost Measles Virus Eradication?" *Expert Opinion on Drug Discovery* 9:2 (February 2014): 201–214.

Plowright, W. "The Application of Monolayer Tissue Culture Techniques in Rinderpest Research. II. The Use of Attenuated Culture Virus as a Vaccine for Cattle." *Bulletin de l'OIE* 57 (1962c): 253–276.

Plowright, W., C. S. Rampton, W. P. Taylor, and K. A. J. Herniman, "Studies on Rinderpest Culture Vaccine, III. Stability of the Lyophilised Product." *Research in Veterinary Science* 11:1 (1970): 71–8.

Plowright, W. and R.D. Ferris. "Cytopathogenicity of Rinderpest Virus in Tissue Culture." *Nature* 179 (February 9, 1957): 316.

"Studies with Rinderpest Virus in Tissue Culture, I. Growth and Cytopathogenicity." *Journal of Comparative Pathology* 69 (1959): 152–172.

"Studies with Rinderpest Virus in Tissue Culture, II. Pathogenicity for Cattle of Culture-Passaged Virus." *Journal of Comparative Pathology* 69 (1959): 173–184.

"Studies with Rinderpest Virus in Tissue Culture: The Use of Attenuated Culture Virus as a Vaccine for Cattle." *Revue Scientifique et Technique (International Office of Epizootics)* 3 (1962): 172–182.

Pomeroy, Laura W., Ottar N. Bjørnstad, and Edward C. Holmes. "The Evolutionary and Epidemiological Dynamics of the Paramyxoviridae." *Journal of Molecular Evolution* 66 (2008): 98–106.

Pottevin, TH.Madsen, and R. Norman White. "Typhus and Cholera in Poland: The Action of the League of Nations." *The Lancet* (December 4, 1920): 1159–1160.

Powell, John W. "A Hidden Chapter in History." *The Bulletin of the Atomic Scientists* 37:8 (October 1981): 44–52.

Preston, Andrew. "Monsters Everywhere: A Genealogy of National Security." *Diplomatic History* 38:3 (June 2014): 492–499.

Reader, John. *Africa: The Biography of a Continent*. New York: Vintage, 1997.

Reinhardt, Bob H. *The End of a Global Pox: America and the Eradication of Smallpox in the Cold War Era*. Chapel Hill: University of North Carolina Press, 2015.

"The Global Great Society and the US Commitment to Smallpox Eradication." *Endeavour* 34:4 (December 2010): 164–172.

Reinisch, Jessica. "'Auntie UNRRA' at the Crossroads." *Past and Present*, Supplement 8 (2013): 70–97.

"Internationalism in Relief: The Birth (and Death) of UNRRA." *Past and Present*, Supplement 6 (2011): 258–289.

"'We Shall Rebuild Anew a Powerful Nation': UNRRA, Internationalism and National Reconstruction in Poland." *Journal of Contemporary History* 43:3 (July 2008): 451–476.

Reisinger, Robert C., et al. "Use of Rabbit-Passaged Strains of the Nakamura LA Rinderpest Virus for Immunizing Korean Cattle." *American Journal of Veterinary Research* 15:57 (October 1954): 554–560.

Renaud, Anne. *Island of Hope and Sorrow: The Story of Grosse Île*. Montreal: Lobster Press, 2007.

Rich, Karl M., David Roland-Holst, and Joachim Otte. "An Assessment of the Socio-Economic Impacts of Global Rinderpest Eradication." FAO Animal Production and Health Working Paper, Rome, 2012. http://www.fao.org /docrep/015/i2584e/i2584e00.htm.

Rimmington, Anthony. "The Soviet Union's Offensive Program: The Implications for Contemporary Arms Control," in *Biological Warfare and Disarmament: New Problems/New Perspectives*. ed. Susan Wright. Lanham, MD: Rowman & Littlefield, 2002: 103–148.

Roeder, Peter. "Making a Global Impact: Challenges for the Future." *Veterinary Record* 169 (2011): 671–674.

Roeder, Peter, Jeffrey Mariner, and Richard Kock. "Rinderpest: The Veterinary Perspective on Eradication." *Philosophical Transactions of the Royal Society of London* 368:1623 (August 5, 2013). DOI: https://dx.doi.org/10.1098%2Frstb.2012.0139.

Roeder, Peter and Karl Rich. "The Global Effort to Eradicate Rinderpest." International Food Policy Research Institute Discussion Paper 00923 (November 2009) available at http://ebrary.ifpri.org/utils/getfile/collection/p15738coll2/id/29876/filename/29877.pdf.

Rogers, Leonard. "Prophylactic Inoculations against Animal Diseases in the British Empire." *The British Medical Journal* 1:4080 (March 18, 1939): 565–566.

Romano, Terrie M. "The Cattle Plague of 1865 and the Reception of 'The Germ Theory' in Mid-Victorian Britain." *Journal of the History of Medicine* 52 (January 1997): 51–80.

Rosebury, Theodor and Elvin A. Kabat, with the Assistance of Martin H. Boldt. "Bacterial Warfare." *The Journal of Immunology* 56 (1947): 7–96.

Rosen, S. McKee. *The Combined Boards of the Second World War.* New York: Columbia University Press, 1951.

Rosenberg, Emily S. *Spreading the American Dream: American Economic and Cultural Expansion, 1890–1945.* New York: Hill and Wang, 1982.

Rusk, Dean. "The Bases of United States Foreign Policy." *Proceedings of the Academy of Political Science* 27:2 (January 1962): 98–110.

Russell, Edmund. *War and Nature.* Cambridge: Cambridge University Press, 2001.

Russell, Paul F. "International Preventive Medicine." *The Scientific Monthly* 71:6 (December 1950): 393–400.

Sasaki, Masao, et al. "Global Rinderpest Eradication and the South Asia Rinderpest Eradication Campaign." *EMPRES: Transboundary Animal Disease Bulletin* 38 (2011): 32–40.

Sasamoto, Yukuo. "The Scientific Intelligence Survey: The Compton Survey," in *A Social History of Science and Technology in Contemporary Japan.* Vol. I, Ed. Shigeru Nakayama. Melbourne: Trans Pacific Press, 2001.

Saunders, P. T. and Rao Sahib K. Kylasam Ayyar. "An Experimental Study of Rinderpest Virus in Goats in a Series of 150 Direct Passages." *The Indian Journal of Veterinary Science and Animal Husbandry* 6 (1936): 1–86.

Schmelzer, Matthias. *The Hegemony of Growth: The OECD and the Making of the Economic Growth Paradigm.* Cambridge: Cambridge, 2016.

Schein, H. "Expériences sur la Peste Bovine." *Bulletin de la Societe de Pathologie Exotique et de ses Filiales* 19 (December 8, 1926): 915–928.

Scott, Gordon R. "Adverse Reactions in Cattle after Vaccination with Lapinized Rinderpest Virus," *Journal of Hygiene* 61 (1963): 193–203.

"Rinderpest," in *Advances in Veterinary Science*, V. 9, ed. C. A. Bradley and E. L. Jungherr. New York: Academic Press, 1964, 113–224.

"Rinderpest Virus" in *Virus Infections of Ruminants*, ed. Z. Dinter and B. Morein. Amsterdam: Elsevier, 1990, 341–354.

Sealey, Anne. "Globalizing the 1926 International Sanitary Convention." *Journal of Global History* 6 (2011): 431–455.

Sharp, Walter R. "The Institutional Framework for Technical Assistance." *International Organization* 7:3 (August 1953): 342–379.

Shen, Tsung-han. *The Sino-American Joint Commission on Rural Reconstruction.* Ithaca: Cornell University Press, 1970.

Shephard, Ben. "'Becoming Planning Minded': The Theory and Practice of Relief 1940–1945." *Journal of Contemporary History* 43:3 (July 2008): 405–419.

Shishido, A., K. Yamanouchi, M. Hikita, T. Sako, A. Fukuda, and F. Kobune. "Development of a Cell Culture System Susceptible to Measles, Canine Distemper, and Rinderpest Viruses." *Archiv für die gesamte Virusforschung* 22:3–4 (1967): 364–380.

Shope, Richard E. "Experimental Wartime Studies on Rinderpest." *Journal of the American Veterinary Medical Association* 110 (April 1947): 216–218.

"Influenza: History, Epidemiology, and Speculation." *Public Health Reports* 73:2 (February 1958): 165–178.

Shope, Richard E., Henry J. Griffiths, and Dubois L. Jenkins. "I: The Cultivation of Rinderpest Virus in the Developing Hen's Egg." *American Journal of Veterinary Research* 7:23 (April 1946): 135–141.

Shope, Richard E., Fred D. Maurer, Dubois L. Jenkins, Henry J. Griffiths, and James A. Baker. "IV: Infection of the Embryos and the Fluids of Developing Hens' Eggs." *American Journal of Veterinary Research* 6:23 (April 1946): 152–163.

Simpson, Bradley R. *Economists with Guns: Authoritarian Development and U.S.–Indonesian Relations, 1960–1968.* Stanford: Stanford University Press, 2008.

Skinner, Robert P. *Abyssinia of Today.* New York: Longmans, Green & Co., 1906.

Skloot, Rebecca. *The Immortal Life of Henrietta Lacks.* New York: Broadway Books, 2010.

Slavin, Philip. "The Great Bovine Pestilence and Its Economic and Environmental Consequences in England and Wales, 1318–50." *The Economic History Review* 65:4 (2012): 1239–1266.

Sluga, Glenda. *Internationalism in the Age of Nationalism.* Philadelphia: University of Pennsylvania Press, 2013.

"Turning International: Foundations of Modern International Thought and New Paradigms for Intellectual History." *History of European Ideas* 41:1 (2015): 103–115.

"UNESCO and the (One) World of Julian Huxley." *Journal of World History* 21:3 (September 2010): 393–418.

Sluga, Glenda and Patricia Clavin, ed. *Internationalisms: A Twentieth-Century History.* Cambridge: Cambridge, 2017.

Smith, Major-General Sir Frederick. *The Early History of Veterinary Literature and Its British Development*, Vol. II. London: Ballière, Tindall and Cox, 1924.

Solomon Hailemariam. *The Diary of an African Veterinary Doctor.* Bloomington, IN: Xlibris, 2010.

Soloman Hailemariam, Rene Besin, and Datsun Kariuki. *Wiping the Tears of African Cattle Owners.* Bloomington, IN: Xlibris, 2010.

Sonoda, Akiro. "Production of Rinderpest Tissue Culture Live Vaccine." *Japan Agricultural Research Quarterly* 17:3 (1983): 191–198.

Speich, Daniel. "The Use of Global Abstractions: National Income Accounting in the Period of Imperial Decline." *Journal of Global History* 6:1 (March 2011): 7–28.

Spinage, C. A. *Cattle Plague: A History.* New York: Kluwer Academic/Plenum Publishers, 2003.

Stafford, Jane. "Vaccine Can Increase Food." *The Science News-Letter* 49:11 (March 16, 1946): 174–175.

Staley, Eugene. "Relief and Rehabilitation in China." *Far Eastern Survey* 13:20 (October 4, 1944): 183–185.

———. *World Economy in Transition.* Reissue. Port Washington, NY: Kennikat Press, 1971.

Staples, Amy L. S. *The Birth of Development: How the World Bank, Food and Agriculture Organization, and World Health Organization Changed the World, 1945–1965.* Kent: Kent State Press, 2006.

Steed, Wickham. "Ariel Warfare: Secret German Plans." *The Nineteenth Century and After* 116 (July 1934): 1–14.

———. "The Future of Warfare." *The Nineteenth Century and After* 116 (August 1934): 129–140.

Stepan, Nancy Leys. *Eradication: Ridding the World of Diseases Forever?* Ithaca, NY: Cornell University Press, 2011.

Stirling, R. F. "Some Experiments in Rinderpest Vaccination: Active Immunisation of Indian Plains Cattle by Inoculation with Goat-Adapted Virus Alone in Field Conditions." *The Veterinary Journal* 88 (1932): 192–204.

Sun Yat-sen. *The International Development of China.* 2nd Ed. New York: G. P. Putnam's Sons, 1929.

Sunseri, Thaddeus. "The Entangled History of *Sadoka* (Rinderpest) and Veterinary Science in Tanzania and the Wider World, 1891–1901." *Bulletin of the History of Medicine* 89 (2015): 92–121.

Talbot, Ross B. "The International Fund for Agricultural Development." *Political Science Quarterly* 95:2 (Summer 1980): 261–276.

Taylor, William P. "Epidemiology and Control of Rinderpest." *Revue Scientifique et Technique (International Office of Epizootics)* 5:2 (1986): 407–410.

Taylor, William P., P. N. Bhat, and Y. P. Nanda. "The Principles and Practice of Rinderpest Eradication." *Veterinary Microbiology* 44 (1995): 359–367.

Taylor, W. P., and W. Plowright. "Studies on the Pathogenesis of Rinderpest in Experimental Cattle. III. Proliferation of an Attenuated Strain in Various Tissues Following Subcutaneous Inoculation." *Journal of Hygiene* 63:2 (June 1965): 263–275.

Thistle, Mel, Ed. *The Mackenzie-McNaughton Wartime Letters.* Toronto: University of Toronto Press, 1975.

Thornton, Philip K. "Livestock Production: Recent Trends, Future Prospects." *Philosophical Transactions of the Royal Society* 365 (2010): 2853–2867.

Tilly, Helen. *Africa as a Living Laboratory: Empire, Development, and the Problem of Scientific Knowledge, 1870–1950.* Chicago: University of Chicago Press, 2011.

Tounkara, Karim, Nick Nwankpa, and Charles Bodjo. "The Role of the African Union Pan African Veterinary Vaccine Centre (AU-PANVAC) in Rinderpest Eradication." *EMPRES: Transboundary Animal Disease Bulletin,* 38 (2011): 43–45.

Toynbee, Arnold J. "Not the Age of Atoms But of Welfare for All." *New York Times* (October 21, 1951): 168.

UNRRA. *The Story of UNRRA.* Washington, DC, 1948.

USDA. *The Plum Island Animal Disease Laboratory,* Miscellaneous Publication No. 730. Washington, DC: The United States Government Printing Office, September, 1956.

Veen, Tjaart W. Schillhorn van. "One Medicine: The Dynamic Relationship Between Animal and Human Medicine in History and at Present." *Agriculture and Human Values* 15 (1998): 115–120.

Vernon, James. *Hunger: A Modern History.* Cambridge: Harvard University Press, 2007.

Vogel, Kathleen M. "Pathogen Proliferation: Threats from the Former Soviet Bioweapons Complex." *Politics and the Life Sciences* 19:1 (March 2000): 3–16.

Wadman, Meredith. *The Vaccine Race: Science, Politics, and the Human Costs of Defeating Disease.* New York: Viking, 2017.

Walker, R. V. L., Henry J. Griffiths, Richard E. Shope, Fred D. Maurer, and Dubois L. Jenkins. "III: Immunization Experiments with Inactivated Bovine Tissue Vaccines." *American Journal of Veterinary Research* 6:23 (April 1946): 145–151.

Waller, Richard. "'Clean' and 'Dirty': Cattle Disease and Control Policy in Colonial Kenya, 1900–40." *The Journal of African History* 45:1 (2004): 45–80.

Walsh, John. "War on Cattle Disease Divides the Tropics." *Science* 237:4820 (September 11, 1987): 1289–1291.

Wang, Jessica. "Colonial Crossings: Social Science, Social Knowledge, and American Power from the Nineteenth Century to the Cold War," in *Cold War Science and the Transatlantic Circulation of Knowledge, History of Science and Medicine Library,* V. 51, ed. Jeroen van Dongen. Leiden: Brill, 2015, 184–213.

Webb, James L. A. *Humanity's Burden: A Global History of Malaria.* Cambridge: Cambridge University Press, 2008.

Wei, C.X. George. *Sino-American Relations, 1944–1949.* Westport, CT: Greenwood Press, 1997.

Weindling, Paul. "Between Bacteriology and Virology: The Development of Typhus Vaccines between the First and Second World Wars." *History and Philosophy of the Life Sciences* 17:1 (1995): 81–90.

"Philanthropy and World Health: The Rockefeller Foundation and the League of Nations Health Organization." *Minerva* 35 (1997): 269–281.

Weller, Thomas H. *Growing Pathogens in Tissue Cultures*. Canton, MI: Science History Publications, 2004.

Wertheim, Joel O. and Sergei L. Kosakovsky Pond. "Purifying Selection Can Obscure the Age of Viral Lineages." *Molecular Biology and Evolution* 28:12 (December 2011): 3355–3365.

Westad, Odd Arne. *The Global Cold War: Third World Interventions and the Making of Our Times*. Cambridge: Cambridge University Press, 2005.

Wheelis, Mark, Lajos Rózsa, and Malcolm Dando, ed. Deadly Cultures. Cambridge, MA: Harvard University Press, 2006.

WHO. "Proceedings of the Global Technical Consultation to Assess the Feasibility of Measles Eradication, 28–30 June 2010." *The Journal of Infectious Diseases* 204: Supplement 1 (July 15, 2011).

Wilcox, Francis O. "The United Nations Program for Technical Assistance." *Annals of the American Academy of Political and Social Science* 268 (March 1950) 45–53.

Wilkinson, Lise. "Rinderpest and Mainstream Infectious Disease Concepts in the Eighteenth Century." *Medical History* 28 (1984): 129–150.

Williams, Andrew J. "'Reconstruction' Before the Marshall Plan." *Review of International Studies* 31:3 (July 2005): 541–558.

Wolton, Suke. *Lord Hailey, the Colonial Office and the Politics of Race and Empire in the Second World War*. New York: St. Martin's Press, 2000.

Woodbridge, George. *The History of the United Nations Relief and Rehabilitation Administration*. V. I. New York: Columbia University Press, 1950.

The History of the United Nations Relief and Rehabilitation Administration. V. II. New York: Columbia University Press, 1950.

The History of the United Nations Relief and Rehabilitation Administration. V. III. New York: Columbia University Press, 1950.

Woodward, Ellen S. "UNRRA and War's Aftermath." *Social Security Bulletin* (November 1945): 10–14.

Worboys, Michael. "Was There a Bacteriological Revolution in Late Nineteenth-Century Medicine?" *Studies in History and Philosophy of Biological and Biomedical Sciences* 38:1 (2007): 20–42.

Wright, Susan, ed. *Preventing a Biological Arms Race*. Cambridge, MA: MIT Press, 1990.

Yamanouchi, Kazuya. "Scientific Background to the Global Eradication of Rinderpest." *Veterinary Immunology and Immunopathology* 148 (July 15, 2012): 12–15.

Sizyo Saidai no Densenbyo, Gyueki. Tokyo: Iwanami Shoten, 2009.

Yamanouchi, K. and T. Barrett. "Progress in the Development of a Heat-Stable Recombinant Rinderpest Vaccine Using an Attenuated Vaccinia Virus Vector." *Scientific and Technical Review of the Office International des Epizooties* 13:3 (1994): 721–735.

Yamanouchi, K., T. Barrett and C. Kai. "New Approaches to the Development of Virus Vaccines for Veterinary Use." *Scientific and Technical Review of the Office International des Epizooties* 17:3 (1998): 641–653.

Yasumara, Y. and Y. Kawatika. "Studies on SV40 Virus in Tissue Cultures." *Nihon Rinsho* 21 (1963): 1201–1215.

Yates, P. Lamartine. *So Bold an Aim: Ten Years of International Co-operation Toward Freedom from Want*. Rome: FAO, 1955.

Youde, Jeremy. "Cattle Scourge No More: The Eradication of Rinderpest and Its Lessons for Global Health Campaigns." *Politics and the Life Sciences* 32:1 (Spring 2013): 43–57.

Zanasi, Margherita. "Exporting Development: The League of Nations and Republican China." *Comparative Studies in Society and History* 49:1 (2007): 143–169.

Saving the Nation: Economic Modernity in Republican China. Chicago: University of Chicago Press, 2006.

Zeiler, Thomas W. "Opening Doors in the World Economy," in *Global Interdependence*. ed. Akira Iriye. Cambridge, MA: Harvard University Press, 2014, 203–361.

Free Trade, Free World: The Advent of GATT. Chapel Hill, NC: University of North Carolina Press, 1999.

Zinsser, Hans. *As I Remember Him: The Biography of R.S.* Boston: Little, Brown and Company, 1940.

Rats, Lice and History. 1935, New Brunswick, NJ: *Transaction* Publishers, 2008.

Index

70 Years of FAO (FAO), 246

A system of complete medical police (Frank), 20
ABC warfare. *See* Atomic, Biological, and Chemical warfare
Abyssinia, 54
accreditation process, 218
Acheson, Dean, 90, 92, 114–115, 151, 196
Addis Ababa, 24, 215
administrative issues, 66
Advances in Veterinary Science, 207
advice, expert, 91, 158
Advisory Council on Scientific Research and Technical Development, 183
aerosol experiments, 187
Afghanistan, 152–153
Africa. *See also* individual countries in Africa
 eradication campaigns in, 212–216, 222–225
 independent countries of, 134
 laboratories in, 230
 outbreaks of rinderpest in, 5, 23–30, 221
 raising living standards in, 148
 regional control efforts in, 131–136
 testing biological weapons in, 78
African buffalos, 235
African Rinderpest Organization, 134
African Scientific Committee, 134
African swine fever, 201
African warthog fever, 201

agitation, Communist, 194
agriculture
 exports, 30, 113, 159, 221, 239
 in post-war planning, 90
 research stations, 31
 revolution, agricultural, 118
 supplies and equipment, 109
Alexandria, 24
Alien Shadows (Simonov), 192
ALSOS Mission, 166–168, 175
American Association of Scientific Workers, 164
American Journal of Veterinary Research, 81
Amrith, Sunil, 10
Ancient Order of Hibernians, 2
Anglo-American Food Committee, 50
animal contact, transmission of virus through, 197
animal disease control, 150, 161
animal health, 84, 126
Animal Health Research Institute, 48
Animal Health Subcommittee, 128
Annual Message to Congress 1941, 88
anthrax, 52, 190
antigens, 36, 230
Antwerp, 13, 33
Anyang, 48
approved facilities for storing rinderpest, 252
arid regions, 239
Army Medical Corps, 171
Asia. *See also* individual countries in Asia
 as source of disease, 21
 eradication efforts in, 155

Assembly of Lower Canada, 1
assessment systems, 226
Associated Press, 191
Atlantic Charter, 51
Atomic, Biological, and Chemical (ABC)
    warfare, 182
attenuated vaccines, 37–38, 138
Attlee, Clement, 123
attrition, economic, 179
Austria, 21
Austrian-Hungarian Empire, 20
Avery, Donald, 189
avianized vaccine, 70, 78, 82–83, 100,
    138–141, 143, 180, 251

"Bacterial Warfare" (Rosebury and Kabat),
    164, 190
Bacteriological Warfare Subcommittee, UK,
    53
balloon bombs, 75–77, 172
Bangkok, 142, 160, 180
Banting, Frederick, 53, 55
basic reproductive number (R_o), 234, 246
beef trade, 221
Belgium, 13, 131, 134, 135
Belshaw, H., 110
Berlin, 168, 170
Bhutan, 227
bilateral government agencies, 122, 146,
    245
bile inoculation, 28
Bill and Melinda Gates Foundation, 243
Bini Hissar laboratory, 153
biological revolution, 248
biological warfare
    accusations of attacks in North Korea,
        189, 194
    ad hoc committee on, 178
    Allied research on, 168
    Canadian research on, 55, 188
    cost of research, 176
    danger from, 59
    defense against, 80, 167, 200
    delivery systems for, 173
    false allegations of, 180
    first use of, 196
    food supply attacked by, 168
    French research on, 53
    German research on, 165–170
    horror of, 178
    infrastructure attacked, 53

international relations and, 180
Japanese research on, 48, 74, 165
offensive research on, 60, 71–77, 85, 166,
    170, 187, 189, 200
possibility of attacks, 53
research on pathogens, 80
Soviet research on, 190
stop order on references to, 177
subversive potential of, 179
testing weapons, 74, 78, 187
UK research on, 61, 188
US research on, 60, 71–77, 85, 187–189,
    200
use in World War I, 52
Biological Weapons Research Panel, 197
Black Death. *See* plague
Black Leg, 168
Blitzarbeiter Committee, 167, 170
blockades, 50
Blome, Kurt, 169–171, 203
borders, political, 236
Botswana Vaccine Institute, 230
botulin, 190
bovine tissue vaccines, 68
Bretton Woods, 112
British Food Mission, 133
British Ministry for Overseas Development,
    219
Brookings Institution, 119
buffalo meat, 239
buffalo spleen vaccine, 141
Bulletin (OIE), 34
Bulletin of the World Health Organization,
    225
Bundy, Harvey Hollister, 71
Bureau of Agriculture, Philippines, 36
Bureau of Animal Industry, US, 30
Burkina Faso, 223
Burma, 157
Burnet, Frank Macfarlane, 69
Busan Institute, 31, 47, 74
Bush, Vannevar, 171

C.1. commission, 61–62, 72
Cabot, Daniel, 55, 136, 150–151
Cairo, 101, 105, 184
Cambodia, 160, 208
Cameroon, 214, 230
Camp Detrick, 183, 197
campaign against infectious animal diseases,
    33

Canada, 1–4, 7, 51, 56, 62, 181, 188, 219, 252
Canadian Journal of Comparative Medicine, 202
cancer, cure for, 191
canine distemper (CDV), 229
Canton, 117
Cape Colony, 26, 28
caprinized vaccine, 68, 138, 141, 144, 155, *See also* goats, passaging in
Caroll, Michael Christopher, 203
Cattle Disease Research Institute, 32
CBAHWs. *See* community-based animal health workers
CCTA. *See* Commission for Technical Cooperation in Africa South of the Sahara
CDV. *See* canine distemper
Center for the Study of Rinderpest, 127
Central Intelligence Agency (CIA), 186
centralization, 242
certifications of eradication, 226
Chad, 238
Chemical Corps, 177, 182–183, 187–190, 198
Chemical Defense Experimental Station, 54
Chemical Warfare Service (CWS), 56, 77
Cheng, S.C., 102, 136, 137–140
Cheysson, Claude, 213
Chiefs of Staff, Canada, 72
China
  elimination of rinderpest, 208
  FAO efforts in, 129
  fighting rinderpest in, 84
  Japanese bacteriological warfare in, 55
  JCRR efforts in, 117
  Korean border, 31
  program of vaccination, 99
  rinderpest rampant in, 20
  storing rinderpest virus, 252
  UNRRA efforts in, 95–111
  US aid to, 52
China National Relief and Rehabilitation Administration (CNRRA), 95
Chisholm, Brock, 204
chloroform-inactivated vaccine, 36, 68
cholera, 1, 123, 168
Chow, T.C., 102
Chunking, 95

Churchill, Winston, 50, 51, 54, 86, 89
CIA. *See* Central Intelligence Agency
civilian agencies, 60
civilizing mission, 111
CLARA. *See* Communist Liberated Areas Relief Administration
Clavin, Patricia, 87
Clayton, William L., 117
Clement XI, 18
CNRAA. *See* China National Relief and Rehabilitation Administration
cold chain from laboratory to field, 228, 240
Cold War, 11, 146–148, 175–178, 190–191, 195, 249
Colombo Plan for Cooperative Economic and Social Development in Asia and the Pacific, 148, 208
colonial development agencies, 11, 122
Colonial Development and Welfare Act, 42
Colonial Office (CO), 131–134
colonialism, 42, 84, 132, 134, 212, 249
Combined Commission for Technical Co-operation in Africa South of the Sahara (CCTA), 149, 212–214
Combined Food Board, 50, 89
commerce networks, 221, *See also* trade
Commission for Technical Cooperation in Africa South of the Sahara (CCTA), 218
Committee on Biological Warfare, 180
commodity divisions, 94
Commonwealth Conference on Foreign Affairs, 148
Commonwealth countries, 199
communal grazing, 29
communication of outbreaks, 21
"Communist Bacteriological Warfare Propaganda," 192
Communist Liberated Areas Relief Administration (CLARA), 95, 100
Communist Party, Japanese, 174
community-based animal health workers (CBAHWs), 240
compensation to livestock owners, 21
competition, commercial, 21
Compton Team, 174
Compton, Karl T., 171
Conference on Food and Agriculture, United Nations, 91

conferences
  1851 Paris International Sanitary Conference, 21
  1863 Hamburg conference of veterinary surgeons, 20, 220, 247
  1871 Vienna conference, 21, 220
  1903 Pan-African Veterinary Conference, 29
  1921 Paris International Conference for the Study of Epizootics, 13, 247
  1942 Oxford conference of Fabian Society, 86
  1948 Nairobi meeting, 132–136, 180
  1949 Bangkok conference, 142
  1950 Colombo Commonwealth Conference on Foreign Affairs, 148
  1950 UN Technical Assistance conference, 146
  1952 Pakistan OIE regional conference, 194
  1955 Bangkok rinderpest conference, 160
  1964 Addis Ababa FAO meeting, 215
  1964 joint OIE-FAO meeting, 207
  1967 joint OIE-FAO meeting, 208
  1992 Rome Expert Consultation on the Strategy for Global Rinderpest Eradication, 231
  2015 Abijan International Conference for the Control and Eradication of *peste des petits ruminants*, 243
  2016 Maintaining Global Freedom from Rinderpest, 252
  international, 129
consciousness, global, 6
contagion, theories of, 16, 18
control of disease, 18, 96, 121, 209, 247
cooperation
  global, 34, 217
  inter-imperial, 29, 38, 149
  international, 6, 38, 44–46, 87–91, 108, 140, 246
  scientific, 38
  technical assistance, 112–114, 118, 123, 129, 142–145, 154, 156–157, 205
  visions of, 208
coordination of eradication efforts, 29, 212
coordinators, designated, 218
cost-benefit analysis of eradication, 238
Côte d'Ivoire, 214, 242
Council of Ministers, Soviet Union, 204
covert warfare, 73

Crop Committee, Advisory Council on Scientific Research and Technical Development, 183
crop production, 220, 223
CWS. *See* Chemical Warfare Service
Czechoslovakia, 116

Dahomey, 214
Dai, Poeliu, 109
Dakar, 131
Danube River, 18
Daubney, R., 41, 83, 127, 136, 137, 184–185
Davies, E. Ll., 184
DDT powder, 123
De Bevoise, Ken, 23
decentralization, 225, 242, 244
Declaration of Global Freedom from Rinderpest, 237
declassification of research, 80
decontamination, 4, 82
deep-freeze units, 143, 144
Defense Research Board (DRB), 177, 182–185, 193, 198–201
delivery systems for biological weapons, 173
democratization of vaccination process, 240
Denmark, 14
Department of Agriculture, Canada, 4, 85, 177
Department of Defense, Canada, 2, 60, 62, 177
Department of Pensions, Canada, 2
Department of Public Works, Canada, 2, 58
Department of War, US, 4, 82
Deressa, Yilma, 112
desiccated vaccine. *See* dried vaccine
destruction of records, 173
development
  as environmental act, 12
  colonialism and, 132
  commitment to, 88
  domestic machinery for, 106
  for Asia, South America, and Africa, 112
  international machinery for, 122, 161, 208, 240, 245, 247, 248
  narrative of, 10
  of crops, 223
  of livestock, 223
  of under-developed areas, 129
  pairings in pursuit of, 242

development (cont.)
 plans for, 113
 problems in, 222
 results from, 219
 rinderpest eradication and, 8
 Sun Yat-sen's views on, 108, 110
 targeted goals for, 216
 technical assistance for, 114, 119
 timeline of, 11
DGV. *See* Dried Goat Virus
diagnostic kits, 236
Diouf, Jacques, 232
diphtheria, 123, 225
disease theory, 16
disinfection, 19, 22
Disney, Walt, 51
dissemination methods for rinderpest, 169,
 187
distribution of food, 91
Doctors' Trial, 170
Dodd, Norris, 129–130, 147–148, 152
Dominion of Animal Pathology, Canada,
 164, 181
donor community, international, 243
DRB. *See* Defense Research Board
Dried Goat Virus (DGV), 215
dried vaccine, 70, 156
duplication of research, 81

East Africa, 232
East African Veterinary Research
 Organization, 202
East Germany, 193
Eastern Europe, 146
ECA. *See* Economic Cooperation
 Administration
Economic and Financial Organization
 (EFO), 34, 43
Economic Cooperation Administration
 (ECA), 117
economic issues
 attrition, 179
 cattle as capital, 26
 development plans, 113
 disruption through biological warfare,
 179
 eradication of rinderpest and, 238–239
 idea of economy, 111
 insecurity, 46
 national income estimates, 44
 post-war planning, 90

progress, 110
 rinderpest and, 39
 visions of the future, 107
EDF. *See* European Development Fund
Edwards, James Thomas, 39, 47, 55, 61, 143
efficiency, 245
eggs, passaging in, 61, 69, 83, 139, 143,
 181, *See also* avianized vaccine
Egypt, 184, 232
Emergency Prevention System for
 Transboundary Animal and Plant Pests
 and Diseases (EMPRES), 232
emergency vaccination campaigns, 222
Enders, John F., 209
England. *See* United Kingdom
eradication of rinderpest. *See also* individual
 countries
 and imperialism, 213
 as national project, 143, 154
 avianized vaccine and, 83
 certification of, 226
 colonialism and, 212
 coordination of efforts, 29, 212
 cost-benefit analysis of, 238
 difficulties of, 246
 economic benefits of, 238–239
 exit strategies for campaigns, 218, 226
 global strategies for, 142, 216, 219,
 226
 in Asia, 155
 in isolated locations, 236
 in South East Asia, 160
 integrated elements of campaigns, 216
 international plans for, 160, 208
 machinery of, 209
 regional campaigns, 231
 structure of programs, 225
 technical and bureaucratic machinery for,
 242
Eritrea, 134
Ethiopia, 24–25, 112–114, 144, 158, 162
EU. *See* European Union
Europe. *See also* specific countries in Europe
 defense from Asian diseases, 21
European Commission, 234
European Development Fund (EDF), 213,
 219, 223
European Union (EU), 246
exchange of knowledge, 44
exit strategies for eradication campaigns,
 218, 226

Expanded Programme on Immunization, 225
experiments, on human beings, 170, 173
Expert Consultation on rinderpest, 226, 232
experts, advice from, 91, 123, 158, 212
exports, 30, 113, 159, 221, 239

Fabian Society, 86
Famine Emergency Committee, 115
FAO. *See* Food and Agriculture Organization
farm implements, in post-war reconstruction, 93
farmers, US, 50
farm-tool shops, 109
fear, freedom from, 86
fecal material, transmission of virus through, 72
Fell, Norbert, 174
fencing of livestock, 29
Ferris, R.D., 210
Feuntun, M., 137
fibromas, 64
financial aid, 112
first use of biological weapons, 196
Fischman, Harvey, 140, 155
Fishman, Yakov, 190
flexibility, 245
flood control, 109
food
    attacks on, 168, 179, 186
    distribution of, 91
    global crisis, 220
    high-energy, 51
    insecurity, 222
    post-war planning and, 89
    production of, 91
    psychological importance of, 51
    security of, 6, 115, 126, 159
    unequal access to, 43
Food and Agriculture Organization (FAO), 121–163
    and internationalism, 248
    beginning budget and mandate, 125
    creation of, 91
    endorsement of Declaration of Global Freedom from Rinderpest, 237
    Expert Consultation on rinderpest, 226, 231
    experts sent by, 208
    funding partners for, 222

global coordination provided by, 233
interference with imperialist campaigns, 132
mission of, 8
monitoring vaccine work, 181
Orr's resignation from, 147
research hidden from, 193
rinderpest chosen as eradication target, 142
role in animal health, 126
seed material requested from, 184
struggles in 1970s, 220
support for, 130
understaffing of, 207
veterinary problems, involvement in, 84
"Food Will Win the War" (Disney), 51
foot and mouth disease, 76, 128, 168, 187, 189
Foreign Aids, 219
foreign policy, US, 118
Forgotten Ally (Mitter), 108
formalin-inactivated vaccine, 68
Forrestal, James, 178
Fort Terry, 198
Foucault, Michel, 15
Four Freedoms, 88
fowl plague, 201
France, 53, 131, 134–137, 146, 149, 151
Frank, Johann Peter, 20
"free world," 148
freeze-drying units, 152, 153, 155, 156
frozen tissue, transmission through, 2
funding
    for biological warfare research, 62
    for eradication campaigns, 142, 146, 208
    for FAO, 130, 158
    for GREP, 234
    for JP 15, 213, 218
    for PARC, 232
future, visions of, 107

G-2, 174
game barriers, 134
Gamgee, John, 247
GDP. *See* gross domestic product
GenBank, 253
General Assembly, UN, 119
Geneva Protocol, 53, 191
germ theory, 27
Germany, 22, 50, 165–171, 219
Ghana, 214

Ghent, 13
Gilfoyle, Daniel, 29
GIN project, 66
GIR-1, 63, 71–72
giraffes, 221
glanders, 52
global consciousness, 6
Global Fund to Fight AIDS, Tuberculosis
   and Malaria, 244
global interdependence, 8
global liberalism, 216
Global Measles and Rubella Strategic Plan,
   245
Global Polio Eradication Initiative, 225
Global Rinderpest Eradication Programme
   (GREP), 232–235
"Global Rinderpest Situation" map, 236
global strategies for eradication, 216, 219
Global Technical Consultation to Assess the
   Feasibility of Measles Eradication, 245
glycerin, use in vaccines, 31
goats, passaging in, 40, 47, *See also*
   caprinized vaccine
Goulele laboratory, 144
governments, purpose of, 15, 30
grain production, 51, 220
grassroots goals and approaches, 157
Great African Rinderpest Panzootic, 23–30
Great Britain. *See* United Kingdom
Great Depression, 43, 111
Great Famine, 17
Greater Horn, 221
GREP. *See* Global Rinderpest Eradication
   Programme
gross domestic product (GDP), 238
Grosse Île
   as quarantine station, 1
   biological warfare research at, 61–62
   chickens left on island, 85
   declared "prohibited place," 164
   end of US support for, 4, 82, 177
   geography of, 58
   joint project at, 184
   made into national park, 250
   meeting of Biological Weapons Research
      Panel, 197
   precautions for, 59
   returned to Defense Research Board, 177
   returned to Department of Agriculture,
      202
   testing stocks of vaccine, 181, 182

turned over to Department of Public
   Works, 58
Gulf War, 232

Hailemariam, Solomon, 215, 222
Hall, H.T.B., 158
Hamblin, Jacob Darwin, 196
Hamburg, 220
Hankey, Maurice, 54
Hannah, John H., 219
Harar, 24
harnessing micro-organisms, 121, 124
Haylu, Alaqa Lamma, 24
Henderson, Donald, 225, 227
Herat, 152
herd immunity, 234, 246
high energy foods, 51
Himmler, Heinrich, 169, 203
Hinduism, 39
history, importance of, 247
Hitler, Adolf, 50, 166
Hodge, Joseph, 42
hog cholera, 98
Hoover, Herbert, 115
hosts, living, 12, 54, 67
House Committee on Foreign Affairs, 92
human welfare, 88, 147, 161
humanitarianism, 80, 108, 200
humans, experiments on, 170, 173
Hungary, 18
hunger, 46, 253
Hutcheon, Duncan, 4
Huxley, Julian, 87–88, 121, 124, 147

IAEA. *See* International Atomic Energy
   Agency
IBAH. *See* Inter-African Bureau for Animal
   Health
immune belt, 31
immunity, duration of, 36, 68, 83, 138, 155,
   211
Imperial Defense Ministry, UK, 53
Imperial Ethiopian Government (IEG), 144
imperialism, 6, 15, 22–23, 84, 131, 135,
   213, 249
improvement, concept of, 110
inactivated vaccines, 36, 39, 137, 160
income estimates, national, 44
India, 39, 143, 207, 221, 227, 232, 238
Indian Veterinary Research Institute, 155
inequality, global, 44

infectivity rates, 246
influenza, 32, 63–64
information, fighting with, 196
infrastructure
    attacks on, 53
    scientific, 45
ingestion, transmission of virus through, 72
insect vectors, transmission of virus
    through, 197
Institute for Animal Health, UK, 234
Institute for Infectious Diseases, 27
Institute for Inter-American Affairs, 123
Institute of Pacific Relations (IPR), 109
integrated elements of eradication
    campaigns, 216
Intensified Smallpox Eradication Program
    (SEP), 217, 224
Inter-African Bureau for Animal Health
    (IBAH), 207, 213
Inter-African Bureau of Animal Resources,
    223
Inter-African Bureau of Epizootic Diseases,
    149
Inter-Allied Committee on Post-war
    Requirements, 89
interconnectedness, 8, 10
interdependence, global, 8
international agencies, 14
International Atomic Energy Agency
    (IAEA), 234
International Bank, 142
International Conference for the Control
    and Eradication of *peste des petits
    ruminants*, 243
International Conference for the Study of
    Epizootics, 247
International Conference of Veterinary
    Surgeons, 247
International Cooperation Year, 208
international development community, 222,
    240, 245, 247
International Health Yearbook (LNHO), 44
International Institute of Agriculture (IIA),
    43, 91
international institutions, 11
International Labor Organization (ILO), 43
international law, 78
international networks for fighting
    rinderpest, 127
international organizations, 142, 208, 216,
    246

international relations, 9, 180, 205
International Sanitary Conference, 21
International Scientific Commission (ISC),
    196
internationalism, 10, 32–33, 48, 147, 178,
    199, 220, 248–249
interviews, with Japanese war researchers,
    172
IPR. *See* Institute of Pacific Relations
Iran, 221, 232, 252
Iraq, 221, 232
Iriye, Akira, 6
Isayama, Ichiro, 73
ISC. *See* International Scientific
    Commission
Ishii Shiro, 173–175
isolated locations, eradication work in, 236
Israel, 221
Italy, 115
Izatangar, 155–156

Japan
    assistance with eradication efforts, 208
    bacterial warfare in China, 55
    biological warfare research, 48, 72–77,
        165, 171–175, 190
    construction of immune belt, 31
    development of vaccine, 31–32
JCRR. *See* Joint Commission on Rural
    Reconstruction
Jiangxi, 99
Johnson, Louis A., 179
joint action, 14
Joint Canada-United States Commission,
    63, 78
Joint Chiefs of Staff, 174, 186
Joint Commission on Rural Reconstruction
    (JCRR), 117–118
Joint FAO/OIE Committee for Global
    Rinderpest Eradication, 235
Joint Project 15 (JP15), 212–219
Joint US-Canada anti-animal program, 201
Journal of Immunology, 164
JP 15. *See* Joint Project 15

Kabat, Elvin A., 164, 190, 193
Kabete Attenuated Goat vaccine (KAG), 41,
    55, 83, 138, 156
Kabete O strain, 41, 67, 69, 106, 197, 210,
    253
Kabete, Kenya, 41, 83, 105, 132, 180

KAG. *See* Kabete Attenuated Goat vaccine
Kaifeng, 95
Kakizaki, Chiharu, 31, 36
Kefu Qan (Evil Days), 25
Keitel, Wilhelm, 167
Kelser, Raymond Alexander, 36, 38, 56, 61, 68, 71, 78
Kenya, 79, 105, 181, 198, 232, 235, 238
Kesteven, Keith V.L., 101–106, 129, 133, 138, 142–145, 150, 184
Khabarovski, 175
Khan, Mirak Shah, 153
Khan, Mohamed Asiam, 153
kidney cells, bovine, 210
Kliewe, Heinrich, 168
Kliueva, Nina, 191
knowledge, exchange of, 44
Koch, Robert, 27–29
Korea, 31, 47, *See also* North Korea; South Korea
Korean War, 186
Krementsov, Nikolai, 192
Kuba, Noboru, 73–76, 173

Lab 257 (Caroll), 203
laboratories. *See also* individual laboratories
    accidents at, 63
    equipment for, 97, 152, 153
    establishment of, 3, 30, 57, 106
    imperial, 7
    in Africa, 230
    updating, 66
Lake Chad Basin, 214
Lake Mweru, 25
Lanchow strain, 101
Lancisi, Giovanni Maria, 18
land, recovery of, 109
lapinized vaccine, 102, 106, 138–139, 144, 185, *See also* passaging, in rabbits
League of Nations. *See also* specific agencies
    agencies of, 87
    end of, 248
    international society created by, 43
    orbit of, 14
League of Nations Health Organization (LNHO), 32, 34, 43
Lebanon, 221
Lebensraum, 50
Leffler, Melvyn, 178
Lehman, Herbert, 93
Leitenberg, Milton, 190

lend-lease aid, 50
Lepissier, H.E., 216, 218
lesser kudus, 221
Lewis, Paul, 63
liberalism, global, 216
Lie, Trygve, 145
Literary Gazette, 193
livestock
    communal grazing of, 29
    compensation to owners, 21
    development of, 223
    fencing of, 29
    importance of, 239
    overstocking and overgrazing of, 42
    potential attack on, 183
    production, 51
    replacement of, 96
LNHO. *See* League of Nations Health Organization
local communities, 235, 240, 245
localized outbreaks, 69
locating rinderpest, 233
London Underground, 53
Lugard, Frederick, 23, 25
lyophilization, 229

M-1000, 52, 56–61
Maass, Otto, 57, 61, 66
MacArthur, Douglas, 194
Macfarlane, I.M., 213
mad itch, 64
Maintaining Global Freedom from Rinderpest meeting, 252
malaria, 123, 125, 204, 216
Mali, 214, 223
Malta, 132
Manchuria, 31, 73
mandatory vaccinations, 32
Mariner, Jeffrey C., 228–229, 240
Marquand, John P., 77
Marshall Plan, 117
Marshall, George C., 114–115, 124, 186
Masai, 25
mass production of vaccines, 186
mass vaccination, 215, 222, 244
Mazower, Mark, 178
McNamara, Robert, 216
McNeill, J.R., 122
measles, 16, 219, 225, 245
measles virus (MeV), 210, 229
meat exports, 159

medical knowledge, 19
Menelik of Ethiopia, 24
MEP, 216
Merck, George W., 63, 66, 71, 77–79, 80–81, 175–177
MeV. *See* measles; measles virus
Mexico, 35
Middle East, 221
military engineering of pathogens, 165
Millett, Piers, 202
Ministry of Agriculture and Fisheries, United Kingdom, 132
Ministry of Agriculture, Soviet Union, 204
Ministry of Agriculture, Thailand, 129, 141
Mitchell, Charles, 4–6, 76, 82, 181–186, 197–201, 203
Mitchell, Timothy, 111
Mitter, Rana, 108
modernization of farming, 31
molecular epidemiology, 16, 231, 234
Molopo River, 26
Molotov, Viacheslav, 191
Moreland, Edward L., 171
Morgenthau, Henry, 112
Mount Meru National Park, 235
mucous membranes, transmission of virus through, 197
Muguga laboratory, 198, 202
Mukden Veterinary Institute, 32, 73
Mukteshwar, 39, 47, 143
Murray, Everitt G.D., 56, 57–58, 60, 66, 72, 76
mutability of rinderpest, 37, 38, 165, 188, 201
mutual help, 44

Nairobi, 101, 132
Nakamura, Junji, 47, 73–76, 102, 139–140
Nakdong River, 74
Nanjing, 100, 138, 140, 180
nasal cavity, transmission of virus through, 74
National Academy of Sciences, 56
national aid programs, 208
national eradication campaigns, 143, 205
National Institute of Health, Japan, 229
National Intelligence Estimate 18 (CIA), 186
National Research Bureau of Animal Husbandry, China, 100, 102, 106

National Research Council, Canada, 53, 55, 56, 164
national security, 176, 180, 199–200, 206
nationalism, 16, 107, 131, 206
Nationalist government, China, 107
NATO. *See* North Atlantic Treaty Organization
natural resources, 50, 113
Near East Panzootic, 221
Nehru, Jawaharlal, 208
Nepal, 221, 227
Netherlands, 252
New York Times, 174, 239
Newfield, Timothy P., 17
NIE 18. See National Intelligence Estimate 18
Niger, 214
Nigeria, 214, 221, 223
Nippon Institute of Biological Science, 139
Noborito Institute, 73, 172
Norfolk, Virginia, 52
North Atlantic Treaty Organization (NATO), 199
North Korea, 189, 194
nuclear weapons, 178, 189
Nuremberg tribunal, 170

OAU. *See* Organization of African Unity
OEEC. *See* Organization for European Economic Co-operation
Office International d'Hygiène Public, 32
Office International des Épizooties (OIE)
  coordination with FAO and Colonial Office, 131–132
  creation of, 7, 33, 247
  FAO's relationship to, 150–152
  freedom from rinderpest status granted by, 226
  Joint FAO/OIE Committee for Global Rinderpest Eradication, 235
  May 1947 meeting, 127
  Pakistan regional conference 1952, 194
  understaffing of, 207
Office of Foreign Relief and Rehabilitation Operations (OFRRA), 89, 92
Office of Scientific Research and Development, 171
Office of Strategic Information, 182
OFRRA. *See* Office of Foreign Relief and Rehabilitation Operations
OIE. *See* Office International des Épizooties

Okinawa, 80
Oman, 221
operational structures, 245
Operations Mission, US, 160
optimal route of infection, 187
Organization for European Economic Co-
    operation (OEEC), 152
Organization of African Unity (OAU), 214,
    218, 223
Orr, John Boyd, 121, 125–126, 130, 133,
    137, 147, 163, 248–249
Ottawa, 52
outbreaks of rinderpest
    14th century panzootic, 17
    1709–1800 Europe, 17
    1742–1760 Europe, 20
    1877 United Kingdom, 22
    1886 Philippines, 22
    1921 Belgium, 13
    569–570 Europe, 17
    986–988 Europe, 17
    communication of, 21
    Great African Rinderpest Panzootic, 5,
        23–30
    localized, 69
    Near East Panzootic, 221
    Second Great African Rinderpest
        Panzootic, 221
    unreported, 233
overstocking of livestock, 135

PACE. *See* Pan-African Programme for the
    Control of Epizootics
packaging vaccines, 184
Packard, Randall, 204, 216
pairings in pursuit of development, 242
Pak Chong laboratory, 129, 141, 154
Pakistan, 155, 194, 221, 227, 238
Pan African Veterinary Vaccine Center
    (PANVAC), 239
Pan-African Programme for the Control of
    Epizootics (PACE), 233
Pan-African Rinderpest Campaign (PARC),
    223–225
Pan-African Veterinary Conference, 29
Pankhurst, Richard, 24
PANVAC. *See* Pan African Veterinary
    Vaccine Center
Papal States, 18
papillomatosis, 64
PARC, 232

Paris, 21
Paris Metro, 53
participatory epidemiology (PE), 240
Participatory Epidemiology Network for
    Animal and Public Health (PENAPH),
    241
partnership organizations, 244
passaging
    attenuation via, 37–38
    in animals, 27
    in bovine kidney tissue, 211
    in eggs, 61, 69, 83, 139, 143, *181*
    in goats, 40, 47
    in pigs, 154
    in rabbits, 39, *102*
    in sheep, 40
Pasteur Institute, 35
Pasteur, Louis, 27
Pasteurella, 168
pasteurellosis, 40
pastoralists, 25
pathogenicity, 37
patriotism, Soviet, 191
Patterson, Robert P., 80
PE. *See* participatory epidemiology
Peace or Pestilence (Rosebury), 190
peace, plans for, 90
peat moss carrier, 72
Pedersen, Susan, 34
Pei-p'ing Radio, 194
PENAPH. *See* Participatory Epidemiology
    Network for Animal and Public Health
persistence of rinderpest, 207
pertussis, 225
peste des petits ruminants (PPR), 242–245
pharmaceutical supplies, 109
Philippines, 22, 36, 38
Pierson, E., 153
pigs, passaging through, 154
Ping Fan, 173
Pirbright Research Station, 55, 61, 72, 183,
    234
plague, 17, 173
"Plan for Eradication of Rinderpest in
    Vietnam" (United States Operations
    Mission), 161
Plowright, Walter, 209, 227
Plum Island, 198, 202, 228
Point Four Program, 118–120, 123, 142,
    145, 157–160
Poland, 14, 32, 34

polio, 225
political success, 217
Porter, William N., 71
Portugal, 131, 134, 135
positive security, 44
post-war planning, 89
potato bugs, 193
poverty
　colonial, 42
　fighting, 224
　global, 111
powder, dried virus, 74
PPR. *See peste des petits ruminants*
Pravda, 193
preservation of vaccines, 5, 82
Preston, Andrew, 176
*Principles for an International Regulation
　for the Extinction of the Cattle Plague*
　(Vienna conference report), 21
print culture, 18
production
　expansion of, 110
　of food, 91
　of necessities, 92
Program of Technical Cooperation, 108
progress, economic, 110
Project 1001, 198
Project 63, 171
propaganda, Soviet, 192–197
prophylaxis, 15, 27, 30
protein famine, 17
Psychological Strategy Board, 195
public health services, 44, 224
public relations, 191, 200, 220
publication of research, 81, 140, 175, 176

quality control for vaccines, 230, 239
quarantine sites, 1, 13, 202, 250
quarantines, 2, 18, 26, 30, 32, 247

rabbits
　as test subjects, 47
　passaging in, 39, 102
　shortage of, 103, 141
racial disparities, 26, 29, 42
Radio Moscow, 193
railroads, 20
Rats, Lice and History (Zinsser), 1, 35, 250
raw materials, 50
Ray, J. Franklin, 107
reassembling of rinderpest virus, 253

recombinant rinderpest vaccines, 230
reconstruction, 87–89, 92, 109–110, 115
Red Cross, International, 196
reestablishment, 110
refrigerators, 98–99
regional eradication campaigns, 231
regional organizations, 246
regulation, 247
rehabilitation, 86–89, 92, 96, 107–109,
　114
Reims, 170
Reisinger, R.C., 144, 159
relief, 86–89, 92, 96, 107, 109, 196
replacement of livestock, 96
repositories for rinderpest virus, 181
reservoirs of rinderpest infection, 217, 221,
　234
results of aid programs, 219
revolution, agricultural, 118
Rhee, Syngman, 194
rice, chemical attacks on, 77
Rich, Karl, 247
rinderpest. *See also* eradication of
　rinderpest; outbreaks of rinderpest;
　strains of rinderpest; transmission of
　rinderpest; vaccines
　2011 eradication of, 12
　and creation of global consciousness, 6
　approved facilities for storing, 252
　courses on, 156
　destruction of stocks, 251
　dissemination methods for, 169, 187
　economic issues of, 39
　eradication efforts, 126
　experimental study of, 15
　grown in bovine kidney cells, 210
　history of, 15–46
　locating, 233
　meaning of word, 2
　molecular clock analysis of, 16
　mutability of, 37, 38, 165, 188, 201
　persistence of, 207
　preservation of virus, 61
　reassembling virus, 253
　repositories of virus, 181, 251
　reservoirs of infection, 217, 221, 234
　routes of infection, 61
　similarity to measles, 16
　spread of, 5
　surveillance for, 235
　symptoms of, 4, 18

rinderpest (cont.)
  testing for presence of, 199, 210
  transparency in reporting, 246
Rinderpest Meeting for Asia and the Far
  East, 142
"Rinderpest: On the Virulence of the
  Attenuated Rabbit Virus for Cattle"
  (Nakamura), 47
Rinderpest Serum Manufacture Institute, 48
Rockefeller Foundation, 242
Rockefeller Institute for Medical Research,
  55, 203
Roeder, Peter, 217, 233, 235, 247
Rome, 156
Roosevelt, Franklin Delano, 51, 58, 59–60,
  86–90, 94, 176
Rosebury, Theodor, 164, 190, 193
Rosenberg, Emily, 111
Roskin, Grigorii, 191
rural industries, 109
Russia, 20, 34, 152, 169, 252, *See also*
  Soviet Union
RVP. *See* rinderpest

sabotage, biological, 169, 179, 189
saliva, transmission of virus through, 72
sanitary measures, 15, 22, 24, 30, 32, 248
SAREC. *See* Sustainable Agriculture
  Research and Extension Center
Saudi Arabia, 221
Schoening, H.W., 150, 151
science
  and development of resources, 248
  and international cooperation, 32, 48,
    209
  confidence in, 163
  defeating want, 130
  dissemination of information, 246
  infrastructure of, 45
  medical and biological, 179
  political aspects of, 165
  sharing knowledge, 201
Scientific Technical and Research
  Commission (STRC), 214
Scott, Gordon R., 207–209
Second Great African Rinderpest Panzootic,
  221
secrecy of research, 60, 191, 199
Secretary of Defense, Ad Hoc Committee,
  182
Security Council, UN, 194, 195

security problems, 232
seed virus, 100, 105, 184–186
seeds, in post-war reconstruction, 93
"Seek, Contain, Eliminate," slogan, 234
Sen, Binay Ranjan, 161
Senegal River Basin, 217, 221
SEP. *See* Intensified Smallpox Eradication
  Program
serosurveillance technology, 234
serum-simultaneous method, 28, 35, 39, 138
Shanghai, 84, 95, 99
sharing
  scientific knowledge, 201
  vaccines, 82, 136, 189
sheep, passaging in, 40
Shope, Richard E., 50, 63–69, 79, 137, 175,
  180–182, 193, 203
Sidibe, Samba, 222
Simonov, Konstantin, 192
simultaneous infections, 39
slaughter of infected animals, 19, 26, 247
Slavin, Phil, 17
small ruminants, 242
smallpox, 205, 217, 219, 224
smart vaccination, 244
Smith, Walter Bedell, 191
So Bold an Aim (FAO), 162
social health and welfare, 43
social strife, 222
Solandt, Omond, 186
solidarity, international, 246
Somali Ecosystem, 234
Somalia, 134
Soper, Frederick, 209
South Africa, 135, 252
South Asia Rinderpest Eradication
  Campaign (SAREC), 227
South East Asia, 160
South Korea, 208
South Manchuria Railway Company, 32
Soviet Union
  and international organizations, 146
  biological warfare research, 182, 186,
    190, 204
  defenses against biological weapons, 201
  eradication of rinderpest in, 35
  fear of attack from, 181
  propaganda, 192–197
  trials of Japanese war criminals, 175
  US aid to, 52
  withdrawal from WHO, 205

Sri Lanka, 221
stabilizing vaccines, 184
Stalin, Joseph, 116, 178, 191–192
"stamping out" policies, 26
standards of living, 42–44, 46, 111, 124, 132, 148
Standing Committee on Animal Health, 126
Stanley, Eugene, 112
Stannard, A.F.B., 200
State Department, 195
Steed, Henry Wickham, 53
Stimson, Henry, 56, 60, 61, 77–79
strains of rinderpest. *See also* vaccines
    African lineages 12, and234
    Asian lineage, 234
    caprine strain, 81
    collection of, 188
    Indian strain, 81
    Kabete O, 41, 67, 69, 106, 197, 210, 228, 253
    L strain, 48
    Nakamura III, 48, 102, 129, 133, 139, 160, 253
    North African strain, 81
    Szechwan strain, 101
    Turkish strain, 197
STRC. *See* Scientific Technical and Research Commission
subcutaneous injections, 197
success, factors of, 225
Sudan, 134, 221, 223, 232
Suebsaeng, Charas, 128
Sueki Kusaba, 172
Sun Yat-sen, 108, 110
surveillance for rinderpest, 222, 226, 235
Sustainable Agriculture Research and Extension Center (SAREC), 232
Sustainable Development Goals, 253
swine influenza, 63–64
Switzerland, 252
Syria, 221
Szechwan strain, 101

Tachikawa laboratory, 160
Taipei, 117
Tanganyika, 25, 42
targeted vaccination, 244
TCRV. *See* Tissue Culture Rinderpest Vaccine
technical advising, 91

Technical Advisory Committee for Animal Disease Control, 159
technical assistance, 112–114, 118, 123, 129, 142–145, 154, 156–157, 205
Technical Assistance Conference, UN, 146
Technical Assistance Fund, United Nations, 157
technology
    and human welfare, 161
    and internationalism, 87
    emphasis on, 122
    field laboratories and, 236
    influence of, 6
    transfers of, 97, 111
    transformation of world through, 119
    value of, 175
Ten reasons for NOT maintaining or storing rinderpest virus (FAO/OIE), 251
testing
    for rinderpest, 210
    of balloon bombs, 76
    of biological weapons, 74, 78, 187
    of vaccines, 182
tetanus, 225
Thailand, 128–129, 141, 154, 208
The Cure (Krementsov), 192
The Future of Food and Agriculture (FAO), 253
The Japanese Journal of Veterinary Science, 47
The Plum Island Animal Disease Laboratory (USDA), 203
"The Preparation of Our Enemies for BW" (Kliewe), 168
The Soviet Biological Weapons Program (Leitenberg and Zilinskas), 190
The Transition from War to Peace Economy (League of Nations), 87, 92
thermostability of vaccines, 211, 228, 235, 240
ThermoVax, 229, 240
Tientsin, 95
Tilly, Helen, 45
Tissue Culture Rinderpest Vaccine (TCRV), 209, 215, 227
tissue cultures, virus grown in, 210
Togo, 214
Tojo, Hideki, 75, 173
Tokyo, 171
Tokyo war crimes trials, 174

top-down interventions, 224
Toynbee, Arnold J., 163, 205
trade
  expansion of, 21
  global, 6
  imperially created networks, 15
  in beef, 221
  in post-war planning, 90
  restriction of, 19
training centers, 239
transmission of rinderpest
  by animal contact, 197
  by fecal material, 72
  by frozen tissue, 2
  by ingestion, 72
  by insect vectors, 197
  by mucous membranes, 197
  by nasal cavity, 74
  by saliva, 72
  by wild animals, 25
transparency in reporting rinderpest, 246
transportation of necessities, 92
Traub, Erich, 170, 203
travel, expansion of, 21
treaties, 21
trials of war criminals, 174, 193
Tripartite Biological and Chemical
    Weapons Agreement, 177, 188, 191
tropical diseases, 124
Truman, Harry S., 96, 115, 118, 125, 145,
    149, 178
Trypanosoma cruz, 191
tsetse flies, 26
tuberculosis, 225
Turkey, 31, 170, 221, 232, 252
typhus, 2, 32, 34, 123

US Canada Joint Defense Board, 58
U.S.A. v. Karl Brandt et al, 170
Uganda, 79, 232
undeclared war, 179
UNESCO. *See* United Nations
    Educational, Scientific and Cultural
    Organization
Unger, Corrina, 122
UNICEF. *See* United Nations Children's
    Fund
Union of Soviet Writers, 191
United Arab Emirates, 221
United Kingdom, 22, 51–53, 61, 131–135,
    146–148, 166, 188

United Nations
  agencies connected to, 121, 149, 162, 249
  Conference on Food and Agriculture, 91
  declaration of International
      Cooperation Year, 208
  development goals, 248
  General Assembly, 119
  help with vaccination programs, 225
  institutions, 245
  Security Council, 194, 195
  struggles of system, 8
  substitutes for bodies set up by, 149
  Sustainable Development Goals, 253
  systems developed around, 205
  Technical Assistance Conference, 146
  technical assistance for economic
      development, 145
United Nations Children's Fund (UNICEF),
    225
United Nations Development Programme,
    227, 232
United Nations Educational, Scientific and
    Cultural Organization (UNESCO),
    121, 124
United Nations Expanded Program of
    Technical Assistance for Economic
    Development of Underdeveloped
    Countries, 120
United Nations Food and Agriculture
    Committee, 84
United Nations Office for the Coordination
    of Humanitarian Affairs, 235
"United Nations Programme for Freedom
    from Want of Food" (Australia), 90
United Nations Relief and Rehabilitation
    Administration (UNRRA), 8, 84,
    91–95, 106–109, 113–115, 123–130,
    144
United States
  assistance to eradication campaigns, 213
  assistance to underdeveloped areas, 142
  biological weapons research, 188
  collection of rinderpest strains, 188
  farmers, 50
  foreign policy, 118
  funding for joint Canada-US research, 62
  involvement in struggle against
      rinderpest, 7
  laboratory established in, 181
  manipulation of internationalism, 248
  reasons for Point Four, 145, 159

relief efforts, 196
stockpiling vaccines, 166
support for FAO, 150
technical assistance program, 156
vaccine production project, 50
working class diet, 52
United States Agency for International
Development (USAID), 213, 219, 223,
234
United States Biological Warfare Committee
(USBWC), 78–80
United States Counter Intelligence Corps,
169
UNRRA. *See* United Nations Relief and
Rehabilitation Administration
Unyanyembe, 25
Upper Volta, 214
US Department of Agriculture (USDA), 30,
51, 82, 125, 198, 202
USAID. *See* United States Agency for
International Development
USBWC. *See* United States Biological
Warfare Committee
USDA. *See* US Department of Agriculture
USSR. *See* Soviet Union

vaccines
attenuated, 37, 138
avianized, 70, 78, 82–83, 100, 138–141,
143, 164, 180, 251
bile inoculation, 28
bovine tissue, 68
caprinized, 68, 128, 138, 141, 144, 155
China provided with, 98
chloroform-inactivated, 36, 68
creation of, 27, 31, 36
dried, 70, 156, 197
Dried Goat Virus (DGV), 215
field apparatus for, 107
formalin-inactivated, 68
freeze-dried, 225, 227
inactivated buffalo spleen, 141
inactivated tissue, 36, 39, 57, 137, 160
joint US-Canadian rinderpest vaccine
production project, 50
KAG, 41, 83, 138
lapinized, 102, 106, 138–139, 144, 185
live, 5
mass production of, 186
packaging, 184
preservation of, 5, 82

production of, 3, 57, 67, 100, 113, 169,
173, 205
quality control for, 230, 239
recombinant rinderpest, 230
sharing, 38, 82, 136, 143, 189
stabilizing, 184
standardized production of, 155
stockpiling, 186
technology of, 175
testing, 182
thermostability of, 211, 228, 235, 240
ThermoVax, 229, 240
Tissue Culture Rinderpest Vaccine
(TCRV), 209, 215
training in administration of, 117
use of multiple, 41
Vero cell rinderpest, 229
viability of, 68
wet tissue, 36, 155
Vavasour, G.R., 202
Venezuela, 123, 151
Vero cells, 228–229
veterinary medicine
colleges of, 20
improvement of services, 239
international community, 226
police measures, 127
services, 15, 30
supplies, 98
surgeons, 21
Veterinary Research Institute, Kenya, 181
Vienna, 21, 220
Vietnam, 161

want, freedom from, 86–89, 95, 130, 253
War Bureau of Consultants Committee
(WBC), 56–60
War Department, US, 60, 62, 176
War Effort Committee, 164
War Food Administration, US, 51
war projects, termination of, 82
War Research Service (WRS), 60, 63
WAREC. *See* West Asia Rinderpest
Eradication Campaign
water buffalo, 23
WBC. *See* War Bureau of Consultants
Committee
Wei, George C.X., 108
West Africa, 232
West Asia Rinderpest Eradication
Campaign (WAREC), 227, 232

wet tissue method, 36, 128
WHA. *See* World Health Assembly
White, Dexter, 112
WHO. *See* World Health Organization
Wilcox, Francis, 145
wild animals, transmission of virus through, 25
Woolf, Leonard, 86
World Bank, 223
World Food Prize, 227
World Health Assembly (WHA), 204, 225
World Health Organization (WHO), 125, 131, 196, 204, 216, 219, 222–225, 245
World Organization for Animal Health, 237
World Reference Laboratory for rinderpest, 234

World War I, 52, 187
World War II, 7, 50–52, 111, 190, 248
WRS. *See* War Research Service

Yamanouchi, Kazuya, 229
yellow fever, 55, 63, 105
Yellow River, 109
Yemen, 221, 232

Zambezi River, 26
zebus, 83
zero incidence goal, 225
Zhdanov, Andrei, 191–192
Zhdanov, Viktor, 204
Zilinskas, Raymond, 190
Zinsser, Hans, 1, 34, 48, 250

Printed in Great Britain
by Amazon

31210015R00179